IMMUNOBIOLOGY OF PROTEINS AND PEPTIDES VIII

Manipulation or Modulation of the
Immune Response

ADVANCES IN EXPERIMENTAL MEDICINE AND BIOLOGY

Recent Volumes in this Series

A Continuation Order Plan is available for this series. A continuation order will bring delivery of each new volume
immediately upon publication. Volumes are billed only upon actual shipment. For further information please contact
the publisher.

IMMUNOBIOLOGY OF PROTEINS AND PEPTIDES VIII

Manipulation or Modulation of the Immune Response

Edited by

M. Zouhair Atassi

Baylor College of Medicine
Houston, Texas

and

Garvin S. Bixler, Jr.

Univax Biologics, Inc.
Rockville, Maryland

SPRINGER SCIENCE+BUSINESS MEDIA, LLC

Library of Congress Cataloging in Publication Data

International Symposium on the Immunobiology of Proteins and Peptides (8th: 1994: Pio Rico, Ariz.)

Immunobiology of proteins and peptides VIII: manipulation or modulation of the immune response / edited by M. Zouhair Atassi and Garvin S. Bixler, Jr.

    p.   cm.—(Advances in experimental medicine and biology; v. 383)

"Proceedings of the Eighth International Symposium on the Immunobiology of Proteins and Peptides, held in November 16-20, 1994, in Pio Rico, Arizona."—T.p. verso.

Includes bibliographical references and index.

ISBN 978-0-306-45125-6   ISBN 978-1-4615-1891-4 (eBook)

DOI 10.1007/978-1-4615-1891-4

1. Immune response—Regulation—Congresses. 2. Proteins—Immunology—Congresses. 3. Peptides—Immunology—Congresses. I. Atassi, M. Z. II. Bixler, Garvin S. III. Title. IV. Series.

QR186.I56   1994   95-36866

616.07'9—dc20   CIP

Proceedings of the Eighth International Symposium on the Immunobiology of Proteins and Peptides, held November 16–20, 1994, in Pio Rico, Arizona

ISBN 978-0-306-45125-6

© 1995 Springer Science+Business Media New York
Originally published by Plenum Press, New York in 1995

10 9 8 7 6 5 4 3 2 1

Scientific Council of the Symposium

M. Zouhair Atassi, *Chairman*
Garvin S. Bixler, Jr., *Secretary*
Colin R. Young, *Secretary*
Howard L. Bachrach
Constantin A. Bona
Chella S. David
John J. Marchalonis
Nicole Suciu-Foca

Symposium Sponsors

Univax Biologics, Inc
Amgen
Smith Kline

PREFACE

This volume summarizes the proceedings of the Eighth International Symposium on the Immunobiology of Proteins and Peptides which was held on November 16-20 in Rio Rico, Arizona.

The articles represent papers by invited speakers as well as papers selected by the Scientific Council, from among those submitted by the participants, on the basis of quality and timeliness. This symposium series was established in 1976 for the purpose of bringing together, once every two or three years, active investigators in the forefront of contemporary immunology, to present their findings, discuss their significance in the light of current concepts and identify important new directions of investigation. The founding of the symposium was stimulated by the achievement of major breakthroughs in the understanding of the immune recognition of proteins and peptides. We believed that these breakthroughs would lead to the creation of a new generation of peptide reagents, which could have enormous potential in biological, therapeutic, and basic applications. This anticipated explosion has since occurred and many applications of these peptides are now being realized.

The eighth symposium focused on the manipulation or modulation of the immune response. This volume broadly covers the areas of adjuvants, cytokines, vaccines, and the use of intravenous immunoglobulins for disease management. There is a clear need to identify methods for improving vaccine efficacy and guiding the host to respond with a particular type of immune response. By directing a response in a desired direction, alternative adjuvants, such as glucans or blocked copolymers, modulation of immunoregulatory pathways with cytokines, alternative vaccine delivery systems, or alternative conjugation chemistries for carrier protein-based vaccines, all hold promise for achieving more efficacious vaccines. Successful vaccine development also benefits from continued efforts to define and understand the critical regions of proteins involved in diseases such as malaria or botulism and from having available appropriate animal models, as is being developed in Lyme disease. In conjunction with the development of new vaccines, passive immunotherapy with intravenous immunoglobulin, a safe and widely used procedure, is being more fully exploited by targeting specific pathogens through active immunization of plasma donor populations.

In the area of autoimmunity, the potential of peptide reagents is clearly evident as insights into the underlying basis for pathogenesis have been revealed and new opportunities for the development of immunotherapeutics identified. In the last few years, there have been significant advances in understanding the molecular basis of several autoimmune diseases. The contact residues involved in the binding of acetylcholine to its receptor have been defined, which has provided insight into myasthenia gravis as well as indicated a potential direction for the development of peptide-based therapeutics. The relationship of band 3 membrane protein to aging and Alzheimer's disease is being dissected with an overlapping peptide strategy.

A major focus of contemporary immunology is an understanding of the roles, relationships, and interactions of antigen, T-cell receptor, and the major histocompatibility complex. State of the art technology has enabled the isolation and identification MHC-specific binding motifs with obvious implications for immunotherapy. Identification of subunit structures on the T-cell receptor itself has lead to the development of TcR peptide vaccines as a possible therapeutic approach to control T cells in rheumatoid arthritis and psoriasis. Clues to the identification of the genetic basis for susceptibility to rheumatoid arthritis are being uncovered through the use of newly developed mouse models congenic for the T-cell receptor. Additional data suggest that apoptosis may be part of a feedback mechanism for regulating T cells, which could potentially provide an avenue for modulating autoimmune activity.

Autoantibodies to the T-cell receptor are now known to exist in normal individuals. Interestingly, these autoantibodies are elevated in patients with autoimmune disease and retroviral infection. The obvious question is what is their role in pathogenesis? The hypothesis that such antibodies are concentrated in immunoglobulin-rich fractions and may beneficially effect immunoregulatory pathways upon passive transfer is being explored. In immunological privilege sites, T cells produce extracellular antigen-specific proteins. The regulatory events in these sites and the roles of T-cell products in the regulatory process are yet another area of investigation.

Symposia of this kind require a considerable amount of funding. We would like to express, on behalf of the symposium organization, our gratitude to our sponsors. In particular, we would like to thank Univax Biologics, Inc., Amgen, and Smith-Kline for their strong support. We would especially like to thank the authors for their efforts in preparing the manuscripts.

M. Zouhair Atassi
Garvin S. Bixler, Jr.

CONTENTS

SILICONE GELS AS ADJUVANTS

Effects on Humoral and Cell-Mediated Immune Responses

John O. Naim[1,2] and Carel J. van Oss[2]

[1] Department of Surgery
Rochester General Hospital
Rochester, New York
[2] Department of Microbiology
State University of New York at Buffalo
Buffalo, New York

INTRODUCTION

Since the safety of silicone gel filled mammary implants has come into question in recent years, our laboratory has begun a program of systematic animal experiments aimed at determining the effects of silicone gel (cross-linked polydimethylsiloxane {PDMS}) on the immune response. Silicone gel is a highly hydrophobic substance [1]. All hydrophilic molecules (e.g. proteins) will therefore adsorb onto the silicone gel surface when immersed in a aqueous media. Conversely, silicone gel particles, immersed in water, will adhere to all other surfaces that are similarly immersed in their immediate vicinity. Therefore silicone gel is a surface active substance and as such may behave as an immunoadjuvant.

Our first experiment at establishing silicone gels immunoadjuvant properties, investigated the humoral adjuvancy of silicone-gel taken from a sterile McGhan mammary implant (McGhan Medical, Santa Barbara, CA) [2]. Silicone-gel was mixed with Dow Corning 360 Medical Fluid (20 cs silicone oil) in a 1:1 ratio using a tissue homogenizer to facilitate mixing. Sixty, 250 gm., male Harlan Sprague Dawley rats were divided into six groups: I- phosphate buffered saline (PBS) only, II- silicone oil, III- 50% silicone-gel in silicone oil, IV- Complete Freund's Adjuvant (CFA), V- Incomplete Freund's Adjuvant (IFA), and VI- 50% silicone oil in IFA. Each adjuvant was mixed or emulsified with an equal volume of 50 μg bovine serum albumin (BSA) in 150 μl of PBS. Each immunization was given intramuscularly in a single injection. Cardiac puncture test bleeds were taken at 12, 22, 40 and 56 days post immunization. The serum was serially diluted and the anti-BSA-antibody was measured by enzyme linked immunosorbent assay (ELISA). The results revealed that BSA in PBS alone generated little antibody response. Silicone oil enhanced the antibody response to BSA slightly as compared to PBS control. However this response was very low in comparison to CFA and IFA. The most remarkable result is the significant enhancement of the antibody response to BSA by silicone-gel. The silicone-gel adjuvancy exceeded that of

Immunobiology of Proteins and Peptides VIII
Edited by M. Z. Atassi and G. S. Bixler, Jr., Plenum Press, New York, 1995

1

CFA at 12 and 22 days post immunization. This trend was continued at 40 and 56 days post immunization, although the difference then was not statistically significant. However, the silicone-gel adjuvancy significantly exceeded that of IFA at all time periods tested.

In our initial experiments we did not test for cell-mediated immunity (CMI) against BSA. Therefore we performed separate experiments in which we measured delayed type hypersensitivity (DTH) in rats that were immunized with either CFA, silicone gel or PBS mixed with BSA [3]. Our results showed that silicone gel elicited a cell-mediated response against BSA that was equivalent to CFA. However, the DTH response seen in both the CFA and silicone gel group of rats was relatively weak as evidenced by less than 50% of the rats responding in either group. From these results we tentatively concluded that silicone-gel is a weak adjuvant for cell-mediated immunity against BSA.

It was apparent from our intial results that silicone gel is a potent humoral adjuvant and as such might be able to cause untoward effects in women with silicone-gel filled mammary implants. To determine the propenstiy of silicone gels to produce autoimmune disease, our next series of experiments were designed to test whether silicone gel could induce thyroiditis and collagen induced arthritis in a rat model.

Methods and Materials

I. The Induction of Antibodies and CMI against Thyroglobulin (Tg) in Rats with Silicone gel as Adjuvant

Experimental details have been described previously [3,4]. Briefly, thirty 250 gm. Wistar female rats (Charles River, Wilmington, MA) were divided among the following treatment groups: CFA-positive control (n = 10), silicone-gel (n = 10), and phosphate buffered saline (PBS)-negative control (n = 10). Each rat was injected intradermally at the base of the tail with 2 mg of purified rat thyroglobulin (mixed with an equal volume of adjuvant) in a total volume of 50 µl. A test bleed was taken at 21 days post-immunization after which all the rats received a second booster injection equivalent to the first injection. A DTH test was performed seven days after the booster immunization according to the method of Fong [5]. Briefly, 10 µg rat Tg was injected subcutaneously in a volume of 250 µl of PBS into the subplantar tissue of the right foot of all the rats by using a 27-gauge needle. An equal volume of PBS was injected into the opposite foot of each rat and served as an internal control. Footpad thickness was measured at 6, 24, and 48 hours post injection using micrometer caliper (MTI Corp., Paramus, NJ) and the data were expressed as the difference in footpad swelling between the experimental foot and the control foot. A final test bleed was taken 9 days after the booster injection after which all the animals were sacrificed and their thyroids harvested for histopathological assessment. All the sera were serially diluted and assayed in duplicate for anti-thyroglobulin antibodies by indirect ELISA. Normal rat serum served as a negative control and was included with each assay.

II. The Induction of Antibodies and CMI against Bovine Collagen II in DA Rats with Silicone gel as Adjuvant

Thirty, 180 gm., female DA [RT^{av1}] rats (Harlan Sprague Dawley, Indianapolis, IN) were equally divided among three groups: **I** -silicone gel, **II**- phosphate buffered saline (PBS)-negative control, and **III**- Incomplete Freund's adjuvant (IFA)-positive control. A stock solution of bovine collagen II (BII) (Sigma Chemical, St. Louis, MO) was prepared in 0.1M acetic acid to a final concentration of 10 mg/ml. Each rat was injected at the base of the tail with 250 µg of BII (mixed with an equal volume of adjuvant) in 50 µl of PBS. Each animal was observed periodically for paw swelling, beginning at 7 days post-primary immunization. The silicone gel and PBS groups of rats were given a booster immunization, identical to the primary immunization, at day 45. Only the non-arthritic rats in the IFA group received a booster immunization. Each animal was test bled via tail amputation at 21, 59 and 89 days post-primary immunization. A DTH test was performed on all the rats beginning on day 18 post-primary immunization according to the method of Larsson [6]. Briefly, 25 µg of BII in 20 µl of PBS was injected intradermally into the pinna of one ear (test) and the opposite ear (control) received 20 µl of PBS. Six, twenty four and forty eight hours after the BII challenge, the change in ear thickness (test ear thickness minus the control ear thickness) was measured using a micrometer caliper (MTI Corp., Paramus, NJ). All rat sera were serially diluted and assayed in duplicate for anti-BII antibodies using an indirect ELISA. The absorbance, at 405 nm, for each well was read on a Bio-Tek EL340 microplate reader (Bio-Tek Instruments, Inc, Winooski, Vt.). The data were analyzed with Kineticalc enzyme immunoassay application software version 2.12 (Bio-Tek Instruments, Inc.) in a titer mode. The best fit for the data was determined with a 4 parameter fit, and the titer threshold was set at 0.5. Under these conditons, the serum dilution producing an optical density at 405 nm, corrected for background, of 0.5 was reported as the endpoint. All the rats were euthanised at 89 days post-primary immunization.

RESULTS

Experiment I

Table 1 shows the results of the experiment in which we tested the propensity of silicone gel to induce antibodies to rat Tg. None of the rats immunized and boosted with Tg in PBS showed anti-Tg antibody formation. These OD values were equivalent to values obtained when assaying normal rat serum for anti-Tg antibodies. All of the rats immunized with Tg mixed with CFA and boosted with Tg mixed with IFA produced significant amounts of anti-Tg antibody. All the rats immunized and boosted with rat Tg mixed with silicone gel showed an anti-Tg antibody response. However, this antibody response is nearly two orders of magnitude lower compared with the CFA treated rats. The histopathology of the thyroid glands from the CFA treated rats showed a mild to moderate degree of thyroiditis in each case. None of the thyroid glands from either the PBS or the silicone gel treated rats showed any signs of thyroiditis.

Table 2 shows the DTH test results from rats immunized with rat Tg. The data indicate that no CMI was elicted against Tg in rats that were treated with either PBS or silicone gel. The CFA group of rats showed a positive CMI response against the homologous Tg.

Table 1. Rat Antibody Response to Tg as measured by ELISA

Groups	21 Days	30 Days[a]
Silicone gel (1:100)[b] n =10	0.36±0.13[c]	0.80±0.19
PBS (1:100) n =10	0.03±0.02	0.07±0.03
CFA (1:4000) n =10	0.81±0.22	1.35±0.21

[a]Nine days after booster administration. [b]Serum dilution factor in parenthesis. [c]Mean OD ± SEM.

Table 2. DTH test against rat Tg beginning seven days post booster immunization

Groups	6 hours	24 hours	48 hours
Silicone gel n = 10	-0.03±0.09[a]	-0.02±0.10	-0.03±0.09
PBS n = 10	0.10±0.09	-0.23±0.13	-0.01±0.09
CFA n = 10	-0.13±0.13	0.46±0.18	0.20±0.06

[a]Mean difference in footpad swelling (mm) ± SEM.

Experiment II

The gross assessement of paw swelling indicated that none of the PBS treated rats showed signs of arthritis, while four rats in the silicone gel group and eight rats in the IFA group had positive signs of arthritis. Figure 1 shows the serum antibody response to BII from all rats tested. At 21 days after the primary injection a slight amount of antibody was detected in the silicone gel and IFA group of rats whereas none of the PBS treated rats had detectable antibody. After administering a booster injection to the rats in the silicone gel and PBS group, the silicone gel treated rats showed a significant rise in their titers whereas the PBS rats continued to show no detectable antibody. We chose not to administer a booster injection, at 45 days after the primary immunization, to the IFA group of rats because eight of nine rats had already developed a moderate to severe arthritis. Since we did not boost these animals their anti-BII antibody titers were barely detectable at 59 and 89 days.

Table 3 shows the DTH test results begining 18 days after the primary immunization. Notice that the ear swelling at six hours was very slight in all of the rats indicating that little anti-BII antibody had formed. By 24 and 48 hours, a clear positive DTH response was seen for all the rats in the silicone gel and IFA group, whereas none of the PBS treated rats showed a positive response. These results indicate that silicone gel is capable of inducing cell-mediated immunity against BII.

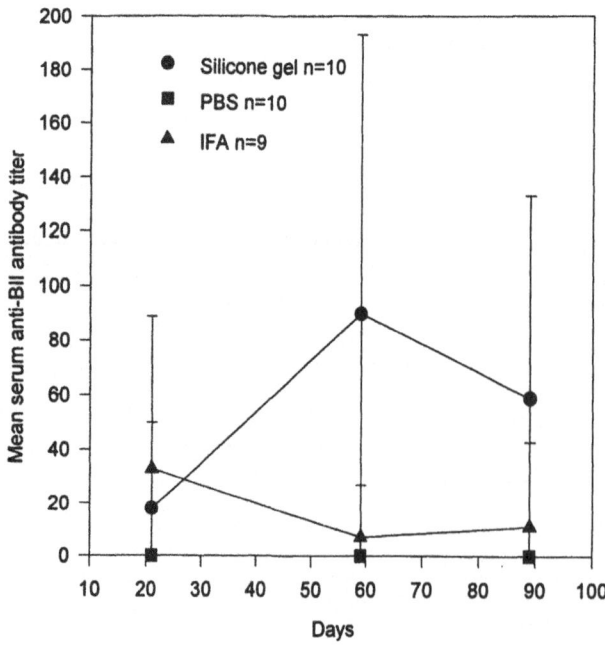

Figure 1. Mean serum antibody titer calculated at an OD of 0.5 ± SEM. Rats in the silicone gel and PBS groups were given a booster injection at 45 days post primary immunization.

Table 3. DTH test against BII beginning 18 days post primary immunization

Groups	6 hrs	24 hrs	48 hrs
Silicone gel n = 10	0.01±0.02[a]	0.29±0.05	0.30±0.06
PBS n = 10	0.02±0.01	0.03±0.01	0.02±0.01
IFA n = 9	0.07±0.02	0.45±0.07	0.45±0.05

[a]Mean difference in ear thickness (mm) ± SEM.

DISCUSSION

Current and previous studies in our laboratory have demonstrated that silicone-gel from commercial breast implants behaves as a potent humoral adjuvant and a relatively weak cellular-mediated immunity adjuvant. Silicone gel was significantly able to heighten the antibody response against both heterologous antigens (i.e., BSA and BII) and a homologous antigen (i.e., rat Tg). However, silicone gel was unable to elicit any substantial CMI against BSA and rat Tg. Although silicone gel was able to elicit CMI against BII in DA rats, previous studies have shown that silicone gel was unable to elicit CMI against BII in Lewis rats [3]. The DA rat has been shown to be highly susceptible to oil induced arthritis which may be T cell mediated [7]. Therefore the propensity of silicone gel to induce CMI against BII in the DA rat is an exception to the observed weak ability of silicone gel to induce CMI against either BII or different antigens in other rat strains.

In our prior studies the low molecular weight silicone oils have shown very little adjuvant properties [1, 9]. Silicone oils, like silicone gels, are made of the same basic chemical of PDMS. The difference between them is that silicone oils are straight chains of PDMS while silicone gels are made of lightly crosslinked PDMS yielding a a sticky cohesive mass without form [8]. During our laboratory preparation of silicone gel, the gel is sheared in a tissue homogenizer, in order to reduce its viscosity so as to be able to to inject the material through a syringe and needle. This preparation produces silicone gel particles which are suspended in a very low molecular weight silicone oil (M.W. 1600) and mixed with an antigen. These silicone gel particles most likely contribute to the overall adjuvant effect of the silicone gel by increasing the effective surface area of the gel. Recent evidence in our laboratory has demostrated that as the molecular weight of silicone oil increases a molecular size is reached whereby humoral adjuvancy is readily apparent [9]. Silicone gel, by virtue of its molecular cross-linking, should have a much higher molecular weight than most of the silicone oils tested. Thus, the humoral aduvancy of silicone oils and silicone gels appears to be dependent on molecular weight. However, we cannot rule out the possibility that other low molecular weight chemical constituents, contained within some of the silicone gels and oils, may also be contributing to their immunoadjuvancy. Further studies are currently being directed toward understanding the adjuvancy difference between silicone oils and silicone gels. The results of these studies may not only help in determining the relative safety of silicone mammary implants, but may also lead to the development of novel immunoadjuvants.

ACKNOWLEDGEMENTS

This work was supported by NIH grant No. AI-34597.

REFERENCES

1. Naim, J. O., and van Oss, C. J., 1992, The effect of hydrophilicity-hydrophobicity and solubility on the immunogenicity of some natural and synthetic polymers, *Immunol. Invest.* 21:649-662.

2. Naim, J.O., Lanzafame, R. J., and van Oss, C. J., 1993, The effect of silicone-gel on antibody formation in rats, *Immunol. Invest.* 22(2):151-161.

3. Naim, J. O., Lanzafame, R. J. , and van Oss, C. J., in press, The effect of silicone-gel on the immune response. *J. Biomaterials Science (Polymer Edition)*.

4. Naim, J. O., van Oss C. J., and Lanzafame, R. J., 1993, The induction of auto-antibodies to thyroglobulin in rats with silicone gel as adjuvant. *Surgical. Forum*, Vol. XLIV.

5. Larsson, P., Kleinau, S., Holmdahl, R., Klareskog, L., 1990,. Homologous type II collagen-induced arthritis in rats: Characterization of the disease and demonstration of clinically distinct forms of arthritis in two strains of rats after immunization with the same collagen preparation. *Arthritis Rheum.*, 33 (5):693-701.

6. Fong, T. A. T., and Mosmann, T. R., 1989, The role of IFN-γ in delayed type hpersensitivity mediated by Th1 clones, *J. Immunol.* 143:2887-2893.

7. Kleinau, S., and Klareskog, L., 1993, Oil-induced arthritis in DA rats: Passive transfer by T cells but not with serum, *J. Autoimmunity,* 6:449-458.

8. Le Vier, R.R., Harrison, M. C., Cook, R. R., and Lane, T. H., 1993, What is silicone?, Plast. Reconstr. Surg. 92:163-167.

9. Naim, J. O., Ippolito, K. M. L., Lanzafame, R. J., and van Oss, C. J., (submitted), The effect of molecular weight and gel preparation on humoral adjuvancy of silicone oils and silicone gels, *Immunol. Invest.*

THE EFFECT OF MOLECULAR WEIGHT AND GEL PREPARATION ON HUMORAL ADJUVANCY OF SILICONE OILS AND SILICONE GELS

John O. Naim[1,2] and Carel J. van Oss[2]

[1] Department of Surgery
Rochester General Hospital
Rochester, New York
[2] Department of Microbiology
State University of New York at Buffalo
Buffalo, New York

INTRODUCTION

The biocompatibilty of medical grade silicones has come into question in recent years, fueled largely by clinical case studies of women who developed unexplained connective tissue type of diseases, i.e., scleroderma, rheumatoid arthritis, polymyositis, systemic lupus erythematosus, *etc.* (1-3) after receiving silicone breast implants for either cosmetic or reconstructive reasons. Recently, our laboratory has shown that silicone gel taken from commercial breast implants possesses humoral adjuvant properties similar to Freund's complete adjuvant (CFA) and is capable of promoting auto-antibodies in a rat model (4-6). Therefore, it is plausible that silicone gel also may be capable of mediating autoimmune disease in humans. On the other hand, our studies indicate that low molecular weight (20 centistoke [cs]) silicone oil (polydimethylsiloxane, PDMS) possesses hardly any adjuvant properties.

Silicone oils are straight chain PDMS of varying molecular weights (7). One physical property that may contribute to the adjuvancy of silicones could be their molecular weight. As the molecular weight of silicone oil increases, the overall size of the molecule increases, which may enhance the adjuvant properties of these materials.

Silicone gel, contained within the outer silicone elastomer shell of a breast implant, is composed of lightly cross-linked PDMS. This polymer network is filled with PDMS fluid (oil) as the liquid phase to yield a sticky, cohesive mass without form (7). During the laboratory preparation of silicone gel, the gel is sheared in a tissue homogenizer, in order to reduce its viscosity so as to be able to inject the material through a syringe and needle. This preparation produces silicone gel particles which are suspended in 20 cs PDMS oil and mixed

Immunobiology of Proteins and Peptides VIII
Edited by M. Z. Atassi and G. S. Bixler, Jr., Plenum Press, New York, 1995

with an antigen. These silicone gel particles most likely contribute to the overall adjuvant effect of the silicone gel by increasing the effective surface area of the gel. However, it is not known to what extent the shearing process influences the adjuvancy of the silicone gel.

This investigation was undertaken to determine the humoral adjuvancy of Dow Corning 360 silicone oils of increasing molecular weight and also aims to determine the effect of the degree of shearing of the silicone gel on its humoral adjuvancy, by using bovine serum albumin (BSA) as the test antigen in rats.

Methods and Materials

Dow Corning Medical 360 Fluids of the following molecular weights were obtained from Dow Corning (Midland, MI): 100 cs (M.W. 5,000), 350 cs (M.W. 10,000), 1000 cs (M.W. 16,500), and 12,500 cs (M.W. 60,000). The very low molecular weight (M.W. 296) octamethylcyclotetrasiloxane (D4) was also obtained from Dow Corning (Midland, MI). The cyclic D4 has been shown by Dow Corning to possess adjuvant properties (8). The D4 compound is the precursor material for silicone oil, gel and elastomer and is present in trace quanities in all three silicone materials (8). Silicone gel, from an unused silicone gel breast implant which was obtained from Dow Corning (Midland, MI.) was prepared as described previously (4).

I. Study of Silicone Oil Adjuvancy as a Function of M.W.

Sixty four, 250 gram, male and female Harlan Sprague Dawley rats (Indianapolis, IN), were equally distributed into eight groups: Group I -Phosphate buffered saline (PBS) only, Group II- D4, Group III- 100 cs oil, Group IV- 350 cs oil, Group V- 1000 cs oil, GroupVI- 12,500 cs oil, Group VII- 50% silicone gel in 20 cs silicone oil, and Group VIII- complete Freund's adjuvant (CFA). Each adjuvant (including PBS) was mixed or emulsified with an equal volume of 50 µg of BSA in 100 µl of PBS. Each immunization was done intramuscularly, in the left thigh, in a single injection. A booster immunization was given intramuscularly, in the right thigh, 71 days post immunization; this consisted of the same antigen and adjuvant dose, except that the CFA group received incomplete Freund's adjuvant. The rats were test bled via tail amputation at 14, 29, 49, 71, 79 and 98 days post immuniza- tion. The blood was allowed to clot and the serum was separated and stored at -70°C until used. After 98 days all of the rats were sacrificed by barbiturate overdose.

Anti-BSA antibodies were measured by the indirect enzyme linked immunosorbent assay (ELISA) as previously described (4). The data were analyzed with Kineticalc enzyme immunoassay application software version 2.12 (Bio-Tek Instruments, Inc.) in a titer mode. The best fit for the data was determined with a 4-parameter fit, and the titer threshold was set at 0.5. Under these conditons, the serum dilution producing an optical density of 0.5 at 405 nm, corrected for background, was reported as the endpoint. A titer of zero was assigned to samples that gave O.D. readings equivalent to naive rat control serum.

II. Silicone Gel Shearing Study

Silicone gel was removed from an unused Dow Corning silicone gel breast implant (Dow Corning, Midland, MI) by drawing the gel through the barrel end a sterile 60 cc syringe by placing a gentle vaccum on the hub of the syringe. An equal volume of 20 cs silicone oil was added to the gel. Using a Brinkman tissue homogenizer (Westbury, NY) fitted with a sterile sawtooth probe, three separate sheared silicone gel preparations were prepared as follows: **Prep. 1**- homogenizer speed was set at 7 for two, 25 sec. cycles, followed by three,

25 sec. cycles and one, 20 sec. cycle at speed 5. **Prep. 2**- homogenizer speed was set at 4 for one, 50 sec. cycle; the speed was then set at 5 for one, 30 sec. cycle followed by two, 15 sec. cycles at speed 8. **Prep. 3**- homogenizer speed was set at 8, for six, 15 sec. cycles. During the preparation the silicone gel was kept on ice constantly. Based on visual observation, Prep. 1 was the most viscous followed by Prep. 2 and the least viscous was Prep. 3 which was prepared similarly as the silicone gel in experiment I.

Twenty eight 250 gram female Harlan Sprague Dawley rats were divided into the following groups: I- Prep. 1, II- Prep. 2, III- Prep. 3 and IV- Silicone gel from experiment I. Each preparation of silicone gel was mixed with an equal volume of 50 μg of BSA in 150 μl of PBS. Each immunization was done intramuscularly in a single injection. The animals were test bled via tail amputation at 18, 40, 61, and 104 days post immunization. The blood was treated and the serum assayed for anti-BSA antibodies as described previously in experiment I. After 104 days all of the rats were sacrificed by barbiturate overdose.

RESULTS

Experiment I

Figure 1A shows the antibody response to BSA when mixed with various molecular weight silicone oils. No antibody response difference was observed between male and female rats, thus the antibody data for both sexes were pooled together for each group. Little antibody response was detected at 14 days post immunization in rats from any of the test groups. Serum samples at 29, 49, and 71 days post immunization from all the rats injected

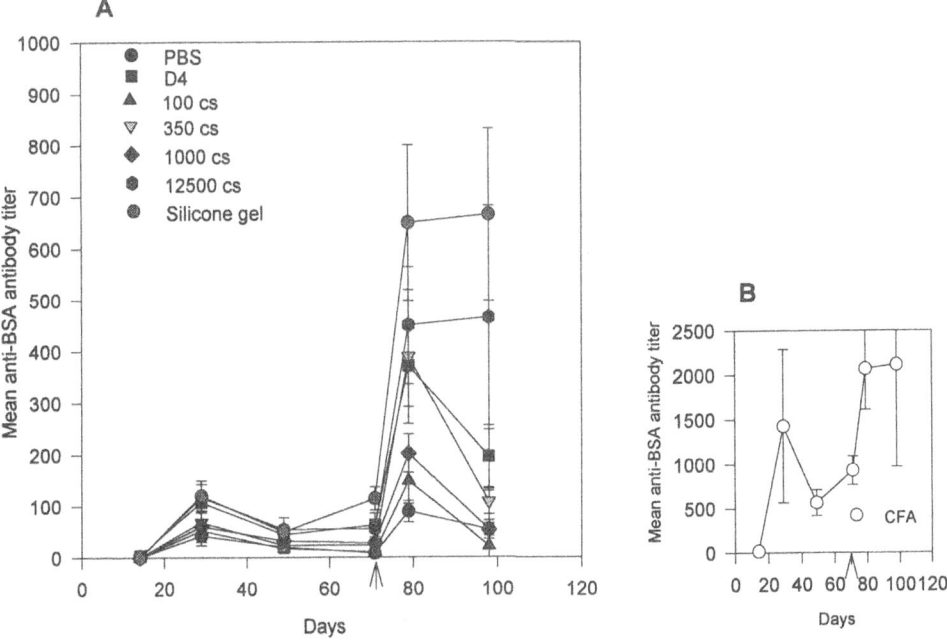

Figure 1. A. Rat antibody response to BSA, as a function of the M.W. of silicone oil. B. Rat antibody response to BSA mixed with CFA. The data are expressed as the mean serum titer at 0.5 OD ±SEM, n = 8. The arrows indicate booster immunization.

with the silicone compounds showed minimal antibody response as compared to the CFA group of rats (Figure 1B). These samples still had anti-BSA titers that were slighly higher than the PBS control rats. The silicone gel and the 12,500 cs silicone oil groups of rats showed a marginally higher antibody response as compared to the 100 cs, 350 cs and 1000 cs groups of rats. However, after boosting this increase in antibody response became significant at 79 and 98 days post immunization. The D4 compound elicited a moderate antibody response as compared to the PBS control.

Experiment II

Figure 2 presents the antibody response to BSA when mixed with different silicone gel preparations. There is no significant difference in antibody response between the silicone gel preparations.

Discussion

The results these experiments suggest that there is a correlation between the molecular weight of silicone oil and silicone gel and their potency as humoral adjuvants. The humoral adjuvancy these various silicone gel preparations did not appear to be dependent on the method of shearing.

Although all of the silicone gel preparations showed a substantially lower humoral adjuvancy response as compared with CFA in these experiments, the silicone gel consistently produced a significantly heightened antibody response when compared with PBS and the lower molecular weight silicone oils i.e., 100 cs, 350 cs, 1000 cs oils and 12,500 cs. In experiment I, silicone gel (after boosting) produced a mean antibody response statistically comparable to CFA at 98 days post immunization. The low molecular weight silicone oils (M.W. <16,500) showed little humoral adjuvancy, while the silicone oil with a molecular weight of 60,000 showed a moderate humoral adjuvant effect. Therefore, it is reasonable to

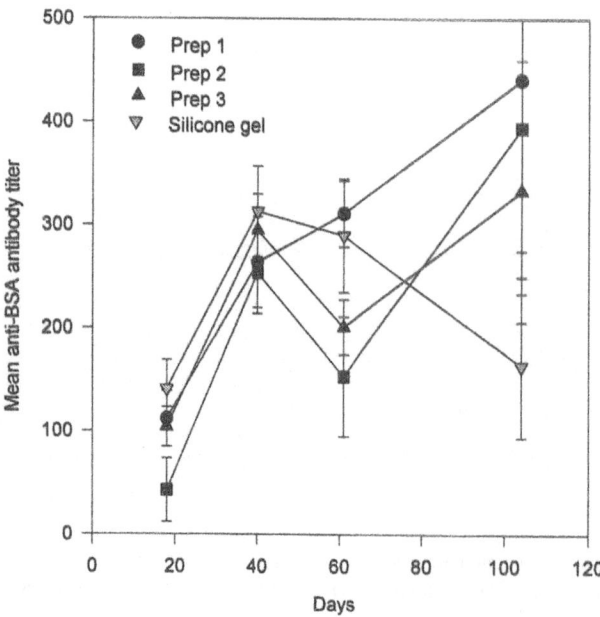

Figure 2. Rat antibody response to BSA, as a function of shear stress applied to silicone gel. The data are expressed as the mean serum titer at 0.5 OD ±SEM, n = 7.

conclude that as the linear chain of PDMS becomes longer, a molecular size is reached whereby the immunological adjuvancy can be readily demonstrated. Because our studies tested only four commercially available silicone oils, we cannot determine precisely at which molecular weight a significant adjuvant effect is seen, but we can situate it between 16,500 and 60,000.

The lowest molecular weight cyclic siloxane material tested (i.e., D4) proved to have a humoral adjuvant effect as has been previously reported (8). However, in contrast to the linear silicone oils, the histopathology of the cyclic D4 injection site showed an intense acute inflammatory response which was still active at 98 days post immunization (9). This acute inflammatory response is no doubt contributing to the humoral adjuvant effect as evidenced by a moderate antibody response to BSA. The D4 compound, by eliciting acute inflammation, is causing significant tissue damage, which is a response that has never been observed for silicone oils and silicone gels. Thus, the humoral adjuvant effect of the D4 compound appears not to be dependent on molecular weight but rather is dependent on its ability to create and maintain an acute inflammatory response.

The process of shearing the silicone gel produces particles of silicone gel suspended in silicone oil. If the particle size of the silicone gel is a critical factor in conferring adjuvant properties to the gel, then testing various particle sizes of silicone gel for humoral adjuvancy would reveal at least some quantitative antibody differences between the different silicone gel particle sizes. Our study crudely prepared three different silicone gel samples which were visually different in terms of their viscosity and thus presumably in their particle sizes. Our results showed no difference in the humoral adjuvancy of these three silicone gel preparations. It is possible that in addition to molecular weight and particle size there may be other chemical constituents within the silicone gel that may also contribute to its humoral adjuvancy, but these are unlikely to be changed by shearing.

In conclusion, our results show that the humoral adjuvancy of silicone oil is dependent on molecular weight and that differential shearing of the silicone gel does not alter its humoral adjuvancy.

ACKNOWLEDGMENTS

This work was supported by NIH grant No. AI-34597

REFERENCES

1. K.C. Shestak, R.J. Restifo, M.J. Wheatley, and D.J. Smith, 1992,The Silicone-gel implant controversy in: *Perspectives in General Surgery*, vol. 3, p. 26, B.A. Levine (Ed.). Quality Medical Publishing, St.Louis, MO.
2. Sergott, T. J., Limoli, J.P., Baldwin jr, C.M., and Laub, D.R., 1986, Human adjuvant disease, possible autoimmune disease after silicone implantation: a review of the literature case studies, and speculation for the future. *Plast. Recontr. Surg.*, 78:104-114.
3. Spiera, H., 1988, Scleroderma after silicone augmentation mammoplasty. *JAMA*, 260:236-238
4. Naim, J.O., Lanzafame, R. J., and van Oss, C. J., 1993, The effect of silicone-gel on antibody formation in rats, *Immunol. Invest.* 22(2):151-161.
5. Naim, J. O., van Oss C. J., and Lanzafame, R. J., 1993, The induction of auto-antibodies to thyroglobulin in rats with silicone gel as adjuvant. *Surgical. Forum*, Vol. XLIV.
6. Naim, J. O., Lanzafame, R. J. , and van Oss, C. J., in press, The effect of silicone-gel on the immune response. *J. Biomaterials Science (Polymer Edition)*.
7. Le Vier, R.R., Harrison, M. C., Cook, R. R., and Lane, T. H., 1993, What is silicone?, *Plast. Reconstr. Surg.* 92:163-167.

8. General and Plastic Surgery Devices FDA Panel Meeting, Bethesda, MD, Feb. 18-20, 1992.
9. Naim, J. O., Ippolito, K. M. L., Lanzafame, R. J., and van Oss, C. J., (submitted), The effect of molecular weight and gel preparation on humoral adjuvancy of silicone oils and silicone gels, *Immunol. Invest.*

3

GLUCANS AS IMMUNOLOGICAL ADJUVANTS

N. Mohagheghpour,[1] M. Dawson,[1] P. Hobbs,[1] A. Judd, R. Winant,[1]
L. Dousman,[1] N. Waldeck,[1] L. Hokama,[1] D. Tusé,[1] F. Kos,[2] C. Benike,[2]
and E. Engleman[2]

[1] Life Sciences Division
SRI International
Menlo Park, California 94025-3493
[2] Stanford Medical School Blood Center
Stanford University School of Medicine
Stanford, California

BACKGROUND

β-1,3-Linked glucopyranose (β-glucan), a major structural component of yeast, fungi, and algae [1], has a wide range of biological activities. Systemic administration of β-glucan (1) provides nonspecific resistance in experimental animals against a variety of pathogenic challenges [2–7]; (2) prolongs the survival time of tumor-bearing animals [8,9]; (3) enhances bone marrow recovery and survival of lethally irradiated mice [10]; (4) promotes wound healing when applied topically [11,12]; and (5) displays an adjuvant effect when coadministered with either bacterial, fungal, protozoan, or viral agents [13–18].

β-Glucan exerts its immunostimulatory activities by targeting macrophages [19]. *In vitro* studies have demonstrated that human and murine monocytes/macrophages express cell surface receptors specific for the β-(1,3)-linked-D-glucopyranosyl oligosaccharides [20]. These receptors, which are functionally distinct from fibronectin, F_c receptors for IgG, and complement type 1 and complement type 3 receptors [21–22], mediate the phagocytosis of β-glucan particles in the absence of opsonic proteins [23] and mediate the production of proinflammatory factors [24–29].

Using either β-glucan particles or particulate β-glucan conjugated to viral protein, we have examined the ability of β-glucan to enhance humoral and cell-mediated immune responses to viral proteins. The viral proteins used in these investigations were herpes simplex virus (HSV) glycoprotein D (gD2), kindly provided by Dr. Rae Lyn Burke of Chiron Corp., Emeryville, CA; recombinant-derived hepatitis B surface antigen (HBsAg) (Cortex Biochem, San Leandro, CA); HIV-1 SF2 gp120 (NIH AIDS Research and Reference Reagent Program); and an HIV-1 envelope-encoded synthetic peptide (amino acid residues 254–274). Administration of a protein–β-glucan conjugate promotes the uptake

Immunobiology of Proteins and Peptides VIII
Edited by M. Z. Atassi and G. S. Bixler, Jr., Plenum Press, New York, 1995

13

of immunogen by the β-glucan-receptor–bearing macrophages, which are the antigen-presenting cells.

Our studies in mice and rabbits demonstrated that coadministering viral protein with β-glucan produces immune responses of a higher magnitude than those elicited by the immunogens alone. In our preliminary experiments in mice, antibody titers in animals injected with the gD2–β-glucan conjugate (the immunoconjugate) were significantly higher than those in animals immunized by the coadministration of viral protein and particulate β-glucan.

PRECLINICAL STUDIES

Algal β-Glucan (AG)

We isolated glucan from the cytoplasm of an adapted strain of *Euglena gracilis* (SRI Strain D86-G). *E. gracilis* was grown heterophically in darkness by the method described by Tusé et al. [30]. Glucan constitutes as much as 50% of the dry cell weight of cells cultivated under these conditions. Following the collection of the organisms and two 30- to 60-min methanol-chloroform extractions to remove cellular debris, the algal glucan (AG) was depyrogenized in hot 1 N HCl. The depyrogenized material was washed extensively with pyrogen-free water. The remaining solid material (AG) consisted of a pure, nonpyrogenic linear β-1,3-glucan, which at the doses tested (5 mg/kg) was well tolerated by mice and rabbits. Analysis of the elemental composition of several batches of particulate Ag has shown that the purification procedure results in preparations which are consistent in composition (Table 1).

Adjuvant Activity of Algal β-Glucan

Five-week-old female mice were immunized by intraperitoneal (i.p.) injection of 5 μg of HSV gD2 and the predetermined optimal dose of AG (100 μg/dose) on Days 0 and 14. Control animals received identical doses of either immunogen alone, immunogen adjuvanted with 200 μg alum (aluminum hydroxide gel, 2% Al_2O_3), or 0.2 ml of pyrogen-free phosphate-buffered saline (PBS). Anti-HSV response was evaluated by measuring serum antibody levels by enzyme-linked immunosorbent assay (ELISA) and gD2-induced T cell response by incorporation of [^{3}H]thymidine and production of interleukin 2 (IL-2).

BALB/c mice immunized by coadministration of gD2 and AG produced antibody levels significantly higher than those found in mice injected with gD2 alone, in both titer (p < 0.01) (Fig. 1) and persistence (p < 0.05) (Fig. 2).

The ability of β-glucan to enhance humoral response was also evaluated in rabbits. Rabbits immunized by coadministration of AG and an HIV-1 envelope-encoded synthetic peptide (amino acid residues 254–274) conjugated to bovine serum albumin (BSA) produced

Table 1. Algal glucan profile

Structure	β-1,3-Glucan (as a crystalline long-chain triple helix)
Molecular weight	Highest measured MW 500,000 (determined by light scattering)
Particle size	Median particle size 3.7 to 4.6 μm [by centrifugal (laser) particle analyzer]
Specific gravity	1.86 ± 0.02 to 2.0 ± 0.03 g/cm^3 (by helium pycnometer)
Purity	0.0001–0.35% phosphorus (by plasma emission spectroscopy)
	0.12–0.27% nitrogen (by Kjedahl)
Pyrogenicity	Negative (by USP rabbit pyrogen test)

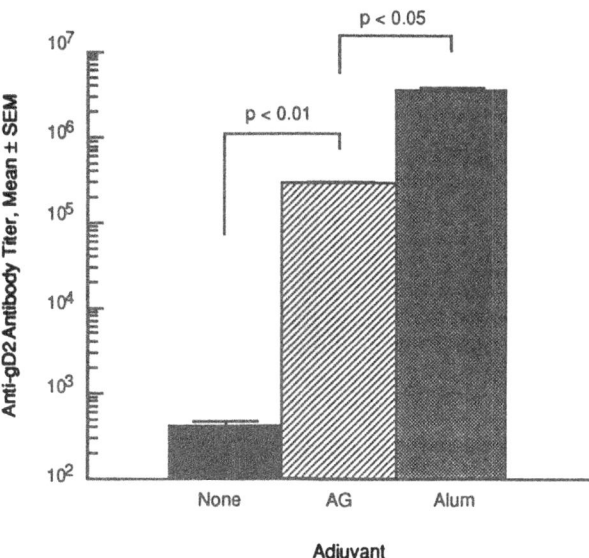

Figure 1. Anti-gD2 antibody response of BALB/c mice immunized by coadministration with linear β-glucan. Anti-gD2 IgG antibody titers in two groups (total of nine BALB/c mice), 14 days after the last injection were measured by ELISA. Titers are expressed as the reciprocal of the serum dilution in which the $A_{490} \geq 0.1$. The anti-gD2 antibody titer of the placebo group injected with PBS was $< 10^{-1}$. Data were evaluated by Student's t-test.

specific antibodies at higher titers than the response exhibited by rabbits injected with the immunogen alone (Fig. 3).

The gD2-specific T cell proliferation (Fig. 4, Panel A) and IL-2 production (data not shown) induced by immunization of BALB/c mice with gD2 adjuvanted with AG were greater than the responses exhibited by animals immunized with gD2 alone. In agreement with these findings in the HSV model, the coadministration of HIV-1 gp120 and AG heightened the antigen-specific response of splenic lymphocytes from C3H/HeJ mice (Fig. 4, Panel B) without affecting the mitogen-induced proliferation response (data not shown). We used C3H/JeH mice to avoid the possible immunoenhancing effect of endotoxin in the preparations. Because of a genetic defect, the response of these mice to lipopolysaccharide (LPS) is markedly reduced [31]. However, C3H/HeJ mice respond normally to particulate glucan [32].

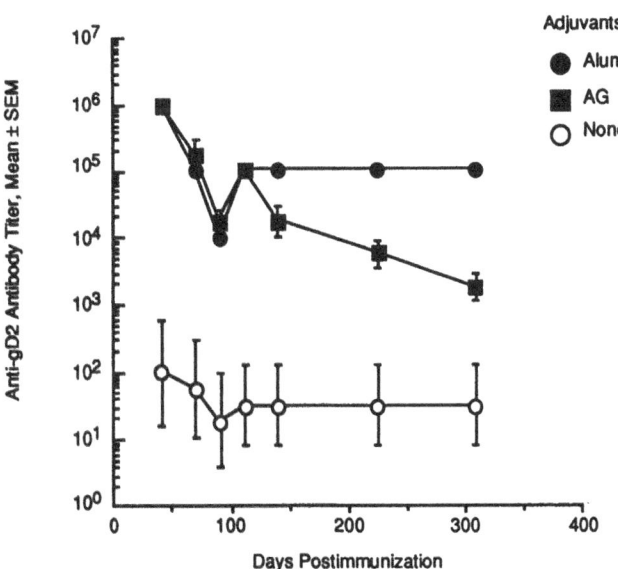

Figure 2. Persistence of antibody response. BALB/c mice were immunized by three biweekly i.p. injections of gD2 and indicated adjuvants. Anti-gD2 IgG antibody titers of sera collected at various intervals were measured by ELISA.

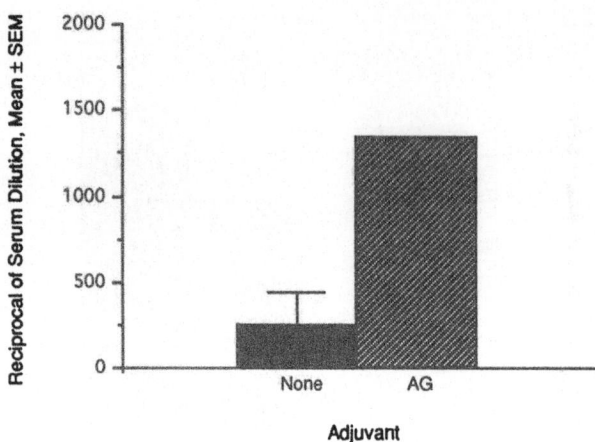

Figure 3. Response to HIV-1 enve-lope-encoded synthetic peptide. Rab-bits were immunized with 1 mg of an HIV-1 envelope-encoded synthetic peptide (amino acid residues 254–274) conjugated to BSA (0.4 mg pep-tide/0.6 mg BSA) mixed with 5 mg AG/kg body weight on Days 0, 14, and 28. Control rabbits were injected with HIV-1 peptide–BSA conjugate alone. Antibody titers in sera col-lected two weeks after the last injec-tion were determined by ELISA using HIV-1 peptide conjugated to keyhole limpet hemocyanin (KLH) as the im-mobilized antigen.

The ability of linear algal glucan to enhance humoral immune responses was also compared to the adjuvant activity of the branched β-glucan, Pleuran. This glucan is a β-1,3-linked glucose polymer having β-D-glucosyl side chains attached by alternating β(1,6) or β(1,4) bonds at the *O*-6 position of every anhydroglucose unit. Pleuran was isolated from the fruit-body of the oyster fungus *Pleurotus ostreatus* [33,34] by alkali extraction at 95–100°C, followed by bleaching with sodium chlorite (pH 3.5–4.5) at 50–60°C. The bleached products were washed in water, dehydrated in organic solvent, and dried by vacuum at 60°C.

In this preliminary study, C3H/HeJ mice immunized with HBsAg mixed with 100 μg of Pleuran produced anti-HBsAg antibodies at titers comparable to those in animals injected with HBsAg adjuvanted with alum (9.11×10^3 and 8.96×10^3, respectively) (Fig. 5). C3H/HeJ mice immunized with HBsAg adjuvanted with AG produced antibody titers that were significantly higher in magnitude than the responses elicited by alum (p < 0.05) (Fig. 5). The difference in the mean antibody titer of mice injected with Pleuran and those

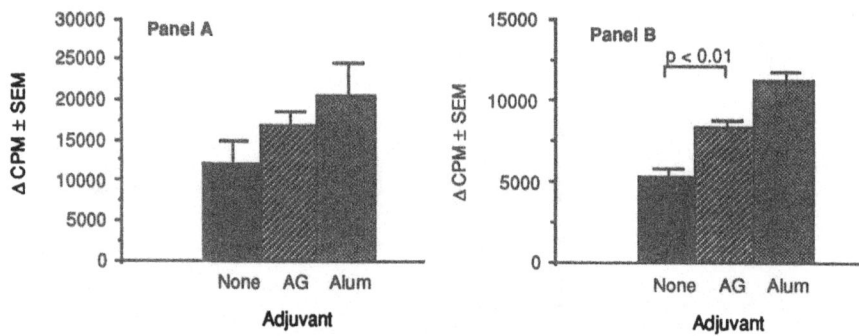

Figure 4. Antigen driven T cell response. BALB/c mice (nine per group) were immunized by two biweekly i.p. injections of HSV gD2 (5 μg/dose) and indicated adjuvants (Panel A). Control animals received PBS. C3H/HeJ mice, five per group, were immunized with two biweekly i.p. injections of HIV-1 gp120 (1 μg/dose) and indicated adjuvants (Panel B). Control groups received either PBS, alum, or AG. Uptake of [³H]thymidine by 2×10^5 splenic lymphocytes cultured in quadruplicate for 72 h in the absence or presence of either 20 μg/ml HSV gD2 (Panel A) or 2.5 μg/ml HIV-1 gp120 (Panel B) was measured during the last 6 h. Data are presented as mean counts per minute (cpm) of stimulated cultures minus the cpm of equivalent control cultures without stimulus (Δcpm). Data were evaluated by Student's *t*-test.

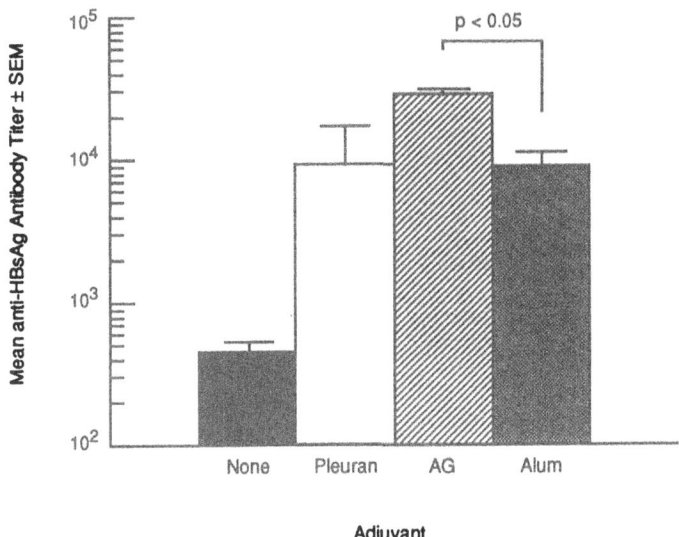

Figure 5. Anti-HBsAg antibody response of C3H/HeJ mice immunized by coadministration with β-glucan. Groups of five mice were immunized intraperitonally with 5 μg HBsAg mixed with 100 μg of each glucan preparation on Days 0 and 14. Control mice were injected with either HBsAg alone or HBsAg adsorbed onto 200 μg of alum. Antibody titers in sera collected two weeks after the last injection were determined by a commercially available radioimmunoassay (RIA) (Ausab assay, Abbott Laboratories, North Chicago, IL). Data obtained from the RIA analysis of individual serum samples were used to calculate the anti-HBsAg antibody titer (estimated Ausab units) for each mouse. The Microsoft Excel program MISMACRO was used for calculating antibody titers. Preimmunization sera from these animals did not contain detectable anti-HBsAg antibody activity. Data were evaluated by Student's t-test.

injected with a comparable dose of algal glucan (2.79×10^4) was not statistically significant because of the large variation in the response of individual animals in the Pleuran group.

In summary, although only a small number of mice were tested, animals immunized with HBsAg adjuvanted with 100 μg of either linear or branched β-glucan produced antibody titers comparable in magnitude to the responses elicited by alum.

Enhancement of the Immune Response by gD2–β-Glucan Conjugation

The efficiency of antigen capture and presentation by "professional" antigen-presenting cells (e.g., macrophages) is enhanced if the antigen can bind to and be internalized via a cell surface receptor [35]. We exploited β-glucan receptors to target antigens to mouse macrophages. Because immunoconjugates taken up by macrophages are delivered to the lysosomes [36], it is reasonable to take advantage of the acidic environment of the lysosome to achieve site-selective release of antigen from an antigen-glucan conjugate. In these studies, we used an acid-labile hydrazone linkage to tether gD2 to β-glucan [37] (Fig. 6). AG was first functionalized with masked aldehyde groups by a two-step procedure: The hydroxyl groups were first activated with 1,1'-carbonyldiimidazole in dioxane to form the imidazolyl carbamate derivative. This derivative, after overnight reaction with 1.3 M 4-aminobutyraldehyde diethyl acetal at pH 10.2, produced *O*-carboxamidobutyraldehyde diethyl acetal glucan. A 2-h treatment with excess 2.1 M ethanolamine at pH 10.2 was used to remove unreacted imidazolyl carbamate groups.

The hydrazide derivative of gD2 was prepared by a stepwise procedure involving thiolation followed by reaction with a linker bearing a hydrazide group and a maleimide

Synthesis of gD2-glucan

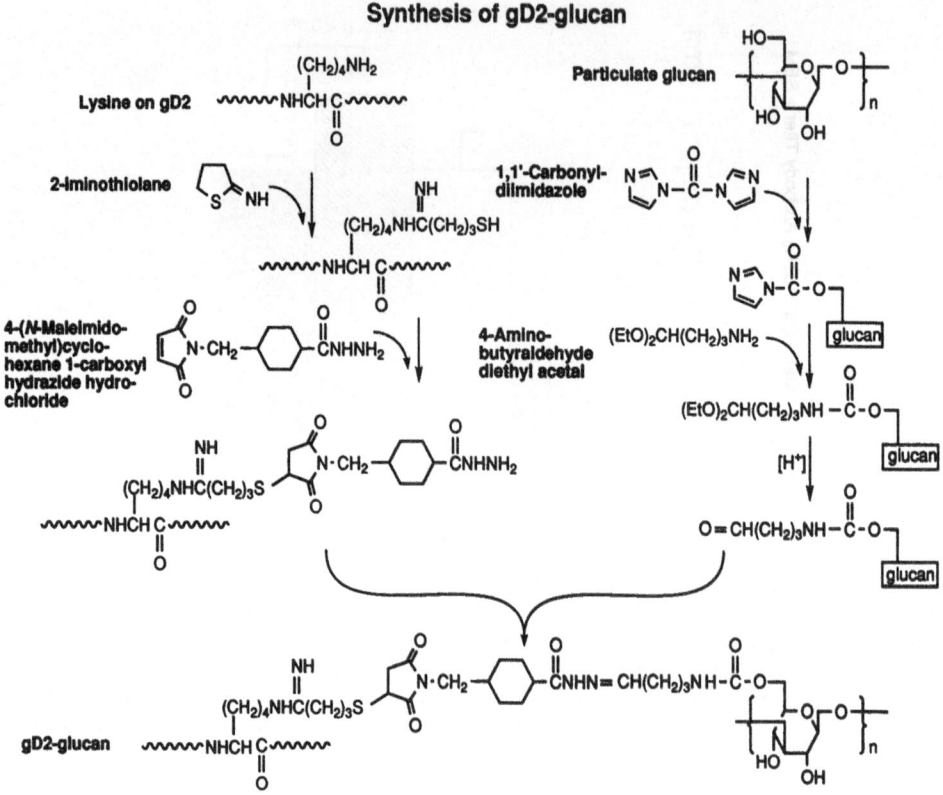

Figure 6. Synthesis of gD2–β-glucan.

group. Briefly, gD2 was thiolated with 4 mM 2-iminothiolane at pH 7.8. Subsequently, 4-(*N*-maleimidomethyl)cyclohexane 1-carboxyl hydrazide hydrochloride (2.3 mM) was added at pH 5.6 for introduction of the carboxyl hydrazide group, which occurred by a Michael reaction of the maleimide group of this reagent with the thiol groups that had been introduced onto gD2.

Just before conjugation, the aldehyde groups (0.12 CHO/mol of glucose) of the particulate *O*-carboxamidobutyraldehyde diethyl acetal glucan suspension were unmasked by hydrolysis of the acetal groups by washing with 1 M hydrochloric acid. Conjugation of the hydrazide derivative of gD2 (2.8 mg) to the *O*-carboxamidobutyraldehyde (CAB) derivative of AG (10 mg) was conducted at pH 5.5 at ambient temperature. Conjugate formation was monitored by periodically removing aliquots of the particles, washing them with PBS to remove untethered derivatized protein, and then determining the protein and glucan contents of the particles by using the bicinchoninic acid method [38] and anthrone reaction [39], respectively. When protein loading was sufficient (≥ 5 μg per 100 μg AG), particles were exhaustively washed with PBS until no protein was detected in the wash by sodium dodecyl sulfate-polyacrylamide gel electrophoresis (SDS-PAGE) and silver staining.

The acid-lability of the hydrazone bond was determined by incubating the gD2-AG conjugate for 7 h at 37°C at both pH 5.0 and 7.2. The release of protein from the conjugate was 4.8-fold greater at pH 5.0 than at pH 7.2 (data not shown).

The gD2-AG conjugate was highly stable on storage. Less than 0.1% of the gD2 was released from the conjugate after 14 days at -80°C, as determined by SDS-PAGE and silver staining (data not shown).

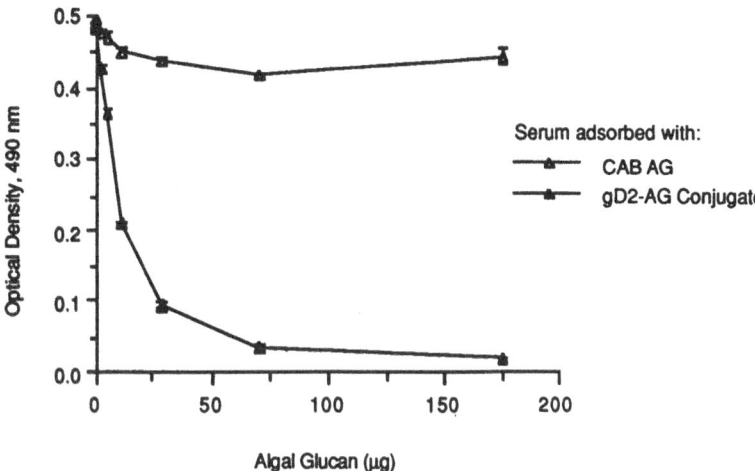

Figure 7. Adsorption of anti-gD2 antibodies with gD2-AG conjugate. Pooled serum (1:20 dilution) from three C3H/HeJ mice immunized twice with 8.4 μg of gD2 mixed with 100 μg of AG was incubated at ambient temperature with increasing quantities of either gD2-AG conjugate (8.4 μg of gD2 per 100 μg AG) or CAB AG. After a 2-h incubation, serum removed by centrifugation was tested for anti-gD2 activity by ELISA against gD2.

The conjugation chemistry and storage at -80°C did not appear to alter the antigenicity of gD2 because the gD2-AG conjugate adsorbed comparable levels (> 90%) of anti-gD2 antibody from the sera of mice immunized either by coadministration of gD2 and AG (Fig. 7) or by gD2-AG conjugate (data not shown).

Adjuvant Activity of gD2-AG Conjugate

Female C3H/HeJ mice (five per group) were immunized on Days 0 and 14 by i.p. injection of 8.4 μg of gD2 conjugated to 100 μg of Ag. Control mice received identical doses of gD2 (8.4 μg/dose) either alone or mixed with 100 μg of Ag. The placebo group received 0.2 ml of PBS. Serum anti-gD2 antibody titers were measured 14 days after the final injection.

Mice immunized with the gD2-AG conjugate had significantly ($p < 0.05$) higher anti-gD2 serum antibody titers than mice immunized by coadministration of gD2 and AG (Fig. 8).

CONCLUSIONS

In summary, these findings confirm the usefulness of β-glucan as an adjuvant and highlight the potential application of β-glucan for delivering an immunogen to macrophages. These observations in conjunction with previously cited studies on the use of glucan for treatment of humans [40–44] indicate the feasibility of administering glucan to humans without toxicity and make glucan a potential adjuvant for immunoprophylactic trials.

ACKNOWLEDGMENTS

This work was supported by NIH Grant AI30939 (N.M.)1.

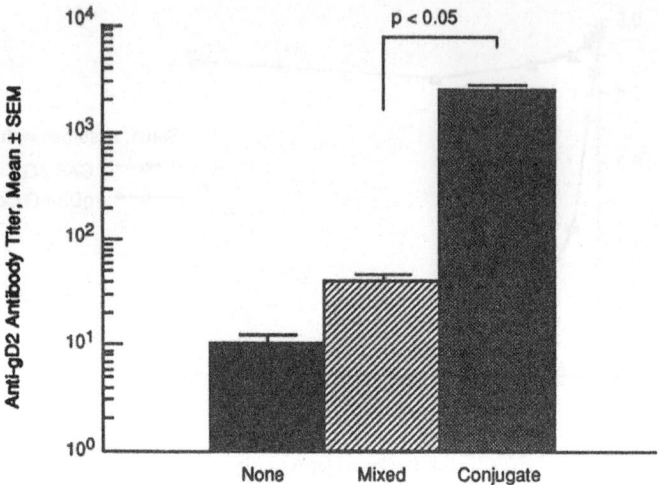

Administration of Algal Glucan with gD2

Figure 8. Anti-gD2 antibody response of mice immunized with gD2-Ag conjugate. Titers are expressed as the mean ± SEM of the reciprocal serum dilution in which the $A_{490} \geq 0.1$. The anti-gD2 antibody titer of the placebo group injected with PBS was < 10^{-1}. Data were evaluated by Student's t-test.

REFERENCES

1. S. Bartnicki-Garcia. Cell wall chemistry; morphogenesis and taxonomy of fungi. *Annu. Rev. Microbiol.* 22:87–108, 1988.

2. J. Delville and P. J. Jacques. Therapeutic effects of yeast glucan in mice infected with *Mycobacterium leprae. Arch. Int. Phys. Biochem.* 85:965–966, 1977.

3. P. L. Kokoshis, D. L. Williams, J. A. Cook, and N. R. DiLuzio. Increased resistance to *Staphylococcus aureus* infection and enhancement of serum lysozyme activity by glucan. *Science* 199:1340–1342, 1978.

4. D. L. Williams, J. A Cook, E. O. Hoffman, and N. R. DiLuzio. Protective effect of glucan in experimentally induced candidiasis. *J. Reticuloendothel. Soc.* 23:479–490, 1978.

5. J. A. Cook, T. W. Hobrook, and B. W. Parker. Visceral leishmaniasis in mice: Protective effect of glucan. *J. Reticuloendothel. Soc.* 27:567–573, 1980.

6. D. L. Williams, H. A. Pretus, R. B. McNamee, E. L. Jones, H. E. Ensley, I. W. Broder, and N. R. DiLuzio. Development, physicochemical characterization and preclinical efficacy evaluation of water-soluble glucan sulfate derived from *Saccharomyces cerevisiae. Immunopharmacology* 22:139–155, 1991.

7. S. James, D. D. Easson, Jr., G. R. Ostroff, and A. B. Onderdonk. PGG-Glucans: Novel class of macrophage activating immunomodulators. In *Polymeric Drugs and Drug Delivery Systems* (R. L. Dunn and R. M. Ottenbrite, Eds.). American Chemical Society, Washington, D.C., 1991, pp. 44–51.

8. N. R. DiLuzio, R. McNamee, W. Browder, and D. Williams. Glucan-inhibition of tumor growth and enhancement of survival in four syngeneic murine tumor models. *Cancer Treat. Rep.* 62:1857–1866, 1978.

9. K. Gomaa, J. Kraus, F. Robkoph, H. Röper, and Z. C. Fran. Antitumour and immunological activity of a β 1→3/1→6 glucan from *Glomerella cingulata. J. Cancer Res. Clin. Oncol.* 118:136–140, 1992.

10. M. L. Patchen and T. J. MacVittie. Stimulated hemopoiesis and enhanced survival following glucan treatment in sublethally and lethally irradiated mice. *Int. J. Immunopharmacol.* 7:923–932, 1985.

11. W. Browder, D. Williams, P. Lucore, H. Pretus, E. Jones, and R. McNamee. Effect of enhanced macrophage function on early wound healing. *Surgery* 4:224–230, 1988.

12. S. J. Leibovich and D. Danon. Promotion of wound repair in mice by application of glucan. *J. Reticuloendothel. Soc.* 27:1–11, 1980.

13. J. A. Reynolds, M. D. Kastello, D. G. Harrington, C. L. Crabbs, C. J. Peters, J. V. Jemski, G. H. Scott, and N. R. DiLuzio. Glucan-induced enhancement of host resistance to selected infectious diseases. *Infect. Immun.* 30:51–57, 1980.

14. T. W. Holbrook, J. A. Cook, and B. W. Parker. Glucan-enhanced immunogenicity of killed erythrocytic stages of *Plasmodium berghei. Infect. Immun.* 32:542–546, 1981.

15. J. L. Benach, G. S. Habicht, T. W. Holbrook, and J. A. Cook. Glucan as an adjuvant for a murine *Babesia microti* immunization trail. *Infect. Immun.* 35:947–951, 1982.

16. J. A. Cook and T. W. Holbrook. Immunogenicity of soluble and particulate antigens from *Leishmania donovani*: Effect of glucan as an adjuvant. *Infect. Immun.* 40:1038–1043, 1983.

17. A. Sharma, A. U. Haq, M. U. Siddique, and S. Ahmad. Immunization of guinea pigs against *Entamoeba histolytica* using glucan as an adjuvant. *Int. J. Immunopharmacol.* 6:483–491, 1984.

18. D. L. Williams, R. G. Yaeger, H. A. Pretus, W. I. Browder, R. B. McNamee, and E. L. Jones. Immunization against *Trypanosoma cruzi*: Adjuvant effect of glucan. *Int. J. Immunopharmacol.* 11:403–410, 1989.

19. N. R. DiLuzio, J. C. Pisano, and T. M. Saba. Evaluation of the mechanism of glucan-induced stimulation of the reticuloendothelial system. *J. Reticuloendothel.* Soc. 7:731–742, 1979.

20. J. K. Czop and K. F. Austen. A β-glucan inhibitable receptor on human monocytes: Its identity with the phagocytic receptor for particulate activators of the alternative complement pathway. *J. Immunol.* 134:2588–2593, 1985.

21. J. K. Czop. Phagocytosis of particulate activators of the alternative complement pathway: Effects of fibronectin. *Ad. Immunol.* 38:361–398, 1986.

22. J. K. Czop. The role of β-glucan receptors on blood and tissue leukocytes in phagocytosis and metabolic activation. *Pathol. Immunopathol. Res.* 5:286–296, 1986.

23. J. L. Kadish, C. C. Choi, and J. K. Czop. Phagocytosis of unopsonized zymosan particles by trypsin-sensitive and β-glucan inhibitable receptors on bone marrow-derived murine macrophages. *Immunol. Res.* 5:129–134, 1986.

24. J. Bagwald, E. Johnson, J. Hoffmann, and R. Seljelid. Lysosomal glycosidases in mouse peritoneal macrophages stimulated *in vitro* with soluble and insoluble glucans. *J. Leukocyte Biol.* 35:357–371, 1984.

25. M. J. Janusz, K. F. Austen, and J. K. Czop. Lysosomal enzyme release from human monocytes by particulate activators is mediated by β-glucan inhibitable receptors. *J. Immunol.* 138:3897–3901, 1987.

26. J. K. Czop and K. F. Austen. Generation of leukotrienes by human monocytes upon stimulation of their β-glucan receptor during phagocytosis. *Proc. Natl. Acad. Sci. USA* 82:2751–2755, 1985.

27. M. Doita, L. T. Rasmussan, R. Seljelid, and P. E. Lipsky. Effect of soluble aminated β-1,3-D-polyglucose on human monocytes: Stimulation of cytokine and prostaglandin E production but not antigen-presenting function. *J. Leuk. Biol.* 49:342–351, 1991.

28. G. Abel and J. K. Czop. Stimulation of human monocyte β-glucan receptors by glucan particles induces production of TNFα and IL-1β. *Int. J. Immunopharmacol.* 14:1363–1373, 1992.

29. N. Mohagheghpour, S. Prohaska, L. Dousman, J. Tefft, L. Hokama, and D. Tusé. Adjuvant activity of an algal glucan. VIII International Conference on AIDS/III STD World Congress, Amsterdam, The Netherlands, 1992.

30. D. Tusé, L. Marquez, and L. Hokama. Production of β-1,3-glucan in *Euglena*. U.S. Patent 5,084,386, 1992.

31. B. M. Sultzer. Genetic control of leukocyte response to endotoxin. *Nature* 219:1253–1254, 1968.

32. E. K. Gallin, S. W. Green, and M. L. Patchen. Comparative effects of particulate and soluble glucan on macrophages of C3H/H3N and C3H/HeJ mice. *Int. J. Immunopharmacol.* 14:173–183, 1992.

33. L. Kuniake, S. Karacsonyi, J. Augusti, A. Ginterova, S. Szechenyl, D. Krvarik, J. Dubaj, and J. Varju. A new fungal glucan and its preparation. W.I.P.O. Patent No. WO93/12243, 1993.

34. S. Karacsonyi and L. Kuniak. Polysaccharides of *Pleurotus ostreatus*: Isolation and structure of Pleuran, an alkali-insoluble β-D-glucan. *J. Biopolymers*, 1994 (in press).

35. A. Lanzavecchia. Receptor-mediated antigen uptake and its effect on antigen presentation to class II-restricted T lymphocytes. *Annu. Rev. Immunol.* 8:773–793, 1990.

36. C. deDuve. Lysosomes revisited. *Eur. J. Biochem.* 137:391–397, 1983.

37. T. Kaneko, D. Willner, I. Monkovic, J. O. Knipe, G. R. Braslawsky, R. S. Greenfield, and D. V. Vyas. New hydrazone derivatives of adriamycin and their immunoconjugates: A correlation between acid stability and cytotoxicity. *Bioconjugate Chem.* 2:133–141, 1991.

38. Pierce Chemical Co. BCA protein assay reagent instructions, Protocol 23225x, 1991.

39. T. A. Scott, J., and E. H. Melvin. Determination of dextran in anthrone. *Anal. Chem.* 25:1656–1661, 1953.

40. P.W.A. Mansell, H. Ichinose, R. J. Reed, E. T. Krementz, R. McMamee, and N. R. DiLuzio. Macrophage-mediated destruction of human malignant cells *in vivo. J. Natl. Cancer Inst.* 54:571–580, 1975.

41. H. Furue. Biological characteristics and clinical effect of schizophyllan (SPG). *Drugs Today* 23:335–346, 1987.
42. G. Chihara, Y. Y. Maeda, T. Suga, and J. Hamuno. Lentinan as a host defense potentiator (HDP). *Int. J. Immunother.* 4:145–154, 1989.
43. K. Okamura, M. Suzuki, T. Chihara, et al. Clinical evaluation of schizophyllan combined with irradiation in patients with cervical cancer. A randomized controlled study. *Cancer* 58:865–872, 1986.
44. G. R. Ostroff. Future therapeutic applications of BETAFECTIN, a carbohydrate-based immunomodulator. The Second Annual Conference on Glycotechnology, organized and sponsored by Cambridge Healthtech Institute, Waltham, MA. La Jolla, CA, May 16–18, 1994.

4

COPOLYMER ADJUVANTS

Robert N. Brey

Vaccine Design Group
Alpharetta, Georgia 30202

INTRODUCTION

A number of experimental adjuvants have been developed in recent years to ultimately supplant or complement the use of alum (salts of aluminum) in individual or combination vaccines for human use. Currently, alum salts are the only adjuvant excipient licensed by the FDA for use in vaccines. Although widely used in pediatric and adult human vaccines, alum may not effectively augment immune responses necessary for a number of subunit vaccines under development. For example, the capacity of alum salts to modulate effective immunity to HIV-1 gp120 and HSV gD is limited (Haigwood, 1992; Sanchez-Pescador, 1988). Alum as adjuvant may not be sufficient to aid in stimulating cellular immunity, and may selectively block development of $CD8^+$ cytotoxic T lymphocytes (CTL) by immunization with some antigens (Schirmbeck, 1994).

Adjuvants can be defined simply as agents added to antigens to augment host immune responses. Experimental adjuvants for enhancing or modifying immune responses to antigens include a variety of compounds, formulations, particles, encapsulation techniques, bioactive molecules, and live vector techniques. Some of these have or will soon be used in clinical evaluation of vaccines. Many of these systems also have theoretical and practical applications to oral vaccination strategies as well as traditional parenteral vaccination. The technologies can be classified accordingly: 1) adsorption technology: e.g. alum, some particle technology, amphipathic micelles; 2) encapsulation or entrapment technology: e.g. liposomes, poly-lactide co-glycolide microspheres (PLGA) (Eldridge, 1991); water/oil emulsion technology (Hilleman, 1967); 3) stimulatory molecules: e.g. materials derived from gram negative bacterial cell walls such as non-toxic LPS (Takayama, 1991) and derivatives; saponin derivatives (Kensil, 1991); and 4) regulatory molecules, including certain cytokines (Afonso, 1994; Good, 1988). Although not strictly defined as an adjuvant technology, strategies to alter structure of antigens by, for instance, addition of lipid tails, or to change T-independent saccharide antigens into T-dependent antigens by conjugation, can have the same immunological outcome as modification of antigens by addition of excipient agents. Additionally, live vectors expressing heterologous antigens, e.g BCG (Bacille Calmette Guerin), *Salmonella spp.*, or vaccinia virus as carriers to induce cell-mediated immune or mucosal immunity incorporate those same ideas (Fuerst, 1992). Some recently

Immunobiology of Proteins and Peptides VIII
Edited by M. Z. Atassi and G. S. Bixler, Jr., Plenum Press, New York, 1995

developed adjuvants have been tested in human clinical trials and are candidates for inclusion in commercial vaccines (Livingston, 1994a; Livingston,1994b).

The current view of adjuvants is that they may, in addition to providing for prolonged contact between antigen and APCs, preserve, expose, or stabilize crucial antigenic epitopes, alter subcellular antigen processing pathways, or help to polarize T cell precursors to either Th1 or Th2 subset of T-cells. Adsorption and encapsulation technologies generally allow for a prolonged or controlled release of antigens from a localized depot. Stimulatory molecules may involve specific interactions with cellular receptors, stimulating cytokine production in professional antigen presenting cells (APCs) and influencing proliferation of specific T and B cells. Adjuvants that interact with antigens in forming micellar complexes may allow for enhanced antigen presentation through direct membrane interaction, as is postulated for antigens encapsulated within the internal space of liposomes (Verma , 1992). Endogenous processing of antigens is generally required for stimulation of $CD8^+$ CTL response to viral or parasite infection (Germain, 1993; Lanzavecchia, 1993; Townsend, 1989). Immune deviation of T cells to either Th1 or Th2 subsets has been implicated in control of several parasitic diseases in animal models and humans, and may be involved in resistance to HIV-1 (Clerici, 1993; Ghalib, 1993; Mosmann, 1989; Scott, 1991; Scott, 1989), implying that the control of Th1 or Th2 respsones will be important in the design of protective vaccines for several viral and parasitic diseases. For example, IL-12 combined with *Leishmania* antigen extract, specifically directed immunity towards protective Th1 responses and replaced requirement for a whole cell bacterial adjuvant *Corynebacterium parvum* (Afonso, 1994). Formulation of IL-2 with a malaria sporozoite vaccine overcame genetic restriction in non-responding mice (Good, 1988). Similarly, vaccination with HBsAg followed by treatment with IL-2 induced protective levels of anti-HB antibodies in previously non-responsive hemodialysis patients (Meuer, 1989).

NONIONIC BLOCK COPOLYMERS

One appraoch to develop an effective adjuvant system has been to take chemical synthetic routes towards anventitious compounds that can promote augmentation of immune responses. An example of this approach is the development of nonionic block copolymers as adjuvants in a variety of formulations. Theoretically, synthetic molecules that could be included in vaccines would have several advantages over molecules obtained by fractionation of natural materials (e.g. cell wall derivatives or saponin): reproducible controlled synthesis, simple purification, and low-cost. The central rationale behind the development of nonionic block polymers in vaccine formulations centers around the observation that these molecules have amphipathic properties, similar to many of the natural molecules with known adjuvant effects, are surface active and are apparently capable of binding proteins or other antigens. Secondly, summary of the current immune response data has indicated that differences in polymer structure have differential effects on the immune system (Zigterman, 1987; Hunter, 1991; also see Table 1). This observation has led to the comparative evaluation as adjuvants of a number of molecules differing from each other in the number and orientation of the building blocks.

Nonionic block polymers are constructed synthetically with relatively simple chemistry from hydrophilic blocks of polyoxyethylene (POE) and hydrophobic blocks of polyoxypropylene (POP). Some species of these block polymers are currently used commercially in shampoos, mouthwashes, cosmetics, topical ointments, and contact lens solutions. These polymers are also used as wetting agents, dispersion agents, and thickening agents. A variety of nonionic block copolymer structures have been synthesized, diagrammed schematically in Figure 1. These structures include triblock, reverse triblock, octablock, and

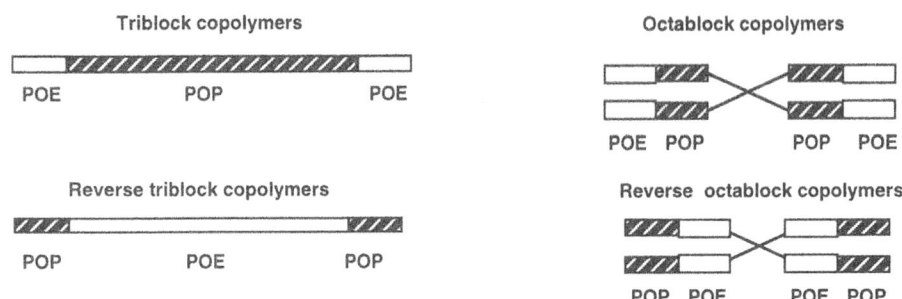

Figure 1. The diagrammatic structure of nonionic block copolymers. POE represents repeating units of polyoxyethylene and POP designates repeating units of polyoxypropylene.

reverse octablock molecules. With respect to each of the four classes of polymer, homologous polymers can be synthesized and defined based on the average (or chromatographic peak) molecular weight and weight percentage or molar content of POE or POP. Many of these polymers are available commercially, but some are available only through custom synthesis.

Polymers with different proportions of POE and POP have unique physical properties. Triblock and octablock copolymer molecules with a high content of hydrophilic POE are more soluble in water, whereas those with a decreasing proportion of POE are less soluble, dissolving only in *cold* water up to 10 g/100 mL. These molecules coalesce in water at "cloud point" temperatures, which are characteristic for each polymeric species. Depending on the particular molecule, the aggregates can be spherical droplets, or can assume unstable fibrous structures in aqueous buffers (Hunter, 1986). Some of the larger, custom synthesized species form stable droplets, a property important for a commercial vaccine formulation. Polymer droplets consist of aggregates of polymers with flexible hydrophobic parts of the chains directed inward and the hydrophilic ethylene oxide side chains interacting with water at the droplet surface. The copolymers having more hydrophobic character are freely soluble in many organic solvents and partition into oils.

Many of the early studies on the physical characteristics of triblock copolymers were performed with a commercially available polymer: L121 (poloxamer 401). L121 binds antigen to the surface of the vehicle oil drops (in emulsions) or presumably to the surface of other hydrophobic vehicles, and promotes the interaction of antigen with cell surfaces (Hunter, 1984; Hunter, 1981). Scanning electron microscopy of oil-in-water emulsions containing L121 and BSA has indicated localization of BSA on the surface of oil droplets. Protein binding promoted by copolymers is thought to be dependent on mild hydrophobic interactions that do not disrupt protein conformation (Hunter, 1986). The ability to bind proteins is thought to be shared amongst the triblock polymers, but has not been strictly examined for each species. Thus, oil-in-water emulsions containing copolymers and aqueous suspensions of antigens and copolymers would fall into the "adsorptive" adjuvant technology. Because of their ability to bind proteins, copolymers are thought to enhance antigen presentation to macrophages or dendritic cells, where proteins could be presented as a condensed matrix with conformational B cell epitopes intact. The mechanism of their action may also include a depot effect in some formulations containing oil. In addition, copolymers may have a more direct role in influencing the production of cytokines and other factors in antigen presenting cells. It has been demonstrated that one of the triblock compounds, L81, in the context of oil-in-water emulsions with squalane, can enhance the expression of MHC class II molecules on the surface of macrophages and may increase complement activation

(Howerton, 1990). The ability of copolymer to influence antigen processing via the class I pathway, inferred by induction of specific CTLs, has not been explored.

Several groups have examined the influence of representatives of the four classes of polymers on immune responses. Most of the evaluation to characterize copolymers and vaccine formulations containing copolymers has been performed with oil-containing vehicles, usually oil-in-water emulsions with metabolizable oils such as squalene (Byars, 1990; Byars, 1991; Hunter, 1991; Millet, 1992; Raychaudhuri, 1992; Takayama, 1991). Consequently, little information is available on the behavior of these molecules in vaccines in oil-free milieu. Early studies conducted with a spectrum of copolymers indicated that these compounds could enhance the antibody titers obtained by injection of formulations with either soluble proteins in oil-in-water emulsion vehicles or haptenated liposomes (Hunter, 1984; Zigterman, 1987). Results obtained with bovine serum albumin (BSA) formulated in oil-in-water emulsions suggested that triblock could enhance anti-BSA antibody responses, depending on the ratio of POP to POE and on the length of the hydrophobic POP moiety. Results have also demonstrated increases in antibody titers correlate to increasing doses of the copolymer. The most effective polymers contained less than 10% POE. On the other hand, early antibody responses (detectable by plaque forming cells 4 days following immunization) to haptenated liposomes demonstrated little correlation to POE content and were associated with higher total molecular weights and the structural class of the copolymer (Zigterman, 1987). Reverse triblock molecules were more effective with haptenated liposomes. Octablock and triblock polymers having less than 10% POE side chains were effective with BSA, whereas reverse block molecules were ineffective. Much of the subsequent work has centered around the development of triblock polymers, because of some of the inflammatory responses observed with reverse triblock and reverse octablock molecules (Hunter, 1984). Those polymers are thought to have ionophoric ability.

An attempt to generalize the effect of triblock polymers on the immune responses to a variety of protein, saccharide/conjugate, and peptide antigens is presented in Table 1. These results have largely been acquired in immunization experiments with mice (although Rhesus monkeys and guinea pig responses have been evaluated in several studies). In some cases, disease challenge models have been included along with antibody studies. The results generally indicate that the structure of copolymers correlates to the magnitude of the immune responses, usually measured by alterations in antibody response. However, this appears to be on an antigen to antigen basis and is probably influenced the oil vehicles, the molar content of polymer, the particular formulation and the method of preparation, the composition of the aqueous buffer components, as well as the route of administration. Most of the experimental vaccines have been formulated on the benchtop and injected minutes to hours after formulation. Some of the vaccines have been made up with lyophilized antigens ground up with oil before reconstitution in buffer; others have been made up with aqueous antigens and then emulsified with oil prior to immunization. The apparent structural effects of polymers on immune responses could be accounted for by different lab to lab methodologies and formulations. One particular formulation containing L121, SAF-m (Syntex Corporation), was developed in part to circumvent some of the instability problems associated with benchtop emulsion preparations. SAF-m has been formulated as an submicron emulsion by microfluidization, a technique that stabilizing oil emulsions for long term storage (Lidgate, 1989). Table 1 does not contain a complete list of antigens evaluated with SAF-m.

Vaccination of young mice with a BSA-*S. pneumoniae* type 3 hexasaccharide conjugate formulated in an oil-in-water emulsion with L121 resulted in significant protection against virulent challenge, which was not achieved by vaccination with the conjugate alone (Zigterman, 1988). Protection could be correlated to elevated levels of more avid IgG2a antibody against the type 3 capsule (Van Dam, 1989). The level of IgG2a was influenced primarily by the polymer; this could be demonstrated whether the vaccines were formulated

Table 1. Influence of formulations containing triblock polymers on the immunogenicity of various antigens

Antigen (formulation)	L81 (2250;10)	L92 (2750;20)	L101 (3200;10)	L121 (4000;10)	SAF-m*	L122 (4000;20)	L141 (4600;10)	L180.5 (5200;5)	L181.5 (5000;15)	p1004 (10,000;5)	CRL1005 (12,500;5)	Ref
BSA (mineral oil/water)	+	–	+++	++		–						19
TNP-HEA (squalene/water)	±	–	++	+		–	+++	++++	–			18
haptenated liposomes	++++	++++	++	+		+						47
Pneumo. type 3 (squalene/water)	–		++	++								42;46
Pneumo. type 14 (squalene/water)				++								43
BSA-(NAGG)5 (squalene/water)				++			++					22
P. yoelii extract†										+++		39
Influenza HA				++								5
HBsAg					++						+++	3;7
Semliki Forest virus T+B cell peptide‡								+++				10
SP66											+++	3
HSV- gD					++							6

Values for molecular weights and proportion of ethylene oxide side chains, with the exception of p1004 and CRL1005, are based on data from (18). Blank spaces in the table imply that no data is available; A (–) indicates no augmentation of the measured immune response(s) relative to control values; (+) to (++++) indicate qualitative increases in measured immune response(s) relative to control values.

*SAF-m (Syntex Corporation) is a submicron squalene/water emulsion, optionally containing threonyl MDP, obtained by microfluidization.

†Evaluated in several formulations: encapsulated in water-in-oil emulsion, oil-in-water emulsions, and in oil-free saline buffer.

‡Evaluated in a multiple oil emulsion format and in oil-free saline buffer.

in oil-in-water emulsions or with the polymer in aqueous suspension. The tendency to induce IgG2a has also been demonstrated with *S. pneumoniae* type 14 polysaccharide-BSA conjugate when L121 was included in oil-in-water vaccines, independent of the presence of co-adjuvant molecules such as MDP or MPL (Van de Wijgert, 1991).

Results obtained with oil-in-water vaccine with TNP-ovalbumin (TNP-OVA) indicated that increased total IgG directed against the TNP hapten correlated to increasing molecular weights of POP, if the content of POE was less than 10% (Hunter, 1991). TNP-OVA vaccines formulated with the highest molecular weight polymers also induced higher proportions of IgG2b subclass antibodies to the hapten. Similarly, IgG2 subclass antibodies have been obtained in mice with a *Plasmodium cynomolgi* (NAGG)$_5$-BSA conjugate anti-sporozoite vaccine formulated with triblock polymers L121 or L141 in oil-in-water emulsions (Kalish, 1991). In other studies, this peptide vaccine was formulated with a custom synthesized copolymer, p1004 (average molecular weight 10,000, 5% POE), and used to immunize Rhesus monkeys (Millet, 1992). Monkeys were immunized up to 3 times with 200 mg of the CS peptide (NAGG)$_5$-diphtheria toxoid in squalene-in-water with either detoxified RaLPS or MTP-PE. Only animals immunized with the conjugate in p1004-RaLPS squalene/water vaccine displayed a significant delay in the onset of parasitemia following challenge. Protection against blood stage parasite challenge following immunization of mice with whole killed *P. yoelii* blood stage parasites has been shown to be adjuvant dependent (ten Hagen, 1993). A whole blood stage parasite vaccine was formulated with p1004 in combination with a variety of emulsions, including water-in-oil, oil-in-water, and aqueous copolymer formulations. Protection against lethal or severe parasitemia appeared to depend on format of the vaccine vehicle. Vaccine formulations of p1004 in squalene-in-water or those with no oil induced protective responses, while all of the water-in-squalene formulations failed. Protection was correlated with the induction of antibody of the IgG2a isotype detected by indirect immunofluorescence. Incorporation of a Semliki Forest virus T and B cell epitope peptide in a multiple (water-in-oil-in-water) emulsion with L180.5 induced equivalent antibody responses to vaccine formulated with Complete Freund's Adjuvant (CFA), whereas peptides formulated with L180.5 in aqueous buffer failed to induce antibodies (Fernandez, 1993). Another custom synthesized molecule (CRL1005; m.w. 12,500;5%) has been evaluated in conjunction with hepatitis B surface antigen (HBsAg) and SPf66, a malaria pepetide vaccine currently undergoing extensive field testing (Teusher , 1994). Both oil-in-water and water-in-oil formulations containing CRL1005 were effective in augmenting antibody responses to these two antigens in comparison to alum-adjuvanted vaccines (Brey, 1995).

A series of immunization experiments has been conducted with vaccines formulated with SAF-m, indicating more rapid antibody response to a recombinant hepatitis B surface antigen (Byars, 1991), and protection against herpes simplex virus type 2 infection in Guinea pigs, modulated by adjuvant (Byars, 1994). Prior evaluation of SAF with inactivated influenza type B virus had included an earlier version of the adjuvant, SAF-1, in which the vehicle emulsions was not microfluidized (Byars, 1990). Usual embodiments of both SAF-1 and SAF-m formulations have contained threonyl MDP, which has been thought to be the essential adjuvant component. Recently, an anti-idiotype melanoma vaccines has been evaluated clinically in conjunction with SAFm, optionally containing threonyl MDP (Livingston, 1994b). In that study, significant systemic side effects, such as profound fatigue and myalgia lasting up to 4 days, were associated with injection of high (1 mg) doses of threonyl MDP. Surprisingly, high rates of seroconversion to the melanoma antigen were elicited by the SAFm vaccine not containing threonyl MDP. These results correspond to human clinical results obtained with another emulsion vehicle (MF59, developed by Chiron Corporation) which has been recently evaluated with a trivalent influenza vaccine (Keitel, 1993; Wintsch, 1991).

Taken together, preclinical and clinical results with emulsion vehicle vaccines containing nonionic block polymers point to an uncertain role of the polymers with respect to the oil emulsion vehicles and other components, with a wide variety of antigens. Moreover, these results tend to point to an underappreciation of the format and role of the vehicle components in modulating immunogenicity. These conclusions tend to support careful reexamination of the role of each of the components of a complex emulsion vehicle vaccine and their role in modulating immunogenicity, before a clinical evaluation of the adjuvant can take place.

With this in mind, we chose to examine the influence on immune responses of aqueous suspension of a single copolymer (CRL1005) in relationship to a standard oil-in-water emulsion vehicle and other controls. CRL1005 was obtained by custom synthesis (CytRx Corporation) and was used in comparison to standard alum-adsorbed HBsAg and in comparison to a commercial influenza vaccine (Fluogen®, Parke-Davis Corporation), which is administered to humans without adjuvant. Historical evaluation of alum-adjuvanted influenza vaccines has indicated that alum is not effective in boosting response rates to HA antigens (Nicholson, 1979). The current trivalent influenza vaccines are effective in protecting against mortality in years where the vaccine strains closely match the strains in circulation. However, the general perception, variably supported by clinical studies, is that influenza vaccine is not very effective in the elderly, the most susceptible population (Beyer, 1989). One of the strategies to boost response rates and protection rates in the elderly is to modify the current vaccine with an adjuvant. Likewise, HBsAg vaccines adsorbed to alum are 90% effective in adults and infants, but require 3 doses to achieve maximum response rates in a human population (Butterly, 1989). Improvement of hepatitis B vaccines with an adjuvant could reduce the required number of doses to two or one and would significantly affect the use of hepatitis B vaccine in developing countries.

An aqueous suspension of CRL1005 and a microfluidized squalene/PBS emulsion were used as a base for comparison of admixtures of either commercial influenza vaccine or a recombinant HBsAg. CRL1005 in aqueous buffers formed uniform submicron droplets that were stable to ambient temperatures, but dissolved when stored at 4 °C. A microfluidized emulsion made with CRL1005 contained homogeneous droplets below 1 mm in diameter and were indistinguishable by particle sizing methodology from emulsions prepared without the polymer. Emulsions with or without the polymer were homogeneous and stable at ambient temperature in terms of droplet distribution, but displayed heterogeneity after warming from a cold state. This suggests that CRL1005 "cold solubilizes" from squalene into the water phase of the emulsions, essentially creating a polymer phase, an oil phase, and an aqueous phase. Balb/c mice were immunized with a trivalent influenza vaccine available during the 1993 season (FLUOGEN, Parke-Davis) mixed directly with several of the adjuvant preparations, or with dilutions mixed with the adjuvant preparations. As shown in Figure 2, flu vaccine at 0.1 the human dose elicited ELISA antibodies and HAI (hemagglutinin inhibiting) antibodies. The levels of those antibodies were enhanced by adjuvants consisting of CRL1005 alone in aqueous buffer, a microfluidized squalene/buffer adjuvant, and a composite adjuvant consisting of both CRL1005 and the microfluidized vehicle. ELISA antibody titers were highest in animals vaccinated with the composite adjuvant up to 70 days following a single immunization. ELISA titers against all three viral types represented in the vaccine correlated well to the HAI titers assayed against closely related subtypes (Table 2), but were highest in animals immunized with the vaccine that was adjuvanted with the microfluidized vehicle without polymer. Balb/c mice responded in a dose dependent fashion that was modulated by the microfluidized adjuvant vehicles (Figure 3). These results suggest that the mixture adjuvant of the polymer and the microfluidized vehicle may have elicited synergistic effects at lower antigen doses.

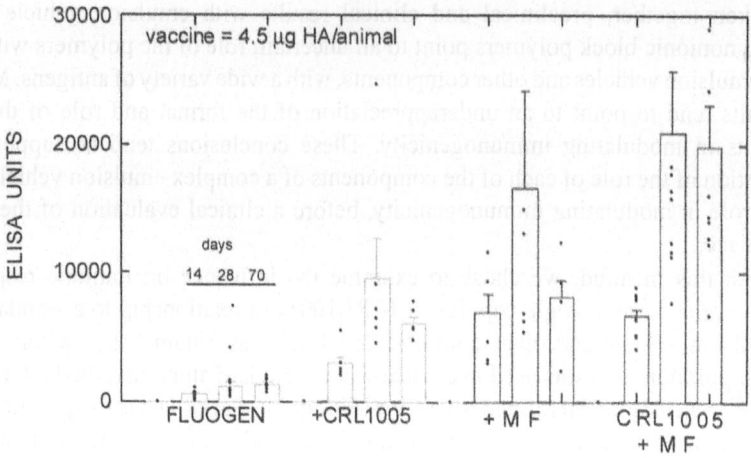

VACCINE FORMULATION

Figure 2. The influence of several types of formulations containing CRL1005 on trivalent influenza vaccine immunogenicity. The vaccine consisted of hemagglutinin (HA) and other virus components from A/Texas/36/91 (H1N1), A/Beijing/32/92 (H3N2), and B/Panama/45/90 and contained 15 μg of each HA antigen. The vaccine was diluted in buffer and mixed 1:1 with either buffer (0.9% NaCl) or copolymer (5% CRL1005, 0.9% NaCl), microfluidized vehicle (MF; 10 squalene, 0.2%tween 80, 0.9% NaCl), or a combination (MF/CRL1005). Mice were injected with 100 μl of the two vaccines. ELISA plates were coated overnight with 0.2 μg/well FLUOGEN® (Parke-Davis, Morris Plains, NJ. 1993 batch containing A/TEXAS/36/91 (H1N1), A/BEIJING/32/92 (H3N2), and B/PANAMA/45/90). Appropriate dilutions of test sera, horseradish peroxidase-conjugated goat anti-mouse IgG, IgA and IgM, and TMB substrate were incubated sequentially in the plate with washing steps between additions. Color intensity of the substrate following development was measured with a VMAX Microplate Reader and the data were analyzed and interpolated with the SoftMAX computer program (both from Molecular Devices, Menlo Park, CA.). Unknown sera were quantitated against a standard curve of an in-house standard serum pool which was designated as having 10,000 Units/ml anti-Fluogen antibody.

Table 2. Serological responses to trivalent influenza HA vaccine (FLUOGEN)

Vaccine	ELISA*	HA Inhibition[†]		
		A/Yamagato 32/89 H1N1	A/Ghizou 54/89 H3N2	B/Aichi 5/89
FLUOGEN 4.5 μg	965	511	303	241
FLUOGEN + CRL1005 micelles	8012	2032	5120	2560
FLUOGEN + CRL1005 MF	17,365	4642	1680	1641
FLUOGEN + MF	11,510	5120	5120	2560

*Against all 3 vaccine antigens
[†]Hemagglutinin inhibiting antibodies were measured in serial dilutions of sera for their ability to neutralize the hemagglutinating activity of 8 units of purified hemagglutinin on chicken red blood cells. Purified hemagglutinins from the following 3 strains were used: A/YAMAGATA/32/89 (H1NI), A/KISHU/54/89 (H3N2), B/AICHI/5/89. Titers are expressed as the highest serum dilution giving complete neutralization of the hemagglutinin.

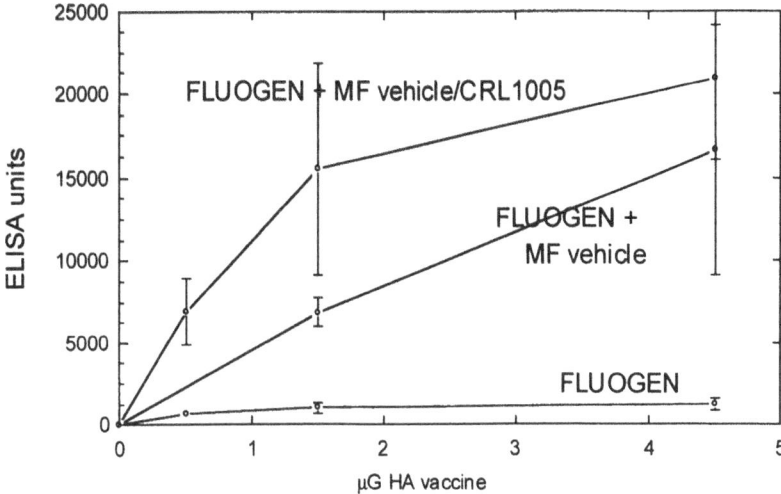

Figure 3. Antigen dose response modulation by adjuvant. CRL1005 was prepared as above with a microfluidized squalene/tween-80 vehicle, and as control the microfluidized vehicle lacking the polymer was also prepared. Mice were immunized subcutaneoulsy with 0.1 ml of formulations containing the same amount of adjuvant as each other, but with various HA doses. Serological response by ELISA were monitored 28 days following immunization.

A variety of HBsAg vaccines were prepared similarly using in-process recombinant HBsAg, but also included alum (as $Al(OH)_3$)adsorption as a control. A dose of 0.5 µg per mouse was chosen because it was judged from prior experiments to be a threshold dose, that when adsorbed to alum elicited less than 10% seroconversion in Balb/c mice 28 days

Table 3. Seroconversion rates of Balb/c mice 28 days following a single immunization with 0.5 µG HBsAg formulated with various adjuvants

Vaccine adjuvant*	α-HBsAg (GMT)	Seroconversion rate (percent)	Increase over control
2.5% CRL1005	127*	3/6 (50)	4
2.5% CRL1005 +alum	26	2/6 (33)	0.8
no additions	2	0/8 (0)	0.06
alum, 0.25 mg/ml	31	2/8 (25)	1
MF/CRL1005	407	6/6 (100)	13
MF	989	6/6 (100)	32
MF/CRL1005 + alum	202	4/6 (66)	6.5
MF + alum	83	3/6 (50)	2.6

*The alum-adsorbed contained 0.25 mgs aluminum hydroxide/ml. Each of the vaccines was made by admixing a recombinant hepatitis B surface antigen (HBsAg) to the adjuvant compositions on ice. The vaccines were warmed to room temperature before injection in animals. The final HBsAg concentration in each vaccine was 5 µg/ml. 0.1 ml of vaccine was administered to each animal via subcutaneous injection at the base of the tail. Murine antibody concentrations against HBsAg were measured with the commercially available AUSAB® EIA (Abbott Laboratories, Abbott Park, IL.) Murine antibody was assayed at a 1:10 dilution or higher and quantitated against a human serum standard curve (also from Abbott Laboratories). Data are expressed as milli international units per ml (mIU), calculated by multiplying the assay value by a factor of 10. Because sera were diluted 1:10 for assay, a value less than 100 was considered below the cutoff of significance.

following a single subcutaneous injection. Data presented in Table 3 demonstrate a more complex relationship of vehicles and polymer in influencing the immunogenicity of HBsAg. The highest AUSAB titers were observed with microfluidized squalene/buffer vehicle as adjuvant, seroconverting 100% of the mice. Adsorbing the MF vehicle vaccine to alum resulted in AUSAB titers similar to alum adjuvant alone. Although CRL1005 addition alone resulted in a 4 fold increase in AUSAB titers over alum control and 60 fold elevation over unadjuvanted HBsAg, adsorption the mixture to alum did not result in additive or synergistic effects. Similarly, the addition of CRL1005 to the microfluidized vehicle, in contrast to its influence on influenza vaccine, reduced AUSAB titers observed with the microfluidized vehicle alone. These results appear to indicate preferential interaction of HBsAg with one of the adjuvant components.

These results with two different complex viral antigens indicate that the nonionic polymer CRL1005 can augment functional antibody responses. This polymer alone in aqueous solution could clearly have advantages over more complex emulsion formulations that would require specialized machinery for manufacture of millions of vaccine doses. Along that line, consistency batches of CRL1005 have been manufactured and formulated for clinical use by sterile filtration techniques. Although CRL1005 can augment immune response to these antigens, it remains a question how this particular adjuvant compares with other polymeric species. However, the comparative studies with microfluidized squalene/buffer adjuvant suggest that CRL1005 will be less potent than clinical formulations such as MF59 (-MTP-PE) or SAF-m (-threonyl MDP). This also could be advantageous for these antigens and others if the oil-containing emulsions fail to meet strict safety criteria upon further human testing.

ACKNOWLEDGMENTS

The synthesis of CRL1005 and evaluation of trivalent influenza and recombinant hepatitis B vaccines was supported by Vaxcel, Inc. I would like to thank C. W. Todd for formulating vaccines and particle analysis, M. Fiorello for developing influenza ELISA and HAI assays and Deborah Godwyn for other serological assays.

REFERENCES

1. Afonso, L.C.C., Scharton, T.M., Vieira, L.Q., Wysocka, M., Trinchieri, G., and Scott, P., 1994, The adjuvant effect of interleukin-12 in a vaccine against *Leishmania major, Science* 263: 235-237.
2. Beyer, W.E.P., Palache, A.M., Baljet, M., and Masurel, N., 1989, Antibody induction by influenza vaccines in the elderly: a review of the literature, *Vaccine* 7: 385-394.
3. Brey, R.N. 1995. "Development of vaccines based on formulations containing nonionic block copolymers." *In* Vaccine Design, M. F. Powell and M. J. Newman ed. Plenum Press. New York.
4. Butterly, L., Watkins, E., and Dienstag, J.L., 1989, Recombinant-yeast derived hepatitis B vaccine in healthy adults: safety and two-year immunogenicity of early investigative lots of vaccine, *J. Med. Virol.* 27: 155-159.
5. Byars, N.E., Allison, A.C., Harmon, M.W., and Kendal, A.P., 1990, Enhancement of antibody responses to influenza B virus haemagglutinin by use of a new adjuvant formulation, *Vaccine* 8: 49-56.
6. Byars, N.E., Fraser-Smith, E.B., Pecyk, R.A., Welch, M., Nakano, G., Burke, R.L., Hayward, A.R., and Allison, A.C., 1994, Vaccinating guinea pigs with recombinant glycoprotein D of herpes simplex virus in an efficacious adjuvant formulation elicits protection against vaginal infection, *Vaccine* 12: 200-209.
7. Byars, N.E., Nakano, G., Welch, M., Lehman, D., and Allison, A.C., 1991, Improvement of hepatitis B vaccine by the use of a new adjuvant, *Vaccine* 9: 309-318.

8. Clerici, M., Lucey, D.R., Berzovsky, J.A., Pinto, L.A., Wynn, T.A., Blatt, S.A., Dolan, M.J., Hendrix, C.W., Wolf, S.W., and Shearer, G.M., 1993, Restoration of HIV-specific cell-mediated immune responses by interleukin-12 in vitro, *Science* 262: 1721-1724.

9. Eldridge, J.H., Staas, J.K., Meulbroek, J.A., Tice, T.R., and Gilley, R.M., 1991, Biodegradable and biocompatible poly(DL-lactide-co-glycolide) microspheres as an adjuvant for staphylococcal enterotoxin B toxoid which enhances the level of toxin neutralizing antibodies, *Infect. Immun.* 59: 2978-2986.

10. Fernandez, I.M., Snijders, A., Benaissa, T.B., Harmsen, M., Snippe, H., and Kraaijeveld, C.A., 1993, Influence of epitope polarity and adjuvants on the immunogenicity and efficacy of a synthetic peptide vaccine against Semliki Forest virus, *J Virol* 67: 5843-5848.

11. Fuerst, T.R., de la Cruz, V., Bansal, G.P., and Stover, C.K., 1992, Development and analysis of recombinant BCG vector systems, *Aids Res. Hum. Retroviruses* 8: 1451-1455.

12. Germain, R.N., and Margulies, D.H., 1993, The biochemistry and cell biology of antigen processing and presentation, *Ann. Rev. Immunol.* 11: 403-440.

13. Ghalib, H.W., Piuvezam, M.R., Skeiky, Y.A., Siddig, M., Hashim, F.A., el Hassan, A., Russo, D.M., and Reed, S.G., 1993, Interleukin 10 production correlates with pathology in human *Leishmania donovani* infections, *J. Clin. Invest.* 92: 324-329.

14. Good, M.F., Pombo, D., Lunde, M.N., Maloy, W.L., Halenbeck, R., Koths, K., Miller, L.H., and Berzofsky, J.A., 1988, Recombinant human IL-2 overcomes genetic nonresponsiveness to malaria sporozoite peptides. Correlation of effect with biologic activity of IL-2, *J. Immunol.* 141: 972-977.

15. Haigwood, N.L., Nara, P.L., Brooks, E., Van Nest, G., Ott, G., Higgins, K.W., Dunlop, N., Scandella, C.J., Eichberg, J.W., and Steimer, K.S., 1992, Native but not denatured recombinant human immunodeficiency virus type 1 gp120 generates broad-spectrum neutralizing antibodies in baboons, *J. Virol.* 66: 172-182.

16. Hilleman, M.R. 1967. "Considerations for safety and application of emulsified oil adjuvants to viral vaccines." *In* International Symposium on Adjuvants of Immunity, Utrect 1966, 13-26. Karger. Basel, New York.

17. Howerton, D.A., Hunter, R.L., Ziegler, H.K., and Check, I.J., 1990, Induction of macrophage Ia expression in vivo by a synthetic block polymer, L81, *J. Immunol.* 144: 1578-1584.

18. Hunter, R., Olsen, M., and Buynitzky, 1991, Adjuvant activity of non-ionic block copolymers. IV. effect of molecular weight and formulation on titre and isotype of antibody, *Vaccine* 9: 250-256.

19. Hunter, R.L., and Bennett, B., 1984, The adjuvant activity of nonionic block polymer surfactants. II. antibody formation and inflammation related ot the structure of triblock and octablock copolymers, *J. Immunol.* 133: 3167-3175.

20. Hunter, R.L., and Bennett, B., 1986, The adjuvant activity of nonionic block polymer surfactants. III. characterizaiton of selected biologically active surfaces, *Scand. J. Immunol.* 23: 287-300.

21. Hunter, R.L., Strickland, F., and Kezdy, F., 1981, The adjuvant activity of nonionic block polymer surfactants. I. the role of hydrophile-lipophile balance, *J. Immunol.* 127: 1244-1250.

22. Kalish, M.L., Check, I.J., and Hunter, R.L., 1991, Murine IgG isotype responses to the *Plasmodium cynomolgi* circumsporozoite peptide (NAGG)5, *J. Immunol.* 146: 3583-3590.

23. Keitel, W., Couch, R., Bond, N., Adair, S., Van Nest, G., and Dekker, C., 1993, Pilot evaluation of influenza virus vaccine (IVV) combined with adjuvant, *Vaccine* 11: 909-913.

24. Kensil, C.R., Patel, U., Lennick, M., and Marciani, D., 1991, Separation and characterization of saponins with adjuvant activity from Quillaja saponaria Molina cortex, *J. Immunol.* 146: 431-437.

25. Lanzavecchia, A., 1993, Identifying strategies for immune intervention, *Science* 260: 937-944.

26. Lidgate, D.M., Fu, R.C., Byars, N.C., Foster, L.C., and Fleitman, J.S., 1989, Formulation of vaccine adjuvant muramyl dipeptides. 3. Processing, optimization, characterization and bioactivity of an emulsion vehicle., *Pharm. Res.* 6: 748.

27. Livingston, P.O., Adluri, S., Helling, F., Yao, T.-J., Kensil, C.R., Newman, M.J., and Marciani, D., 1994a, Phase I trial of immunological adjuvant QS-21 with a GM2 ganglioside-KLH conjugate vaccine in patients with malignant melanoma, *Vaccine* 1275-1280.

28. Livingston, P.O., Adluri, S., Raycauduri, S., Hughes, M.H., Calves, M.J., and Merritt, J.A., 1994b, A phase I trial of the immunological adjuvant SAFm in melanoma patients vaccinated with the anti-idiotype antibody MELIMMUNE_, *Vaccine Res. In press.*

29. Meuer, S.C., 1989, Low dose interleukin-2 induces systemic immune response against HBsAg in immunodeficient non-responders to hepatitis B vaccination in dialysis patients, *Lancet* i: 15-18.

30. Millet, P., Kalish, M.L., Collins, W.E., and Hunter, R.L., 1992, Effect of adjuvant formulations on the selection of B-cell epitopes expressed by a malaria peptide vaccine, *Vaccine* 10: 547-550.

31. Mosmann, T.R., and Coffman, R.L., 1989, TH1 and TH2 cells: different pattern of lymphokine secretion lead to different functional properties, *Ann. Rev. Immunol.* 7: 145-173.

32. Nicholson, K.G., Tyrell, D.A.J., Harrison, P., Potter, C.W., Jennings, R., Clark, A., et al, 1979, Clinical studies of monvalent inactivated whole virus and subunit A/USSR/77 (H1N1) vaccine: serological and clinical reactions, *J. Biol. Stand.* 7: 123.

33. Raychaudhuri, S., Tonks, M., Carbone, F., Ryskamp, T., and Morrow, J.W., 1992, Induction of antigen-specific class I-restricted cytotoxic T cells by soluble protein *in vivo*, *Proc. Natl. Acad. Sci. USA* 89: 8308-8312.

34. Sanchez-Pescador, L., Burke, R.L., Ott, G., and Van Nest, G., 1988, The effect of adjuvants on the efficacy of a recombinant herpes simplex virus glycoprotein vaccine, *J. Immunol.* 141: 1720-1727.

35. Schirmbeck, R., Melber, K., Kuhrober, A., Janowicz, Z.A., and Reimann, J., 1994, Immunization with soluble hepatitis B virus specific surface protein elicits murine H-2 Class-I restricted cytotoxic T lymphocyte respnses in vivo, *J. Immunol.* 152: 1110-1119.

36. Scott, P., 1991, IFN-gamma modulates the early development of Th1 and Th2 responses in a murine model of cutaneous leishmaniasis, *J. Immunol.* 147: 3149-3155.

37. Scott, P., Pearce, E., Cheever, A.W., Coffman, R.L., and Sher, A., 1989, Role of cytokines and CD4+ T-cell subsets in the regulation of parasite immunity and disease, *Immunol. Rev.* 112: 161-182.

38. Takayama, K., Olsen, M., Datta, P., and Hunter, R.L., 1991, Adjuvant activity of non-ionic block copolymers. V. Modulation of antibody isotype by lipopolysaccharides, lipid A and precursors, *Vaccine* 9: 257-265.

39. ten Hagen, T.L.M., Sulzer, A.J., Kidd, M.R., Lal, A.A., and Hunter, R.L., 1993, Role of adjuvants in the modulation of antibody isotype, specificity, and induction of protection by whole blood-stage *Plasmodium yoelii* vaccines, *J. Immunol.* 151: 7077-7085.

40. Teusher, T., Armstrong-Schellenberg, J.R.M., Bastos de Azvedo, I., Hurt, N., Smith, T., Hayes, R., Masanja, H., Silva, Y., Lopez, M.C., Kitua, A., Kilama, W., Tanner, M., and Alonzo, P.L., 1994, SPf66, a chemically synthesized subunit malaria vaccine, is safe and immunogenic in Tanzanians exposed to intense malaria transmission, *Vaccine* 12: 328-336.

41. Townsend, A., and Bodmer, H., 1989, Antigen recognition by class I-restricted T lymphocytes, *Ann. Rev. Immunol.* 7: 601-624.

42. Van Dam, G.J., Verheul, A.F.M., Zigterman, G.J.W.J., De Reuver, M.J., and Snippe, H., 1989, Nonionic block polymer surfactants enhance the avidity of antibodies in polycloncal antisera against *Streptococcus pnuemoniae* type 3 in normal and Xid mice, *J. Immunol.* 143: 3049-3053.

43. Van de Wijgert, J.H.H.M., Verheul, A.F.M., Snippe, H., Check, I.J., and Hunter, R.L., 1991, Immunogenicity of Streptococcus pneumoniae type 14 capsular polysaccharide: influence of carriers and adjuvants on isotype distribution, *Infect. Immun.* 59: 2750-2757.

44. Verma, J.N., Rao, M., Amselem, S., Krzych, U., Alving, C.R., Green, S.J., and Wassef, N.M., 1992, Adjuvant effects of liposomes containing lipid A: enhancement of liposomal antigen presentation and recruitment of macrophages, *Infect. Immun.* 60: 2438-2444.

45. Wintsch, J., Chaignat, C.-L., Braun, D.G., Jeannet, M., Stadler, H., and Abrignanai, S., 1991, Safety and immunogenicity of a genetically engineered human immunodeficiency virus vaccine, *J. Infect. Dis.* 163: 219-225.

46. Zigterman, G.J.W.J., Snippe, H., Jansze, M., Ernste, E.B.H.W., De Reuver, M.J., and Willers, J.M.N., 1988, Nonionic block polymer surfactants enhance immunogenicity of pneumococcal hexasaccharide-conjugate vaccines, *Infect. Immun.* 56: 1391-1393.

47. Zigterman, G.J.W.J., Snippe, H., Jansze, M., and Willers, J.M.N., 1987, Adjuvant effects of nonionic block polymer surfactants on liposome-induced humoral immune repsonse, *J. Immunol.* 138: 220-225.

5

REGULATION OF IL-4 AND IL-5 SECRETION BY HISTAMINE AND PGE$_2$

Manzoor M. Khan

Department of Pharmaceutical Sciences
Creighton University Health Sciences Center
Omaha, Nebraska 68178

SUMMARY

This study was designed to study the effects of autacoids on IL-4 and IL-5 secretion. IL-4 and IL-5 are secreted by TH$_2$ cells. TH$_2$ cells were generated by the culture of peripheral blood lymphocytes from atopic individuals in the presence of ragweed or dustmite antigen. The cloned TH$_2$ lymphocytes were then stimulated with PMA (10 ng/ml) and α-CD3 (50 ng/ml) in the presence and absence of histamine (10^{-4} - 10^{-8}M) and PGE$_2$ (10^{-6} - 10^{-8}M) for 48 hours. Other cAMP elevating agents were used as control. The supernatants were then assayed for the presence of IL-4 and IL-5 by ELISA. Both histamine and PGE$_2$ suppressed the secretion of IL-4 in a dose dependent manner. Other cAMP elevating agents did not affect IL-4 secretion. In contrast, histamine upregulated the secretion of IL-5, whereas the effects of PGE$_2$ on IL-5 secretion were not conclusive. Chloride channels have been implicated in the secretory processes. The effects of a chloride channel blocker, DIDS, were studied on histamine-induced suppression of IL-4 secretion. DIDS (10^{-7} - 10^{-12}M) abrogated the inhibitory effects of histamine on IL-4 secretion. The observations suggest that histamine may inhibit IL-4 secretion via activation of chloride channels.

INTRODUCTION

The cytokines, IL-4 and IL-5, have been increasingly implicated in the pathogenesis of diseases such as allergy, asthma, AIDS and autoimmune diseases. Recently, considerable progress has been made in understanding the molecular basis by which cytokines released from CD4[+] helper T cells contribute to allergic disease (1). A subset of CD4[+] helper T cells, TH$_2$ cells, synthesize and secrete IL-4, IL-5, IL-10 and IL-13 but not IL-2 and γ-IFN. IL-2 and γ-IFN are synthesized and secreted by TH$_1$ cells (2 & 3). IL-4 is important in the regulation of the allergic response, being instrumental in the activation of B cells and promotion of IgE synthesis (4). Differentiation of T-helper cells and potentiation of the antigen-induced proliferation and cytokine secretion by T-helper type 2 (TH$_2$) cells are also

Immunobiology of Proteins and Peptides VIII
Edited by M. Z. Atassi and G. S. Bixler, Jr., Plenum Press, New York, 1995

35

controlled by IL-4 (5). IL-4 also increases endothelial adhesion molecule expression. IL-5 plays an important role in immune modulation of the inflammatory disease. It promotes the IgE synthesis induced by IL-4 (6). Furthermore, IL-5 regulates eosinophil differentiation (7), maturation (8 & 9), activation (10), degranulation (11), and endothelial cell adherence (12). As eosinophil activation is a common feature of both atopic and non-atopic forms of bronchial asthma (13), and both eosinophils and TH_2 cells expressing IL-5 mRNA are found in bronchoalveolar lavage fluid from both nonatopic and atopic asthmatics, IL-5 seems to play an important role in bronchial asthma.

The control pathways which regulate the differentiation of TH_1 and TH_2 cells *in vivo* have not been determined. There are probably a number of different pathways which control the differentiation of a naive T helper cells into TH_1 and TH_2 subsets. It has been suggested that the way and route of immunization influences the type of T cell response. A subcutaneous injection with allergen will result in increased production of IgG (14), whereas bronchial or nasal challenge with the same allergen will result in elevated levels of IgE (15). Second, the differences in the structure of allergens and antigens may influence the type of immune response (16). Finally, the microenvironment of the cells, particularly with reference to endogenous mediators, antigens and cytokines, may control whether a TH_1 or TH_2 response is mounted. Whereas IL-4 potentiates antigen-induced proliferation and cytokine secretion by TH_2 cells, γ-IFN selectively inhibits these responses (5). IL-4 in bulk cultures shifts differentiation of TH_1 cells, causing them to become TH_0 or TH_2 like in their cytokine secretion pattern. TH_0 cells have the capacity to produce virtually all known cytokines. In comparison, γ-IFN has a reciprocal effect, causing TH_2 cells to become TH_0 or even TH_1 like cells (5). Other cytokines which preferentially affect one subset over another have also been suggested (4 & 17).

We have been interested in the immunoregulatory effects of histamine and PGE_2. Autacoids are mediators and modulators of a number of the functions of the immune response (18). For example, histamine inhibits T cell proliferation (19), antibody production (20), generation of cytolytic T cells (21) and the production of cytokines (22). Other immunoregulatory effects of histamine include suppression of complement production by monocytes (23), lysozyme enzyme release by neutrophils and the neutrophil chemotaxis (24). Histamine stimulates thymocyte maturation (25), eosinophil chemotaxis, and the expression of eosinophil complement receptors (26).

Histamine directly inhibits the production of IL-2 and γ-IFN. While it has been demonstrated that histamine regulates TH_1 cells by inhibiting the production of IL-2 and γIFN, the effects of histamine on TH_2 cells have not been reported. As a matter of fact, it has been reported that cAMP elevating agents including PGE_2 suppressed IL-2 and γ-IFN secretion but did not affect TH_2 cells. Based on these reports this study was designed to study the effects of histamine and PGE_2 on IL-4 and IL-5 secretion from aeroallergen-specific TH_2 lymphocytes.

MATERIALS AND METHODS

Materials

Cell culture media RPMI-1640 obtained from Fisher Scientific (Fairlawn, NJ) was supplemented with L-glutamine (1.7 mM), non-essential amino acids (0.087 mM), sodium pyruvate (0.87 mM), streptomycin (100 µg/mL), penicillin (100 U/ml), all obtained from Gibco, and 5% human AB serum purchased from Biocell Laboratories (Rancho Dominguez, CA). Recombinant human IL-2 was a gift from Cetus Corporation (Emeryvelle, CA). Whole mixed ragweed antigen and dustmite antigen was obtained for the Division of Allergy,

Creighton University. Cell culture plates, centrifuge tubes and microtiter plates were obtained from Fisher Scientific (Fairlawn, NJ). Ficoll-Hypaque, phorbol myristate acetate and α-CD3 were purchased from Sigma Corporation (St. Louis, MO). Phytohemagglutinin was obtained from Gibco (Grand Island, NY). The ELISA kits for IL-4 were obtained from R&D Systems (Minneapolis, MN) and the reagents for IL-5 ELISA were purchased from PharMigen (San Diego, CA).

Generation of Antigen-Specific TH$_2$ Cell Lines

Atopic asthmatic patients aged 18 to 40 years, had 50 ml of blood drawn. All allergic patient had a 4$^+$ skin prick test reaction to either ragweed, dustmite, or both. The procedure was done in the Division of Allergy, Department of Medicine, Creighton University under the supervision of Robert G. Townley, M.D. Peripheral blood lymphocytes (PBL) from atopic asthmatic patients were then isolated by Ficoll-Hypaque gradient centrifugation. The number of cells was adjusted to 1 x 10^6/ml and PBL were stimulated with either ragweed or dustmite antigen using a concentration of 10 μg/ml. The cells were cultured at 37°C in a humidified atmosphere of 5.5% CO$_2$ in 24 well-flat bottomed Costar plates. After 7 days human recombinant IL-2 (5 U/ml) was added to all the cell lines and cultures continued for another 7 days. Thereafter, the cells were supplemented with fresh media, fresh antigen, IL-2 and allogeneic irridiated cells (1 x 10^5/ml) every 7 days.

Stimulation of Cytokine Production

The cells were induced to produce cytokines by stimulation with PMA (10 ng/ml) and anti-CD3 (50 ng/ml) or PHA (1%) for 24 hours. The supernatants were then collected to determine the presence of TH$_2$ cells in the cultures.

Determination of the Effects of Histamine and PGE$_2$ on IL-4 and IL-5 Secretion

To determine the effects of histamine and PGE$_2$ on IL-4 and IL-5 secretion, cell lines which had been in continuous culture for more than 4 weeks and were secreting high levels of IL-4 were obtained by centrifugation. After washing, the cells were resuspended at 10^6 cells/ml in RPMI-1640 supplemented with 5% human serum. The cells were stimulated in a 24 well-flat bottomed culture dishes with PMA (10 ng/ml) and α-CD3 (50 ng/ml) in the presence of histamine (10^{-4} - 10^{-8}M) and PGE$_2$ (10^{-6} - 10^{-9}M). The cells were incubated with drugs in the presence of PMA and α-CD3 for 24 hrs at 37°C in an atmosphere of 5% humidified CO$_2$. The cells were also treated with other cAMP elevating agents (PGE$_2$, forskolin, isoproterenol, serotonin) which were used as control. The supernatants were then collected and assayed for IL-4 and IL-5.

Effects of Chloride Channel Blockers on Histamine-Induced Effects on IL-4 Secretion

We explored the effects of agents which interfere with chloride channels on histamine-induced effects on IL-4 secretion by incubating the cells with α-CD3, PMA and histamine, and with and without the chloride channel blocker, stilbene disulfonate derivative, DIDS, (10^{-7} - 10^{-12}M). The supernatants were then collected and quantified for the presence of IL-4.

Measurement of Cytokines

Levels of IL-4 in supernatants were assayed with an ELISA Kit according to the manufacturer's instructions. Calculation of peak IL-4 levels was based on the differences between the amount of cytokine measured in the supernatants from the cells which were stimulated with PMA and α-CD3 and from the cells which were not stimulated. IL-4 production was expressed as picograms/1 x 10^{-6} cells.

IL-5 was quantitated by using ELISA according to Schumacher et. al. (27). Briefly, enhanced protein binding ELISA plates were coated with purified anti-IL-5 monoclonal antibody (2 μg/ml). After blocking with 10% fetal bovine serum in phosphate buffered saline (pH 7.0), standards and samples were applied to the coated wells in duplicate (100 μL/well) and the plates incubated overnight (4°C). After washing the plates with PBS/Tween, 100 μL of biotinylated anti-cytokine monoclonal antibody (2 μg/ml) in PBS/FBS was added to each well. Incubation was done at room temperature for 45 minutes. After washing, plates were incubated at room temperature for 30 minutes with 100 μL of 2.5 μL avidin-peroxidase in PBS/FBS. Immediately after washing 8 times, 100 μL of the ABTS substrate was added to each well and the color reaction allowed to develop for 40 minutes. Optical density was read at 404 nm with a Biorad 3500 microplate reader. IL-5 levels were expressed as Units/10^6 cells.

RESULTS AND DISCUSSION

This study was designed to study the effects of two cAMP elevating agents, histamine and PGE$_2$, on IL-4 and IL-5 production by TH$_2$ lymphocytes. To assess the effects of these cAMP elevating agents on IL-4 and IL-5 secretion, the TH$_2$ cells were incubated with various agonists for various time intervals in complete medium in the presence of PMA and α-CD3. The supernatants were then collected and IL-4 and IL-5 concentrations were measured by ELISA. We have previously established that TH$_2$ cells responded to histamine, PGE$_2$, serotonin, forskolin and isoproterenol by accumulation of cAMP in a dose dependent manner (data not shown). The histamine-induced response was mediated by H$_2$ receptors since it was blocked by H$_2$ antagonists (data not shown).

As shown in Table 1 histamine inhibited IL-4 synthesis/secretion in a dose dependent manner. Incubation with histamine for 4 to 6 hours produced maximum inhibition of IL-4 secretion (data not shown). Shorter or longer incubation periods diminished histamine's effects. We have previously reported similar kinetics for the effects of histamine on the generation of suppressor T cells (28).

We next asked whether other cAMP elevating agents also inhibited IL-4 secretion. We chose serotonin, forskolin, isoproterenol and PGE$_2$ because all of the agonists induced cAMP accumulation in this cell line. With the exception of PGE$_2$, none of the cAMP elevating

Table 1. Effects of histamine on IL-4 secretion

Histamine concentration (M)	Percent suppression
10^{-8}	15 ± 2
10^{-7}	13 ± 4
10^{-6}	24 ± 3
10^{-5}	35 ± 2
10^{-4}	40 ± 1

Table 2. Effects of cAMP elevating agents on IL-4 secretion

Drug concentration (M)	Percent change[*]	Drug concentration (M)	Percent change[*]
PGE$_2$		Forskolin	
10^{-6}	-89 ± 6	10^{-5}	+4 ± 4
10^{-7}	-61 ± 3	10^{-6}	-7 ± 6
10^{-8}	-23 ± 4	10^{-7}	-6 ± 2
Serotonin		Isoproterenol	
10^{-6}	-6 ± 4	10^{-6}	+8 ± 6
10^{-7}	+10 ± 8	10^{-7}	-7 ± 6
10^{-8}	+6 ± 4	10^{-8}	-4 ± 2

[*]Negative sign implies a suppression and a positive sign suggest an increase in cytokine secretion.

agents tested suppressed IL-4 secretion, suggesting that not all cAMP elevating agents act in a similar fashion (Table 2). Some of these results have been confirmed by others. It has been shown (29 & 30) that forskolin, dibutaryl cAMP and other agents that make cAMP in large quantities in TH$_2$ cells, did not inhibit IL-4 secretion from TH$_2$ cells. Since these agents are known to inhibit IL-2 secretion, the data suggest that TH$_1$ and TH$_2$ cytokines may be differentially sensitive to intracellular signals such as cAMP and the T cell subsets may have different signaling pathways. Based on these observations, it is also feasible that the mechanism of suppressive effects of histamine on IL-4 secretion is different than other cAMP elevating agents. Furthermore, this makes histamine and PGE$_2$ unique regulators of the immune response, because they may play an important role in regulating the development of a response dominated by TH$_1$ or TH$_2$ associated cytokines.

We also assessed whether histamine inhibitory effects on IL-4 secretion were mediated by H$_1$ or H$_2$ receptors. As shown in Table 3, the H$_2$ antagonist tiotidine blocked the histamine-mediated inhibition suggesting that the activation of H$_2$ receptors suppressed IL-4 secretion/synthesis. H$_1$ antagonists did not have a similar effect (data not shown).

Chloride channels have been implicated in secretory processes. We explored this possibility by testing the effects of agents that interfered with chloride flux on histamine-mediated regulation of IL-4 secretion. For these studies TH$_2$ cells were incubated with histamine in the presence and absence of stilbene disulfonate derivative DIDS and stimulated with PMA and α-CD3 and the supernatants were collected. The supernatants were assayed for IL-4 by ELISA. As shown in Table 4 DIDS (4,4-diisothiocyano-2,2'-disulfonic acid stilbene) abrogated the inhibitory effects of histamine, suggesting that histamine may inhibit cytokine secretion via activation of chloride channels.

We next analyzed the effects of histamine on IL-5 secretion. The cells were incubated with histamine (10^{-3} - 10^{-6}M) or PGE$_2$ (10^{-6} - 10^{-9}M) as described. As shown in Table 5 histamine at concentrations of 10^{-4} - 10^{-6}M enhanced the production of IL-5, but the results with PGE$_2$ were inconclusive (data not shown). Histamine-mediated effects on IL-5 produc-

Table 3. Effects of histamine and H$_2$ antagonist on IL-4 secretion

Drug concentration	Percent suppression
Histamine (10^{-5}M)	36 ± 2
Histamine (10^{-5}M) + Tiotidine (10^{-6}M)	6 ± 1
Histamine (10^{-5}M) + Tiotidine (10^{-8}M)	20 ± 4
Histamine (10^{-5}M) + Tiotidine (10^{-10}M)	38 ± 4

Table 4. Effects of chloride channel blocker on suppression by histamine

	Drug concentration	Percent change
Exp. I	Histamine (10^{-4}M)	40 ± 2
	Histamine (10^{-4}M) + DIDS (10^{-7}M)	35 ± 3
	Histamine (10^{-4}M) + DIDS (10^{-5}M)	16 ± 1
Exp. II	Histamine (10^{-4}M)	34 ± 1
	Histamine (10^{-4}M) + DIDS (10^{-7}M)	21 ± 2
	Histamine (10^{-4}M) + DIDS (10^{-5}M)	16 ± 3
Exp. III	Histamine (10^{-4}M)	42 ± 3
	Histamine (10^{-4}M) + DIDS (10^{-7}M)	28 ± 4
	Histamine (10^{-4}M) + DIDS (10^{-5}M)	10 ± 2

Table 5. Effects of histamine on IL-5 secretion

	Levels of IL-5 (ng/ml)[*]	
Drug concentration	Exp. I	Exp. II
Histamine (0)	41	135
Histamine (10^{-3}M)	37	140
Histamine (10^{-4}M)	89	210
Histamine (10^{-5}M)	102	260
Histamine (10^{-6}M)	105	255

[*]Two representative experiments are shown out of five.

tion were mediated via H_2 receptors because H_2 receptor antagonists abrogated histamine-mediated upregulation of IL-5 production (data not shown).

It has been reported (29) that PGE_2 selectively and dose dependently inhibited IL-2 and γ-IFN production, but did not affect IL-4 production but even upregulated IL-5 production. Furthermore forskolin, 2'-0-dibutaryl cAMP and 3-isobutyl-1-methylxanthine upregulated IL-5 production, whereas IL-2 production was inhibited. Betz and Fox (29) have also reported that PGE_2 enhanced IL-5 production. Our observations of the effects of histamine on IL-5 production are in agreement with other reports in literature regarding the effects of cAMP elevating agents on IL-5 production. Further understanding of the transduction mechanisms associated with the secretory mechanisms of IL-4 and IL-5 will open new avenues for therapeutic manipulations.

ACKNOWLEDGMENTS

We thank Ms. Thelma A. Cornelius for secretarial assistance. This project was supported in part by a grant from Health Future Foundation.

REFERENCES

1. Andersen, G.P. and Coyle, A.J., 1994, TH_1 amd "TH_2 like" cells in allergy and asthma, Trends in Pharmacological Sciences. 15:324-332.

2. Romagnani, S., 1991, Human TH_1 and TH_2: doubt no more, Immunology Today. 12:256-257.

3. Maggi, E., Parronchi, P., Manetti, R., Simonelli, C., Pacccinni, M-P, Rugiu, F.S., De Carli, M., Ricci, M., Romagnani, S., 1991, Reciprocal regulatory effects of IFN alpha and IL-4 on the *in vivo* development of human TH_1 and TH_2 clones, J. Immunol., 148:2142-2147.

4. Romagnani, S., 1992, Human TH$_1$ and TH$_2$ subsets: regulation of differentiation and role in protection and immunopathology, Int. Arch. Allergy Immunology, 98:279-285.

5. Romagnani, S., 1991, Type 1 T helper and Type 2 T helper cells: functions, regulation and role in protection and disease, Int. J. Lab. Clin. Res., 21:152-158.

6. DelPrete, G-F., Maggi, E., Parronchi, P., Chertien, I., Tirl, A., Macchia, D., Ricci, M., Bacchereau, J., DeVries, J., Romagnani, S., 1988, IL-4 is an essential factor for IgE synthesis induced in vitro by human T cell clones, J. Immunol., 140:4193-4198.

7. Campbell, H.D., Tucker, W.Q.J., Hort, Y., Martinson, M.E., Mayo, G., Clutterbuck, E.J., Sanderson, C.J., Young, I.G., 1987, Molecular cloning,nucleotide sequence, and expression of the gene encoding IL-5, Proc. Natl. Acad. Sci. USA, 84:6629-6633.

8. Saito, H., Hatake, K., Dvorak, A.M., Leiferman, K.M., Donenberg, A.D., Arai, N., Ishizaka, K., Ishizaka, T., 1988, Selective differentiation and proliferation of haematopoietic cells induced by recombinant human interleukins, Proc. Natl. Acad. Sci., USA, 85:2288-2292.

9. Sonoda, Y., Arai, N., Agawa, M., 1989, Humoral regulation of eosinophilopoiesis in vitro, Leukemia, 3:14.

10. Lopez, A.F., Sanderson, C.J., Gamble, J.R., Campbell, H.D., Young, I.G., Vadas, M.A., 1988, Recombinant human IL-5 is a selective activator of human eosinophil function, J. Exp. Med., 167:219-224.

11. Fijisawa, T., Abu-Ghazaleh, R., Kita, H., Sanderson, C-J., Gleich, G.J., 1990, Regulatory effect of cytokines on eosinophil degranulation, J. Immunol., 144:642-646.

12. Walsh, G.M., Hartnell, A., Wardlaw, A.J., Kurihara, K., Sanderson, C.J., Kay, A.B., 1990, IL-5 enhances the in vitro adhesion of human eosinophils, but not neutrophils in a leukocyte integrin dependent manner, Immunology, 71:258-265.

13. Robinson, D.S., Durham, S.R., Kay, A.B., 1993, Cytokines in asthma, Thorax, 48:845-853.

14. Norman, P.S., 1980, An overview of immunotherapy, implications for the futrue, J. Allergy Clin. Immunol., 65:87-95.

15. Parronchi, R., Macchia, D., Paccinni, M-P, Biswas, P., Simonelli, C., Maggi, E., Ricci, M., Ansari, A.A., Romagnani, S., 1991, Allergens and bacterial-specific T-cell clones established from atopic donors show a different profile of cytokine production, Proc. Natl. Acad. Sci., 88:4538-4542.

16. DelPrete, G-F., DeCarli, M., Mastromauro, C., Macchia, D., Biagiotti, R., Ricci, M., Romagnani, S., 1991, Purified protein derivative of *Myobacterium tuberculosis* and excretory-secretory antigen(s) of *Toxocara canis* expand in vitro human T cells with stable and opposite profile of cytokine production, J. Clin. Invest., 88:346-350.

17. deWaal Malefyt, R., Yssel, R., deVries, J.E., 1993, Direct effects of IL-10 in subset of human CD4$^+$ T cell clones and resting T cells, J. Immunol, 150:4754-4765.

18. Melmon, K.L., Rocklin, R. and Rosekranz, R.P., 1981, Autacoids as modulators of inflammatory and immune response, Am. J. Med., 71:100-106.

19. Suzuki, S. and Huchet, R., 1981, Mechanisms of histamine-induced inhibition of lymphocyte response to mitogens in mice, Cell Immunol., 62:396-402.

20. Lima, M. and Rocklin, R.E., 1981, Histamine modulates in vitro IgG production by human mononuclear cells, Cell Immunol., 64:324-336.

21. Khan, M.M., Keaney, K.M., Melmon, K.L., Clayberger, C., Krensky, A.M., 1989, Histamine regulates the generation of human cytolytic T lymphocytes, Cell Immunol., 121:60-73.

22. Dohlsten, M., Sjogran, H.O. and Carlsson, R.A., 1987, Histamine acts directly on human T cells to inhibit IL-2 and γ-IFN production, Cell Immunol., 101:65-78.

23. Lappin, D., Moseley, H.L. and Whaley, K., 1980, Effects of histamine on monocytes complement production, Clin. Exp. Immunol., 42:515-523.

24. Busse, W.W. and Sosman, J., 1976, Histamine inhibition of neutrophil lysosomal enzyme release, Science, 194:737-740.

25. Singh, U. and Owen, J.J.T., 1976, Studies on the maturation of thymus stem cells, Eur. J. Immunol., 6:59-64.

26. Anwar, A.R.E. and Kay, A.B., 1977, The ECF-A tetrapeptide and histamine selectively enhance human eosinophil complement receptors, Nature, 269:522-525.

27. Schumacher, J.H., O'Garra, A., Shrader, B., Kimmenade, A., Bond, M.W., Mosman, T.R. and Coffman, R.L., 1988, The characterization of four monoclonal antibodies specific for mouse IL-5 and the development of mouse and human enzyme-linked immunosorbant assay, J. Immunol., 141:1576-1581.

28. Khan, M.M., Sansoni, P., Engelman, E.G. and Melmon, K.L., 1985, The pharmacologic effects of autacoids on subsets of human T cells, J. Clin. Invest., 75:1578-1583.

29. Betz, M. and Fox, B., 1991, Prostaglandin E$_2$ inhibits production of TH$_1$ lymphokines but not TH$_2$ lymphokines, J. Immunol., 146:108-113.

30. Snijdewint, F.-G.M., Kalinski, P. Wierenga, E.A., Bos, J.D. and Kaspensberg, M.L., 1993, Prostaglandin E$_2$ differentially modulates cytokine secretion profiles of human T helper lymphocytes, J. Immunol., 150:5321-5329.

6

IMMUNOGLOBULIN ISOTYPE MODULATION AFTER ADMINISTRATION OF IL-12

Victor Van Cleave, Stan Wolf, Kristin Murray, Anna Wiencis,
Mara Ketchum, Judy Bliss, Theresa Haire, Christine Resmini,
Rich Maylor, and Ed Alderman

Pre-Clinical Biology Department
Genetics Institute
One Burtt Road, Andover, Massachusetts 01810

ABSTRACT

We have begun a series of experiments assessing the role of IL-12 in the humoral immune response. IL-12 is known to enhance cellular immunity causing a shift toward a Th1, as opposed to a Th2, response. IL-12 is also a potent stimulator of IFN-γ production which, among other activities, modulates isotype expression particularly with respect to IgG_{2a}. We have performed a series of experiments involving the concurrent dosing of mice with murine IL-12 and TNP-KLH followed by the monitoring of IgG_1 and IgG_{2a} anti-TNP responses and total IgG_1 and IgG_{2a} levels. Following administration of IL-12, specific anti-TNP titers showed an IgG_{2a} increase while IgG_1 responses were markedly lower than those exhibited by animals which did not receive IL-12. Total IgG_1 levels in IL-12 treated mice remained at or near baseline while untreated mice demonstrated an increase in total IgG_1 levels. In addition, lymph nodes from these mice were removed, stimulated with KLH and assayed for expression of murine IFN-γ and IL-4. Murine IFN-γ levels in supernatants obtained from IL-12 treated mice were elevated over those seen in untreated mice while IL-4 levels were suppressed.

INTRODUCTION

Interleukin-12 (IL-12) is a potent immunomodulatory cytokine produced by macrophages and B cells. Cloning of the factor (originally termed natural killer cell stimulatory factor or NKSF) from an EBV-transformed B cell line revealed that this cytokine is a disulfide-linked heterodimeric protein composed of 35-kDa light chain and 40-kDa heavy chain subunits (Wolf, et al., 1991). IL-12 acts as a stimulatory signal in cell-mediated responses by means of induction of a T helper type-1 (Th1) response. Examples of this

Immunobiology of Proteins and Peptides VIII
Edited by M. Z. Atassi and G. S. Bixler, Jr., Plenum Press, New York, 1995

activity include a) activation of Th1 cells, b) increased cytolytic activities of T cells, macrophages, and NK cells, c) enhanced proliferation of T cells and NK cells, d) upregulated expression of cell surface markers, and e) induction of cytokine production including IFN-γ, GM-CSF, and TNF-α (Chehimi and Trinchieri, 1994; Aste-Amezaga, et al., 1994; Trinchieri and Scott, 1994). A number of studies have shown that IL-12 can induce protection against intracellular organisms such as *Toxoplasma gondii* (Hunter, et al., 1994) and *Leishmania major* (Sypek, et al., 1993), extracellular parasites such as *Schistosoma mansoni* (Oswald, et al., 1994; Trinchieri and Scott, 1994), and gram-positive pyogenic bacteria such as *Staphylococcus aureus* (Gladue, et al., 1994).

Recognizing that IL-12 upregulates IFN-γ, we were interested in ascertaining if alterations in the immunoglobulin isotype profiles would be observed during an immune response to a defined hapten. McKnight et al. (1994) have demonstrated changes in isotype profile after antigen and IL-12 administration while others have suggested that IL-12 might play a role in the development of prophylactic vaccines (Salk, 1993; Afonso, 1994). To answer questions concerning the effect of IL-12 on immunoglobulin profiles and their temporal relationship to cytokine appearance (IFN-γ and IL-4), we developed a mouse model incorporating Balb/c mice administered IL-12 and immunized with TNP-KLH. Serum levels of IgG_1 and IgG_{2a} anti-TNP titers as well as total IgG_1 and IgG_{2a} levels were then monitored. In addition, lymph nodes were removed and cultured to measure secretion of IFN-γ and IL-4.

MATERIALS AND METHODS

Experimental Design

Prebleeds from groups of three Balb/c mice each were obtained on Day -2. Experimental groups 1, 3, and 5 received 1 μg of mIL-12 subcutaneously on Days -1, 0, and +1 while control groups 2, 4, and 6 received a sham injection. On Day 0, groups 1 and 2 were immunized with 100 μg/mouse of TNP-KLH in alum, groups 3 and 4 were given 100 μg/mouse of TNP-KLH (no adjuvant), and groups 5 and 6 were administered PBS with alum. All experimental and control groups were set up in replicates to allow sacrifice of the animals at various time points in the experiment. Bleeds were obtained on Days 7 and 15. Replicate groups of mice (as described above) were exsanguinated on Days 4, 7 and 15. Their lymph nodes were removed and were cultured with either 2C11 conditioned media containing anti-CD3 antibody, 100 μg/mL KLH, 10 μg/mL KLH, or media only. After three days supernatants were analyzed for levels of IFN-γ and IL-4.

Total Mouse IgG_1 and IgG_{2a} Concentration

96-well flat bottom EIA (Costar) plates were coated with 2 μg/mL, 50 μl/well of rabbit anti-mouse Ig (Southern Biotech) at 4°C overnight. Plates were washed and blocked with 5% gelatin for ninety minutes at 37°C. After washing, mouse serum samples were diluted in PBS-Tween, titered onto the capture plate, and incubated for two hours at room temperature. After incubation, plates were thoroughly washed and 50 μl/well rabbit anti-mouse Ig-horseradish peroxidase (isotype specific for IgG_1 or IgG_{2a}; Southern Biotech) diluted 1:1000 in PBS-Tween was added and incubated for ninety minutes. After washing, ABTS substrate (Kirkegaard & Perry) was added and the O.D. at 405 nm recorded. A standard curve using mouse Ig standards of the appropriate isotype was generated on the same assay plate and sample O.D. values were fitted against the standard curve.

TNP-Specific Titer Determination

96-well flat bottom EIA (Costar) plates were coated with 50 µg/mL, 50 µl/well of TNP-BSA at 4°C overnight. Plates were washed and blocked with 5% gelatin for ninety minutes at 37°C. After washing, mouse serum samples were diluted in PBS-Tween, titered onto the capture plate, and incubated for two hours at room temperature. After incubation, plates were thoroughly washed and 50 µl/well rabbit anti-mouse Ig-horseradish peroxidase (isotype specific for IgG_1 or IgG_{2a}; Southern Biotech) diluted 1:1000 in PBS-Tween was added and incubated for ninety minutes. After washing, ABTS substrate (Kirkegaard & Perry) was added and the O.D. at 405 nm recorded.

Murine Interferon-gamma (mIFN-γ) Determination

Lymph nodes from experimental animals were removed and placed in culture for three days with anti-CD3 antibody, 10 µg/mL keyhole limpet hemocyanin (KLH), 100 µg/mL KLH, or media only. Supernatants from these cultures were added to gelatin-blocked 96-well flat bottom EIA (Costar) plates which had been coated with 3 µg/mL, 50 µl/well of R46A2, a rat monoclonal anti-mIFN-γ antibody. After a two hour incubation, the plate was washed, and 50 µl/well biotinylated XGM1.2 (a rat monoclonal anti-mIFN-γ antibody; Scott, 1991) at a 1:1000 dilution in PBS-Tween was added for ninety minutes. 50 µl/well of avidin D-horseradish peroxidase (Vector) at a 1:4000 dilution in PBS-Tween was incubated for one hour, washed and ABTS substrate (Kirkegaard & Perry) was then added. The O.D. was recorded at 405 nm and concentrations were determined by fitting the sample O.D. to a standard curve prepared using mIFN-γ (Genzyme).

IL-4 Determination by CT4.S Bioassay

The CT4.S T cell line was used to determine IL-4 concentrations in lymph node culture supernatants. As described by Hu-Li, et al. (1989), CT4.S cells were washed extensively and placed in 96 well plates with equal volumes of fresh media and supernatant obtained from the mouse lymph node cultures (as summarized above in the experimental design) for 48 hours. Cells were incubated with ^{3}H-thymidine (1 µCi/well) for 16 hours, washed, and harvested. Incorporated counts were plotted against a standard curve to determine IL-4 concentration.

RESULTS

Total Immunoglobulin Concentration

Animals receiving IL-12 showed a drop in IgG_1 concentrations by Day 7 after immunization with TNP-KLH or PBS as seen in Figure 1. IgG_1 levels in IL-12-treated mice rebounded by Day 15 to concentrations exceeding the non-IL-12 treated mice (TNP-KLH immunized) or were equivalent to their own prebleed IgG_1 concentrations (PBS group). IgG_{2a} isotype levels were elevated by four fold in the TNP-KLH immunized groups by Day 15 (Figure 2). Similarly, by Day 15 the PBS-treated animals had a 40% increase in total IgG_{2a}.

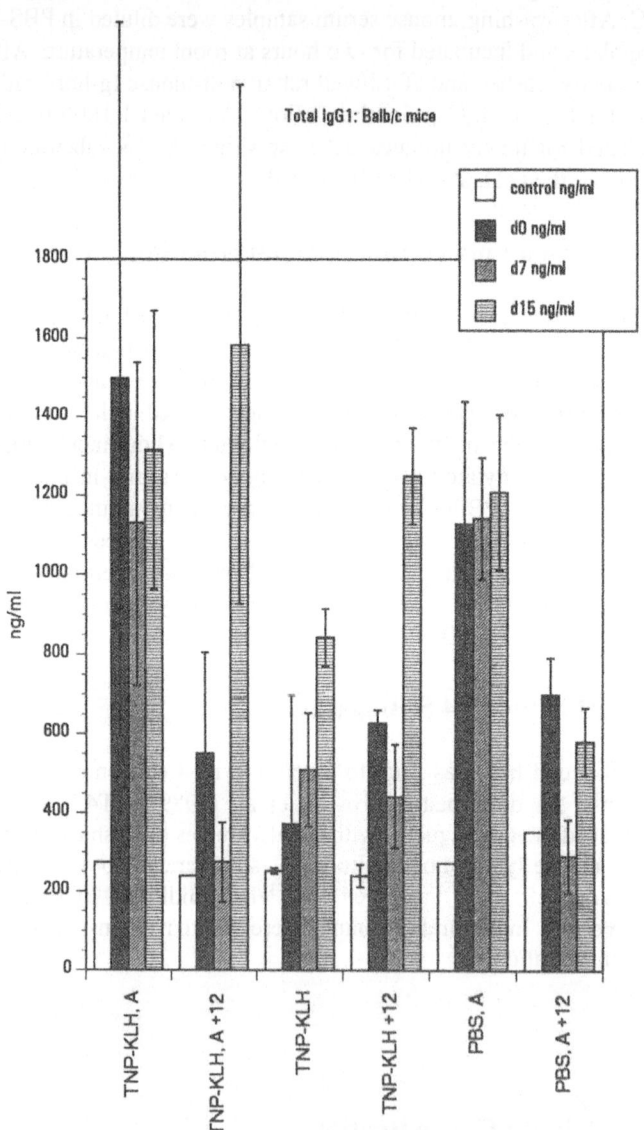

Figure 1. Total IgG1 levels in ng/mL by ELISA.

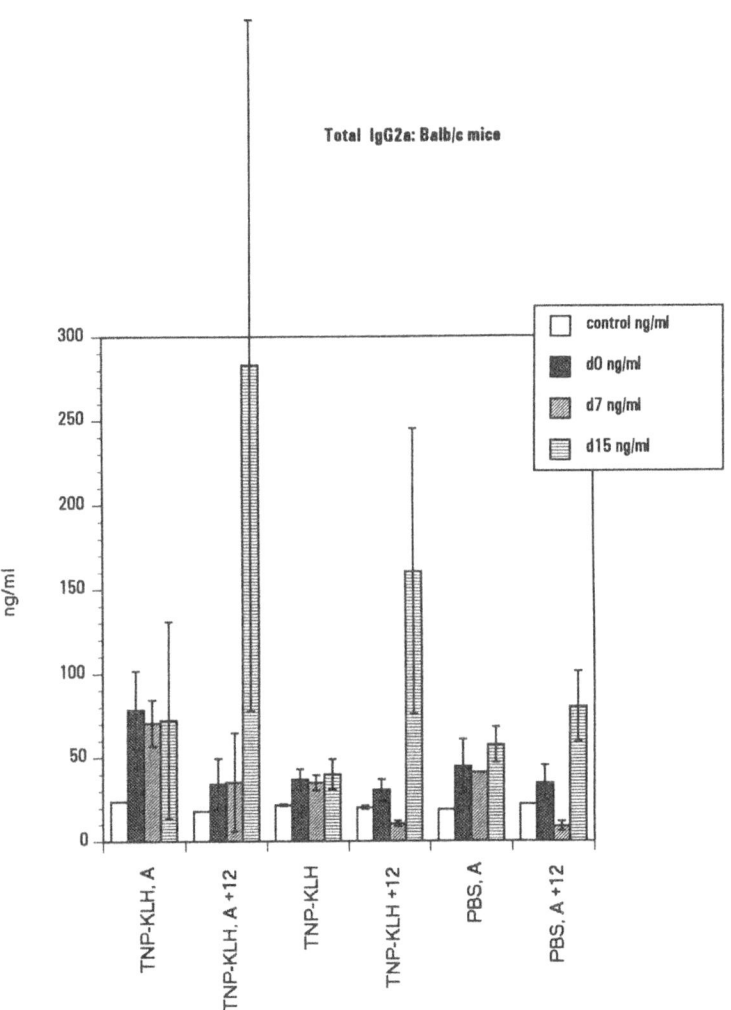

Figure 2. Total IgG2a levels in ng/mL by ELISA.

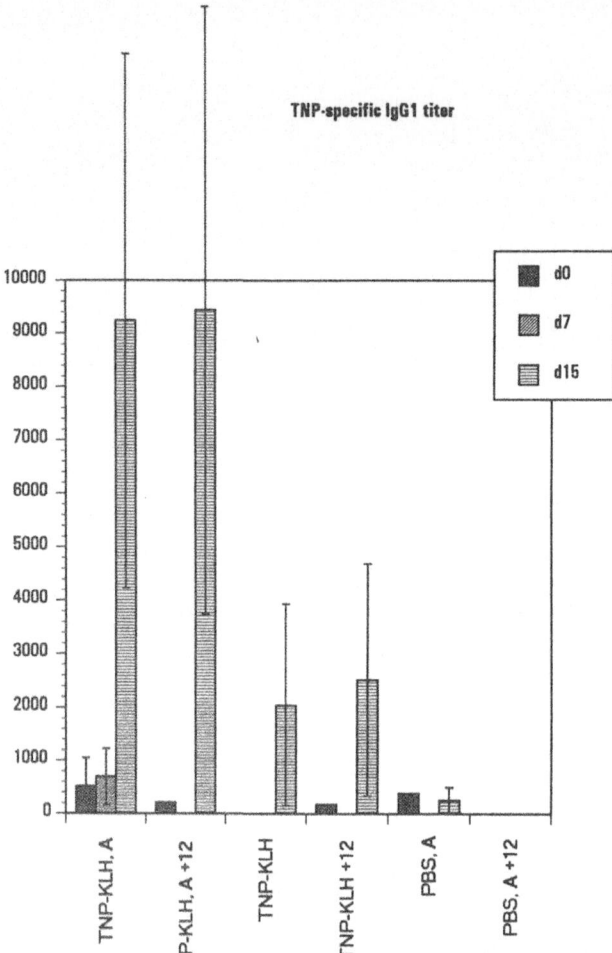

Figure 3. TNP specific IgG1 titers as measured by ELISA. Administration of IL-12 did not significantly alter TNP specific IgG1 titers by Day 15.

Test animals receiving TNP-KLH (+/- Alum) showed a depressed IgG1 concentration on Day 7 as compared to animals receiving only TNP-KLH (+/- Alum). By Day 15 all IL-12 treated groups had rebounded to IgG1 levels similar to those seen in untreated animals.

Test animals receiving TNP-KLH and IL-12 (+/- Alum) exhibited four-fold increases in total IgG2a levels over non-IL-12 treated animals by Day 15.

Anti-TNP Specific Titers

Day 15 bleeds from Balb/c mice administered IL-12 produced IgG_1 titers to TNP that were near identical to those seen in non-IL-12 treated mice (Figure 3). Conversely, IgG_{2a} titers against TNP were highly elevated. Animals receiving TNP-KLH in alum and IL-12 showed a five fold increase in specific anti-TNP titers over that seen in non-IL-12 treated mice. Interestingly, TNP-KLH administered without alum induced a twenty-four fold increase (4254 to 172) in IgG_{2a} anti-TNP titers among mice receiving IL-12 as compared to non-IL-12 treated mice (see Figure 4).

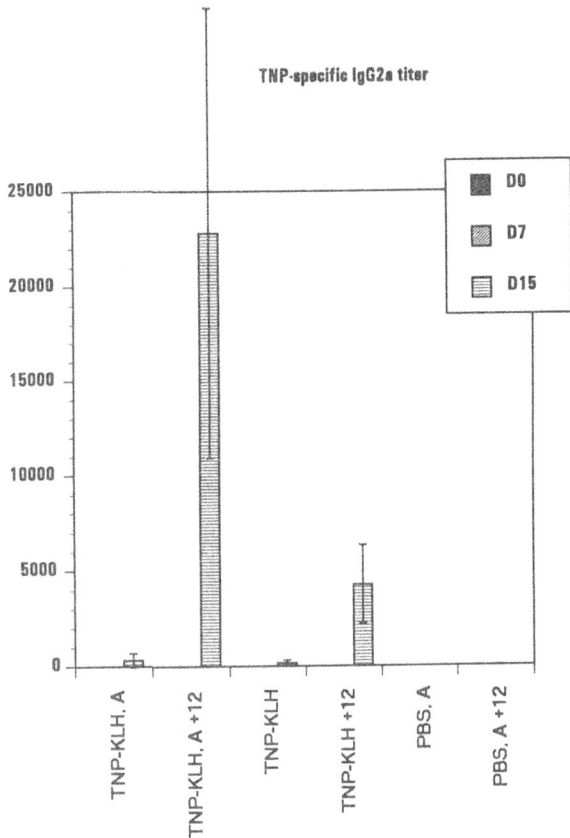

Figure 4. TNP specific IgG2a titers as measured by ELISA. Administration of IL-12 enhanced TNP specific IgG2a titers from five- (+Alum) to 24-fold (-Alum) over non-IL-12 treated animals.

Murine IFN-γ Concentrations

Lymph node cultures from IL-12 treated or non-treated mice were stimulated with anti-CD3, 10 μg/mL KLH, 100 μg/mL KLH, or media only. Murine IFN-γ levels in tissue culture supernatants were examined after three days of culture and the levels were markedly increased in cultures obtained from IL-12 treated mice as compared to non-IL-12 treated mice as shown in Figure 5. These increases were noted by Day 4 in the mouse experimental groups and continued through Day 15.

Lymph node cultures from animals receiving IL-12 showed markedly enhanced levels of IFN-γ by Day 4. IFN-γ levels remained elevated thru Day 15.

IL-4 Determinations

Groups of mice were immunized with TNP-KLH (+/- alum) and treated with or without IL-12 followed by lymph node removal on Day 4, 7, or 15. Lymph nodes cell preparations were cultured for three days in the presence of 10 μg/mL KLH, 100 μg/mL KLH, or media only. Conditioned tissue culture supernatants were then evaluated in the CT4.S bioassay. Supernatants obtained from Day 4 IL-12-treated mouse lymph nodes generated less than one-tenth the stimulatory signal (^{3}H-thymidine incorporation) as compared to non-IL-12-treated mice (average of IL-12 treated versus non-IL-12-treated was

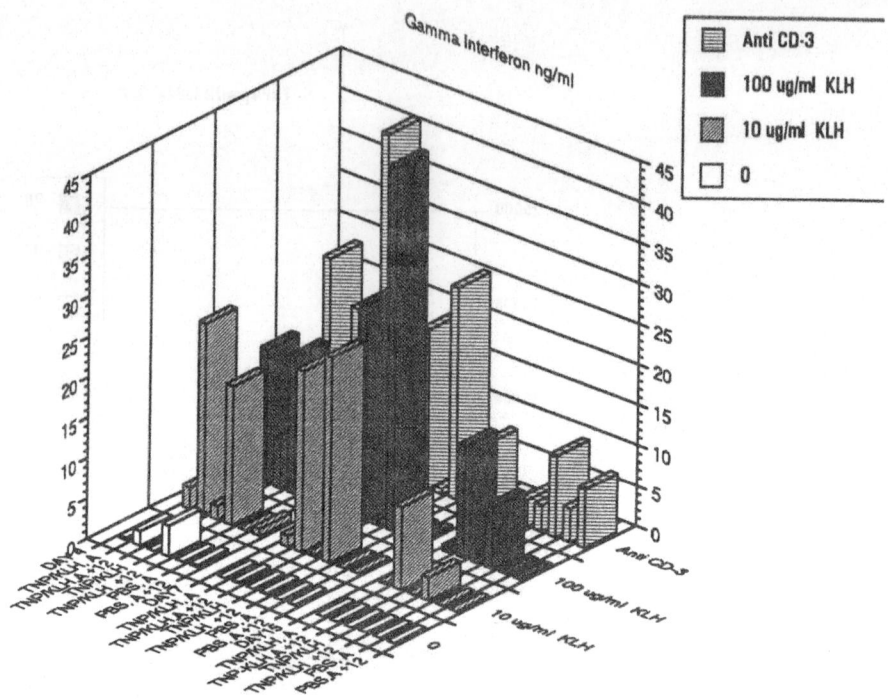

Figure 5. Interferon-gamma (IFN-γ) levels in ng/mL from cultured lymph node supernatants as measured by ELISA.

128,558 to 9911 cpm; see Figure 6). IL-4 signal remained depressed in IL-12-treated mice at Days 7 and 15 although levels were slowly rising (data not shown).

CONCLUSIONS

IL-12 plays an important role in directing the immune response toward a Th1 or cellular immune response (Manetti, et al., 1993; Hsieh, et al., 1993). The mechanism of action responsible for this activity is not fully understood although upregulation of IFN-γ, itself a promoter of Th1 cell development (Maggi, et al., 1992), appears to be important. Pertubations in IFN-γ levels can also alter isotype profiles particularly with regard to IgG$_{2a}$ (Snapper and Paul, 1987). We have demonstrated that Balb/c mice immunized with TNP-KLH mount a strong IgG$_{2a}$ response to the hapten TNP after dosing with IL-12. This response is evidenced by both a four-fold increase in absolute or total IgG$_{2a}$ serum concentrations and increases of up to twenty-four-fold in the specific IgG$_{2a}$ titers against TNP. IgG$_1$ responses appear to be initially depressed (Day 7) but rebound to levels equivalent to those observed in untreated animals by Day 15. Culture supernatants obtained from lymph nodes of IL-12 treated Balb/c mice show striking increases in IFN-γ after stimulation with KLH or anti-CD3. C57bl/6 mice have been observed to exhibit a similar result although of decreased magnitude (data not shown). Conversely, IL-4 levels in culture supernatants obtained from IL-12-treated mice were reduced consistent with the role that IL-4 plays as a promoter of Th2 responses (Coffman, et al., 1991). These results demonstrate that IL-12 can mediate changes in humoral responsiveness and suggest a potential application in the area of vaccine development.

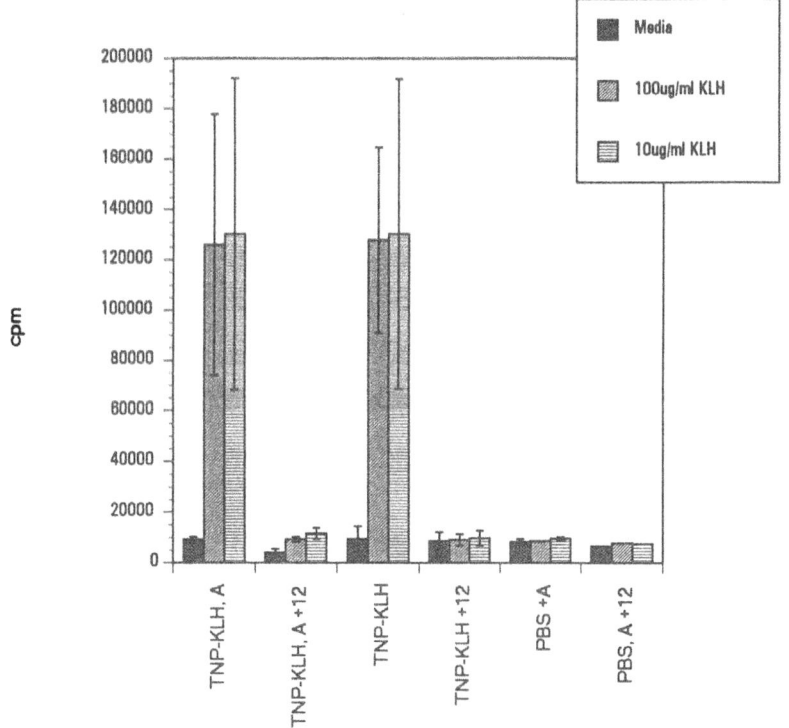

Figure 6. Relative levels of IL-4 in Day 4 culture supernatants. IL-4 levels were substantially depressed in lymph node culture supernatants obtained from IL-12-treated mice as compared to non-IL-12-treated mice. Levels on Days 7 and 15 (data not shown) were similarly depressed.

REFERENCES

Afonso, L., Scharton, T., Vieira, L., Wysocka, M., Trinchieri, G., and Scott, P. 1994. The adjuvant effect of Interleukin-12 in a vaccine against Leishmania major. *Science* 263:235.

Aste-Amezaga, M. D'Andrea, A., Kubin, M., Trinchieri, G. 1994. Cooperation of natural killer cell stimulatory factor/ interleukin-12 with other stimuli in the induction of cytokines and cytotoxic cell-associated molecules in human T and NK cell. *Cell. Immunol.* 156:480.

Chehimi, J. and Trinchieri, G. 1994. Interleukin-12: A bridge between innate resistance and adaptive immunity with a role in infection and acquired immunodeficiency. *J. Clin. Immunol.* 14:149.

Coffman, R., Varkila, K., Scott, P., and Chatelain, R. 1991. Role of cytokines in the differentiation of CD4+ T-cell subsets in vivo. *Immunol. Rev.* 123:189.

Gladue, R., Laquerre, A., Magna, H., Carroll, L., O'Donnell, M., Changelian, P., and Franke, A. 1994. *In vivo* augmentation of IFN-γ with a rIL-12 human/mouse chimera: Pleiotropic effects against infectious agents in mice and rats. *Cytokine* 6:318.

Hsieh, C-S., Macatonio, S., Tripp, C., Wolf, S., O'Garra, A., and Murphy, K. 1993. Development of T_H1 CD4+ T cells through IL-12 produced by *Listeria*-induced macrophages. *Science* 260:547.

Hu-Li, J., O'Hara, J., Watson, C., Tsang, W., and Paul, W. 1989. Derivation of a T cell line that is highly responsive to IL-4 and IL-2 (CT4.R) and of an IL-2 hyporesponsive mutant of that line (CT4.S). *J. Immunol.* 142:800.

Hunter, C., Subauste, C., Van Cleave, V., and Remington, J. 1994. Production of gamma interferon by natural killer cells from *Toxoplasma gondii*-infected SCID mice: Regulation by Interleukin-10, Interleukin-12, and tumor necrosis factor alpha. *Inf. and Imm.* 62:2818.

Maggi, E., Parronchi, Manetti, R., Simonelli, C., Piccinni, M.-P., Rugiu, F., de Carli, M., Ricci, M., and Romagnani, S., 1992. Reciprocal regulatory effects of IFN-γ and IL-4 on the in vitro development of Th1 and Th2 clones. *J. Immunol.* 148:2142.

Manetti, R., Parronchi, P., Giudizi, M., Piccinni, M.-P., Maggi, E., Trinchieri, G., and Romagnani., S. 1993. Natural killer cell stimulatory factor (interleukin-12 (IL-12)) induces T helper type 1 (Th1)-specific immune responses and inhibits the development of IL4-producing Th cells. *J. Exp. Med.* 177:1199.

McKnight, A., Zimmer, G., Fogelman, I., Wolf, S., and Abbas, A. 1994. Effects of IL-12 on helper T cell-dependent immune responses *in vivo*. *J. Immunol.* 152:2172.

Oswald, I., Caspar, P., Jankovic, D., Wynn, T., Pearce, E., and Sher, A. 1994. IL-12 inhibits Th2 cytokine responses induced by eggs of *Schistosoma mansoni*. *J. Immunol.* 153:1707.

Salk, J., Bretscher, P., Salk, P., Clerici, and Shearer, G. 1993. A strategy for prophylactic vaccination against HIV. *Science* 260:1270.

Scott, P. 1991. IFN-γ modulates the early development of Th1 and Th2 responses in a murine model of cutaneous leishmaniasis. *J. Immunol.* 147:3149.

Snapper, C. and Paul, W. 1987. Interferon-gamma and B cell stimulatory factor-1 reciprocally regulate Ig isotype production. *Science* 236:944.

Sypek, J., Chung, C., Mayor, S., Subramanyam, J., Goldman, S., Sieburth, D., Wolf, S., and Schaub, R. 1993. Resolution of cutaneous leishmaniasis: Interleukin 12 initiates a protective T helper type 1 immune response. *J. Exp. Med.* 177:1797.

Trinchieri, G. and Scott, P. 1994. The role of interleukin-12 in the immune response, disease and therapy. *Immunol. Today* 15:460.

Wolf, S., Temple, P., Kobayashi, M., Young, D., Dicig, M., Lowe, L., Dzialo, R., Fitz, L., Ferenz, C., Hewick, R., Kelleher, K., Herrmann, S., Clark, S., Azzoni, L., Chan, S., Trinchieri, G., and Perussia, B. 1991. Cloning of cDNA for natural killer cell stimulatory factor, a heterodimeric cytokine with multiple biologic effects on T and natural killer cells. *J. Immunol.* 146:3074.

7

SUBSTANCE P MEDIATED STIMULATION OF CYTOKINE LEVELS IN CULTURED MURINE BONE MARROW STROMAL CELLS

Jill Marie Manske, Erin L. Sullivan, and Shawna M. Andersen

Department of Biology
University of St. Thomas
St. Paul, Minnesota

ABSTRACT

Substance P (SP) is a neuropeptide which has been reported to have immunomodulatory activity. Most studies on SP have been performed on cells of the peripheral immune system. More recently, SP has been reported to have stimulatory activity on human bone marrow cells *in vitro*, and this activity was dependent on the presence of an adherent layer of cells. The *in vitro* adherent layer represents the stromal cells of the marrow. In this study, we directly addressed the effect of SP on cultured bone marrow stromal cells. Since stromal cells play an important role in the regulation of hematopoiesis, interactions of neuropeptides such as SP with this cell population could lead to an alteration of stem cell development within the bone marrow. Previously we have shown that SP stimulates protein synthesis in this cell population with two waves of protein synthesis activation, after 2 hr and 48 hr of SP incubation. In this study, we asked whether levels of known stromal cell cytokines were altered in response to SP incubation. We assayed the levels of Interleukin-7 (IL-7) and Stem Cell Factor (SCF) associated with the stromal cell surface following 2 hr and 48 hr of SP incubation. Cells were stimulated with SP for 2 hr or 48 hr. Following SP incubation, cells were washed and the levels of cell associated cytokine was determined by ELISA. Following 2 hr of treatment, 0.1 nM of SP significantly increased (p = 0.05) the level of IL-7 as compared to untreated controls. After 48 hr of treatment, 1, 10, and 100 nM SP significantly increased the levels of IL-7 in this cell population. When SCF levels were assayed, SP at all concentrations tested was found to increase significantly the levels of SCF following 2 hr of incubation. Following 48 hr of incubation, 10 and 100 nM of SP significantly increased the levels of SCF. The ability of SP to affect cytokine levels varied with time. Following 2 hr of SP incubation, cytokine levels were enhanced at the lower end of the concentration range as compared to 48 hr of SP treatment. A 48 hr incubation with SP yielded the highest levels of cytokine at the higher end of the concentration range. Taken together, the results of these studies suggest that SP has an immunoregulatory effect on bone marrow stromal

Immunobiology of Proteins and Peptides VIII
Edited by M. Z. Atassi and G. S. Bixler, Jr., Plenum Press, New York, 1995

cells leading to alteration in the production and/or secretion of regulatory cytokines such as IL-7 and SCF.

INTRODUCTION

The possibility that neurophysiologic factors influence immunity was demonstrated initially by the finding of altered lymphocyte and antibody responses in animals with experimentally induced lesions of the central nervous system (1,2). The observation that endorphins and methionine-enkephalins express both stimulatory and suppressive activities for lymphocytes *in vitro* suggested that one mechanism for central nervous system modulation of immunologic responses may be the direct effects of neuropeptides on lymphocytes (3,4). More recently, another neuropeptide, Substance P has been demonstrated to have immunomodulatory activity.

Substance P (SP) is an 11 amino acid neuropeptide belonging to a family of structurally related bioactive peptides known as tachykinins. It was originally isolated from equine brain and intestine and is now known to be distributed throughout the central and peripheral nervous system of many species (5). SP is involved in the regulation of motor control, contraction of smooth muscle, salivation, micturition, vasodilation, and sensory perception (6,7). Numerous studies have shown SP to have immunoregulatory properties as well. These include stimulation of T-cell proliferation (8), augmentation of Ig synthesis by B-cells (9), the induction of mast cell degranulation and histamine release (10), enhancement of PGE_2 and collagenase secretion from synoviocytes (11), macrophage and neutrophil activation (12), modulation of lymphocyte traffic (13), and modulation of cytokine release (11, 14-17).

Most recently, SP has been reported to have stimulatory activity on human bone marrow cells *in vitro* (18). In this study, the authors observed a marked reduction in the ability of SP to stimulate colony formation in adherent cell-depleted bone marrow cultures. Since the adherent cells represent the bone marrow stromal cells, the authors concluded that the stromal cells were the main mediator in the stimulatory effect of SP in their cultures (18). The direct effect of SP on stromal cells, however, was not addressed in this study.

Stromal cells, the supporting cells of the bone marrow, provide the necessary environment for hematopoiesis. In addition to providing the physical "scaffolding" or "meshwork" on which hematopoietic stem cells develop within the marrow, stromal cells produce a number of cytokines. Through their interactions with stem cells, stromal cells are believed to be intimately involved in regulating the production, differentiation, and development of cell types derived from the bone marrow (19).

In vivo, the bone marrow stroma is made up of a variety of cell types. Depending on culture conditions, the phenotype of bone marrow stromal cells *in vitro* has been reported to include endothelial cells, macrophages, and fibroblasts (19-21). Under the culture conditions used in our laboratory, we have observed most of the stromal cells to exhibit morphology and cell markers characteristic of fibroblasts. This is in agreement with other studies which have concluded that the bone marrow fibroblast is the *in vitro* equivalent of the dominant *in vivo* supporting cell of the bone marrow (19). SP has been shown to stimulate the proliferation of fibroblasts (22) and it shares amino acid homology with the brain-derived fibroblast growth factor (23). It is therefore likely that SP will have a stimulatory effect on cultured bone marrow stromal cells.

Since interaction of SP with stromal cells may alter the interactions of stromal cells with stem cell populations, we chose to investigate the direct effects of SP on cultured murine stromal cells. Specifically, we addressed the ability of SP to affect the levels of the cytokines Interleukin-7 (IL-7) and Stem Cell Factor (SCF).

Interleukin-7 was originally isolated from stromal cells (24) and was discovered to be a factor which was capable of inducing proliferation of murine pre-B cells in long term bone marrow culture (24,25). IL-7 receptors are present on murine pre-B cells, but not on mature B cells. IL-7 also functions as a growth and differentiation factor for murine and human T-cells (26-28). IL-7 has shown to be mitogenic for resting early murine (CD4$^-$CD8$^-$) thymocytes (28), and is an important factor in the regulation of B and T cell development. Therefore, the ability of SP to alter IL-7 secretion by stromal cells could have implications relative to lymphocyte development and differentiation.

Stem cell factor (SCF) is a pleiotropic growth factor with diverse hematopoietic target cells including early progenitor (stem) cells (29-31). The cDNA sequences for human, mouse, and rat SCF encode transmembrane proteins which are composed of a signal peptide, a 189 amino acid extracellular domain, a hydrophobic transmembrane domain, and an intracellular domain (32,33). Native SCF can exist either as the membrane bound form or as a soluble form consisting of the first 164 or 165 amino acids of the extracellular domain (32,33). The soluble form is believed to be a proteolytic cleavage product of the transmembrane protein. Both soluble and membrane bound forms have growth factor activities on hematopoeitic precursors purified from mouse and human bone marrow. Stem Cell Factor acts in synergy with other growth factor to induce myeloid, erythroid, and lymphoid lineage colony formation (29).

In this study, we investigated the effect of SP on IL-7 and SCF in cultured murine bone marrow stromal cells. Our data indicate that SP is capable of enhancing the levels of these cytokines in this cell population.

MATERIALS AND METHODS

Mice

BALB/c mice were purchased from Jackson Laboratories (Bar Harbor, Maine). Animals were bred and housed in the mouse colony at the University of St. Thomas. Male or female mice 6 to 8 weeks old were used. All protocols involving animals were reviewed and approved by the University of St. Thomas Institutional Animal Care and Use Committee.

Cells

Mice were killed by cervical dislocation. Femora and tibiae were removed by sterile dissection. The marrow plug was flushed from the bone using a syringe containing RPMI-1640 (Sigma, St. Louis, MO). Cells were washed three times by centrifugation in RPMI-1640 and resuspended in RPMI containing 10% FBS, 1% Penicillin/Streptomycin (PS), and 1% L-Glutamine (LG) (all purchased from Sigma Chemical Co., St. Louis, MO). The cells were counted and viability was determined by trypan blue exclusion. Cells were plated in 96-well flat bottom tissue culture plates at a final concentration of 1 x 10^6 cells per well and incubated at 37° C, 5% CO_2 for 4 days. Following 4 days, the non-adherent cells were removed and media replaced with 200 µl of fresh cell culture media. The media was replaced every 48 hours for approximately two weeks until the cells became confluent.

Reagents

Substance P and the SP antagonist Spantide (D-Arg-Pro-Lys-Pro-Gln-Gln-D-Trp-Phe-D-Trp-Leu-Leu-NH$_2$) were purchased from Sigma Chemical Company (St. Louis, MO). Reagents were diluted into RPMI, 10% FBS, PS, LG from stock solution before use.

Anti-murine-IL-7 was purchased from R & D Systems (Minneapolis, MN). Anti-murine-SCF was obtained from Genzyme (Cambridge, MA). The alkaline phosphatase conjugated secondary antibodies were purchased from Sigma Chemical Company (St. Louis, MO).

Protein Synthesis Assays

To determine the effect of SP on protein synthesis in stromal cells, confluent monolayers of cells were washed and cultured in the presence of various concentrations of SP. Select cultures were incubated in various concentrations of SP and equimolar concentrations of the SP antagonist spantide in RPMI/10% FBS. All cultures were treated in triplicate and incubated for 48 hr at 37°C, 5% CO_2. Controls were incubated in RPMI/10% FBS without SP. After 24 hr, SP containing media was removed and replaced with leucine-free RPMI. Cells were pulsed for 24 hr at 37°C with (^{3}H)leucine (Amersham Corp., Arlington Heights IL). Cells were trypsinized and harvested onto filter paper using an automated harvester (Brandel, Gaithersgurg, MD). (^{3}H)leucine incorporated into cells was determined by standard scintillation counting techniques.

For kinetic analysis of protein synthesis, cells were incubated at 37°C with or without SP in leucine-free RPMI containing 10% heat-inactivated, dialyzed FBS. Cells were incubated in the presence of (^{3}H)leucine for the duration of the SP incubation. At designated intervals, cells were harvested as described above. Data were normalized against controls and expressed as percent difference in protein synthesis above or below untreated controls.

ELISA for IL-7 and Stem Cell Factor Levels

To determine the levels of IL-7 or SCF on stromal cell surfaces following SP incubation, confluent monolayers of cells were washed and cultured in the presence of various concentrations of SP. All cultures were treated in triplicate and incubated for either 2 hr or 48 hr at 37°C, 5% CO_2. Controls were incubated in RPMI/10% FBS without SP. Following SP incubation, the cells were washed three times in Tris-buffered saline (TBS). Primary antibodies, goat-anti-murine IL-7 or rat-anti-murine SCF, were added to the cells. Following one hour of incubation on ice, the plates were washed, and alkaline phosphase conjugated secondary antibodies (Anti-Goat IgG-Alkaline Phosphatase (AP) or Anti-Rat IgG-AP) were added to the cultures. The plates were incubated for an additional hour. Following three washes in TBS, p-Nitrophenyl Phosphate (Sigma, St. Louis, MO) was added as substrate. Since stromal cells have been reported to have endogenous AP activity (20), incubation with the AP labeled secondary Ab was performed in the presence of Levamisole which inactivates endogenous AP activity. The plates were read in an automatic plate reader at 405 nm. Results were expressed as percent control cytokine level.

Statistics

Student's t-test analysis of unpaired normally distributed samples and ANOVA were performed using the Macintosh StatView computer program (Redmond, WA).

RESULTS

Effect of Substance P on Protein Synthesis

To determine the effect of SP on protein synthesis in cultured stromal cells, confluent monolayers of cells were incubated in the presence of SP and protein synthesis was measured

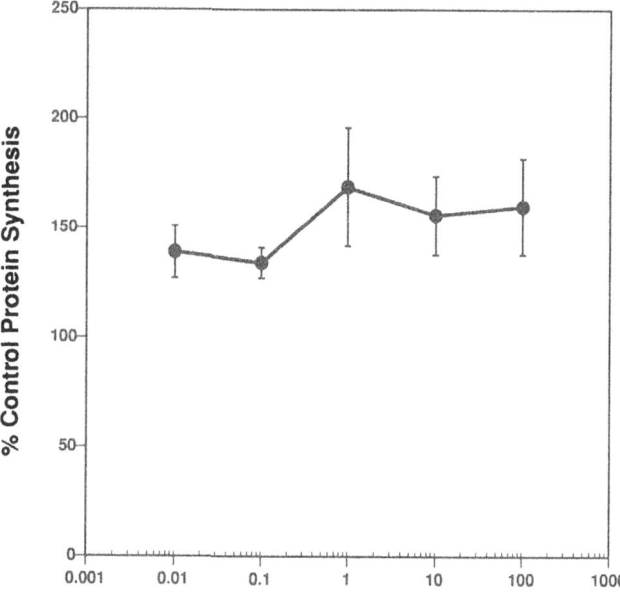

Figure 1. The effect of Substance P on protein synthesis. Confluent monolayers of stromal cells were incubated in the presence of SP for 48 hr. Protein synthesis was measured by (^{3}H)leucine incorporation. Data are presented as percent untreated control. Each point represents the mean ± SE of six experiments. Percent control protein synthesis is plotted against increasing Substance P concentration.

Substance P Concentration (nM)

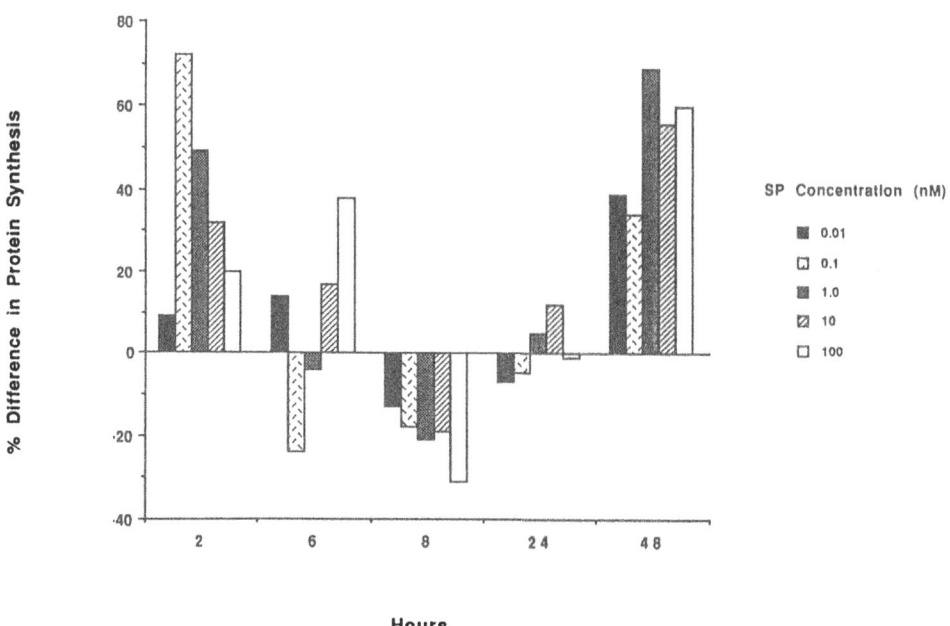

Hours

Figure 2. The kinetics of protein synthesis enhancement by Substance P. To determine the kinetics of protein synthesis induction by SP, cells were incubated in the presence of SP for various times and protein synthesis was measured. Data were normalized against controls and expressed as percent difference in protein synthesis above or below untreated controls. The figure represents the results of three experiments.

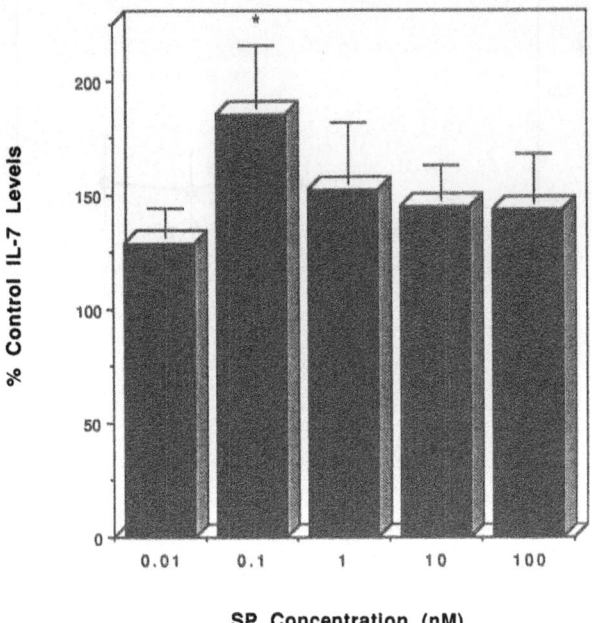

Figure 3. Effect of Substance P on Interleukin-7 Levels: 2 hr incubation. Stromal cells were incubated in various concentrations of SP for 2 hr. Following incubation, stromal cell layers were washed and levels of cell-associated IL-7 were determined by ELISA. Results are expressed as percent control IL-7 level and represent the mean ± SE of 4 experiments.

by (^{3}H)leucine incorporation. Substance P significantly enhanced (p < 0.05) incorporation of (^{3}H)leucine at all concentrations tested.

Substance P at concentrations of 0.01, 0.1, 1, 10, and 100 nM enhanced protein synthesis in stromal cells 39, 34, 69, 56, and 60% respectively, as compared to untreated controls. In contrast, incubation of cells in the presence of SP plus equimolar concentrations of the SP antagonist spantide did not induce significant enhancement of protein synthesis (data not shown).

To determine the kinetics of protein synthesis induction by SP, cells were incubated in the presence of SP for various times. At all concentrations tested, (^{3}H)leucine incorporation was above untreated controls following 2 hr and 48 hr of SP incubation. At all other time points, protein synthesis was either not significantly enhanced or was inhibited as compared to untreated controls. At most time points, there was no apparent correlation between SP concentration and protein synthesis. As in previous experiments, incubation of cells in the presence of SP plus equimolar concentrations of spantide abrogated the effect of SP (data not shown).

Effect of Substance P on Interleukin-7 Levels

The levels of IL-7 on the surface of SP stimulated stromal cells were monitored by ELISA. Since previous data had shown that SP induced protein synthesis at both 2 and 48 hr after incubation, we assayed SP stimulated cells for IL-7 at these time points. After two hr of incubation, all concentrations of SP induced increased levels of IL-7 as compared to untreated controls.

Treated cells expressed IL-7 levels that were 29% - 85% above the levels of control cells. At 0.1 nM, the level of IL-7 was significantly increased (p = 0.007) as compared to controls.

When stromal cells were incubated in various concentrations of SP for 48 hr, SP induced increased levels of IL-7 as compared to untreated controls at all concentrations tested. Treated cells expressed IL-7 levels that were 18% - 78% above the levels of control

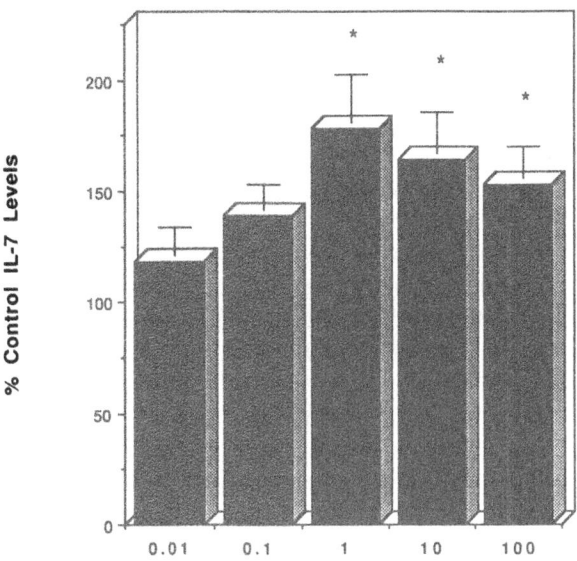

Figure 4. Effect of Substance P on Interleukin-7 Levels: 48 hr incubation. Stromal cells were incubated in various concentrations of SP for 48 hr. Following incubation, stromal cell layers were washed and levels of cell-associated IL-7 were determined by ELISA. Results are expressed as percent control IL-7 level and represent the mean ± SE of 7 or 8 experiments.

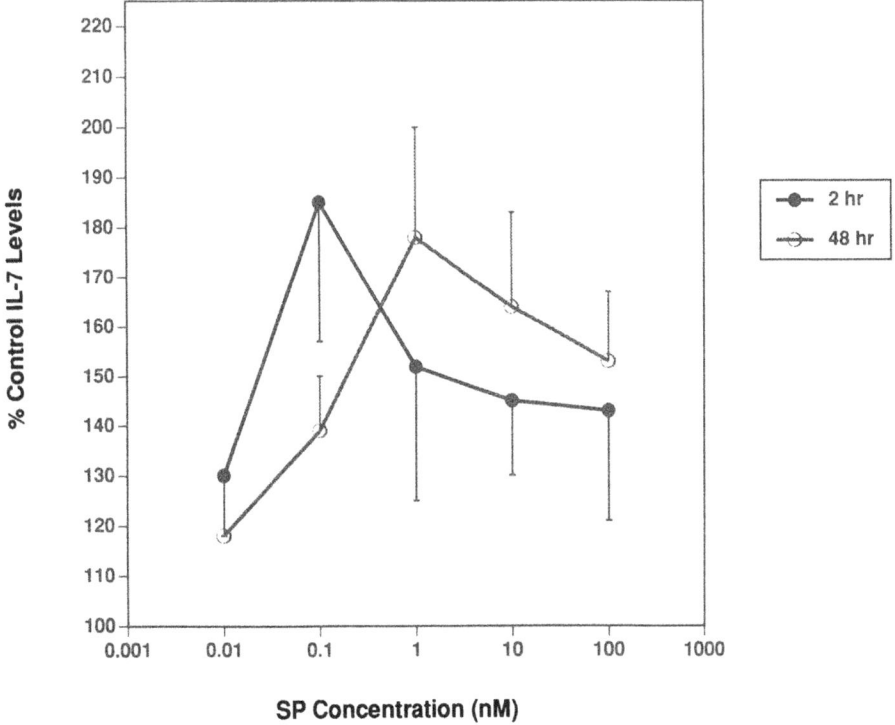

Figure 5. Effect of Substance P on Interleukin-7 Levels: 2 hr vs 48 hr. The effect of SP on IL-7 levels following 2 hr of incubation was compared to the effect following 48 hr of incubation. The levels of IL-7, as determined by ELISA, were compared by ANOVA.

cells. At 1 nM (p = 0.0006), 10 nM (p = 0.003), and 100 nM (p = 0.013), the level of IL-7 was significantly increased as compared to controls.

The effect of SP on IL-7 levels following 2 hr of incubation was compared to the effect following 48 hr of incubation. Although a shift of the dose response curve was noted at 48 hr, the difference between 2 hr and 48 hr of SP incubation was not significant.

Effect of Substance P on Stem Cell Factor Levels

Stromal cells were incubated in various concentrations of SP for 2 hr or 48 hr. Following incubation, stromal cell layers were washed and levels of cell-associated SCF were determined by ELISA. Following 2 hr of SP exposure, treated cells expressed SCF levels that were 40% - 60% above the levels of untreated control cells. At all concentrations tested, the level of SCF was significantly increased as compared to controls (p < 0.03). There was no dose response curve observed at these concentrations of SP.

When stromal cells were incubated in various concentrations of SP for 48 hr, treated cells expressed SCF levels that were 7% - 46% above the levels of untreated control cells. Only at the highest concentrations tested, 10 nM (p = 0.0008) and 100 nM (p = 0.03), were the levels of SCF significantly increased as compared to controls.

When SCF levels following 2 hr of SP incubation were directly compared to the levels following 48 hr of incubation, 2 hr of SP exposure led to significantly higher levels of SCF on the stromal cell surface than did 48 hr of SP exposure (p = 0.001).

DISCUSSION

Through their interactions with stem cells, stromal cells are believed to be intimately involved in regulating the production, differentiation, and development of cell types derived

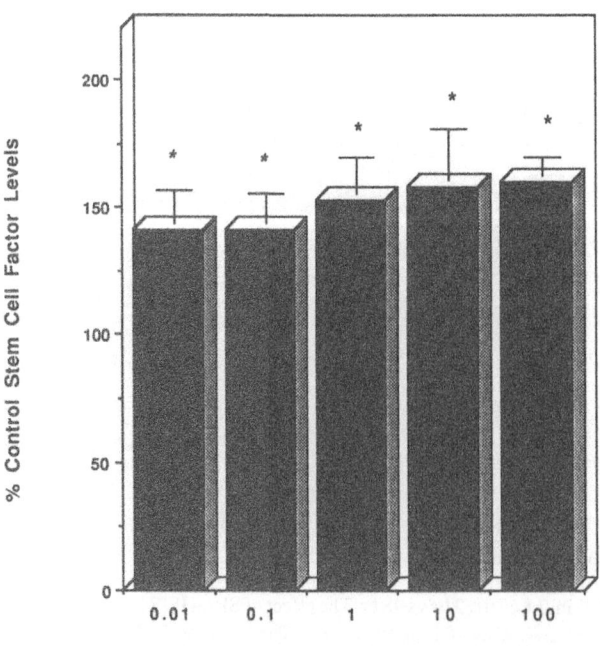

Figure 6. Effect of Substance P on Stem Cell Factor Levels: 2 hr incubation. Stromal cells were incubated in various concentrations of SP for 2 hr. Following incubation, stromal cell layers were washed and levels of cell-associated SCF were determined by ELISA. Results are expressed as percent control SCF level and represent the mean ± SE of 5-6 experiments.

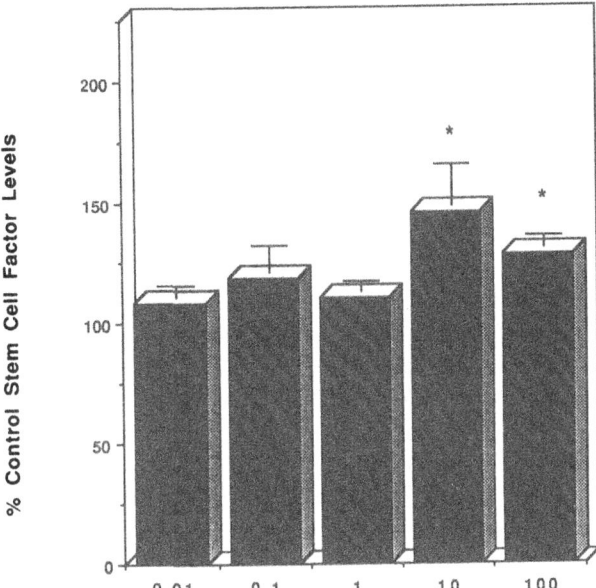

Figure 7. Effect of Substance P on Stem Cell Factor Levels: 48 hr incubation. Stromal cells were incubated in various concentrations of SP for 48 hr. Following incubation, stromal cell layers were washed and the levels of cell-associated SCF were determined by ELISA. Results are expressed as percent control SCF level and represent the mean ± SE of 3-8 experiments.

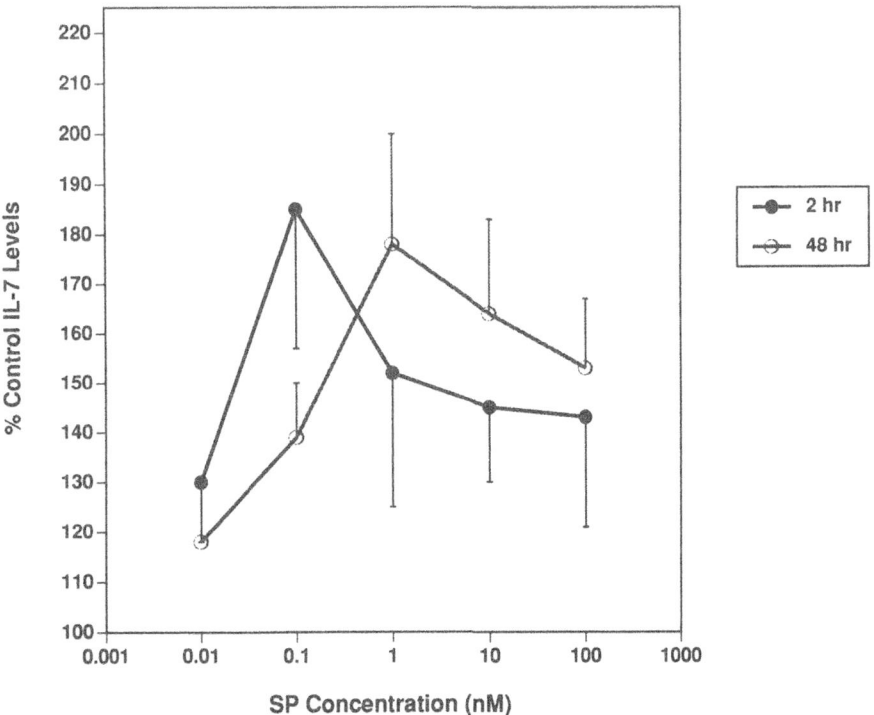

Figure 8. Effect of Substance P on Stem Cell Factor Levels: 2 hr vs 48 hr. The effect of SP on SCF levels following 2 hr of incubation was compared to the effect following 48 hr of incubation. The levels of SCF, as determined by ELISA, were compared by ANOVA.

from the bone marrow (34). If neuropeptides such as SP are capable of stimulating stromal cell processes, they could indirectly effect the development of stem cells in the bone marrow. An effect of SP on stem cell development has been reported by Rameshwar et al. who found that SP had potent *in vitro* stimulatory effects on hematopoiesis for both erythroid and granulocytic progenitors in short-term bone marrow cultures (18). These researchers also reported that the stimulatory effect of SP observed in their study was mediated by the adherent, or stromal, cell population. In the present study, we directly addressed the ability of SP to effect cytokine levels in cultured stromal cells.

Prior to addressing the effect of SP on cytokine levels, we examined the ability of SP to affect protein synthesis. We found that SP increased protein synthesis in cultured stromal cells. We observed two waves of protein synthesis stimulation at two hours and 48 hours of SP incubation. These results suggest that SP induces the synthesis of some proteins early while other, perhaps different proteins are produced at a later time. We have no explanation for the decrease in protein synthesis observed after 8 hours of SP incubation. It is possible that at time points following 2 hr the SP receptors are down modulated or desensitized. Modulation of the SP receptors on stromal cells has not been studied; however, it has been reported that SP treatment of IM-9 lymphoblastoid cells leads to a decrease in the number of SP receptors (35). If down modulation had occurred, the receptors could be re-expressed at later time points resulting in re-stimulation of the cells.

Since SP was observed to induce protein synthesis, we wanted to determine which proteins were being stimulated. Stromal cells serve as regulatory cells within the bone marrow and produce a number of cytokines including IL-7 and Stem Cell Factor. Although bone marrow stromal cells are known to produce a variety of cytokines, it has been reported that supernatant media collected from stromal cells may not contain detectable levels of cytokine (36). This is thought to be due to cytokines, synthesized by stromal cells, remaining bound to the stromal cell layer extracellular matrix (36) or to the cell membrane itself as in the case of SCF which exists in a membrane bound form. For this reason, we chose to investigate the effect of SP on cell surface associated IL-7 and SCF. Based on the results of the protein synthesis assays, we monitored cytokine levels following 2 hr or 48 hr of SP exposure.

Interleukin-7 levels were increased at both 2 hr and 48 hr. However, a 10-fold shift in the dose response curve was observed following 48 hr of incubation. The reason for this shift remains to be investigated. As mentioned above, it is possible that 48 hr of SP exposure leads to desensitization or turn over of receptors on the cell surface. Receptor recycling could lead to changing levels of SP sensitivity or reactivity in the cells over time. This is quite feasible since the biologic effects of SP are mediated through three types of receptors: neurokinin-1 (NK-1R), NK-2R, and NK-3R (37,38). In terms of binding affinity, SP exhibits a preference for NK-1R and lesser affinity for NK-2R and NK-3R (37,38). It is possible that following 48 hr of SP exposure, the NK-1 receptors are desensitized or down modulated. It would then require higher concentrations of SP to mediate effects via these lower affinity receptors.

When Stem Cell Factor levels were assayed following 2 hr of SP incubation, all treated samples expressed higher levels of SCF than was found on untreated control cells. No dose response curve was observed at the concentrations tested. Further experiments will be performed assessing both higher and lower concentrations of SP in order to determine the range of the dose response. Following 48 hr of induction, less SCF was observed associated with the cell surface than was seen after 2 hr. Since we were assaying the levels of cell associated SCF, it is likely that following 48 hr, SCF is secreted into the cell supernatant and is no longer detectable on the cell surface. This will be addressed in future studies.

The release of cytokines such as IL-7 and SCF is essential in providing signals for blood cell progenitors to proliferate and differentiate. Therefore, the results of this study; 1) suggest that SP may play a role in the regulation of hematopoiesis; 2) support the theory that SP mediates its effect on bone marrow stem cells via interactions with the stromal cell population; and 3) further implicate the nervous system and neuropeptides in hematopoietic regulation.

REFERENCES

1. Korneva, E.A., and L.M. Khai. 1964. Effect of destruction of hypothalmic areas on immunogenesis. Fed. Proc. Trans. Suppl. 23: T88.
2. Macris, N.T., R.C. Schiavi, M.S., Camerino, and M. Stein. 1970 Effect of hypothalmic lesions on immune processes in the guinea pig. Am. J. Physiol. 219: 1205.
3. McCain, H.W., I.B. Lamster, J.M. Bozzone, and J.T. Gribic. 1982. B-endorphin modulates human immune activity via nonopiate receptor mechanisms. Life Sci. 31: 1619.
4. Johnson, H.M., E.M. Smith, B.A. Torres, and J.E. Blalock. 1982. Regulation of the in vitro response by neuroendocrine hormones. Proc. Natl. Acad. Sci. USA 79: 4171.
5. Pernow, B. 1983. Substance P. Pharmacol. Rev. 35: 85.
6. Payan, D.G. 1989. Substance P: A modulator of neuroendocrine-immune function. Hosp. Pract. 24: 67.
7. Strand, F.L., J. Kenneth, R. Lisa, A.Z. Lisa, K. June, E.A. Stephan, J.A. Francis, and Y.G. Lara. 1991. Neuropeptide hormones as neurotrophic factors. Pharmacol. Rev. 71: 1017.
8. Payan, D.G., and E.J. Goetzl. 1985. Modulation of lymphocyte function by sensory neuropeptides. J. Immunol. 135: 783.
9. Pascual, D., J. Xu-Amano, H. Kiyono, J.R. McGhee, and K.L. Bost. 1991. Substance P acts directly on cloned B lymphoma cells to enhance IgA and IgM production. J. Immunol. 146: 2130.
10. Shanahan, F., J.A. Denburg, J. Fox, J. Bienenstock, and D. Befus. 1985. Mast cell heterogeneity: effects of neuroenteric peptides on histamine release. J. Immunol. 135: 1331.
11. Lotz, M., D.A. Carson, and J.H. Vaughan. 1987. Substance P activation of rheumatoid synoviocytes: neural pathway in pathogenesis of arthritis. Science 235: 893.
12. Hartung, H.P, K. Wolters, and K.V. Toyker. 1986. Substance P: binding properties and studies on cellular responses in guinea pig macrophages. J. Immunol. 136: 3856.
13. Moore, T.C., J.L. Lami,. and C.H. Spruck. 1989. Substance P increases lymphocytes traffic and lymph flow through peripheral lymph nodes of sheep. J. Immunol. 67: 109.
14. Calvo, C.F., G. Cavanel, and A. Senik. 1992. Substance P enhances IL-2 expression in activated human T-cells. J. Immunol. 148: 3498.
15. Laurenzi, A.M., M.A.A. Person, C. Dalsgaard, and A. Hoegerstrand. 1990. The neuropeptide substance P stimulates production of interleukin-1 in human blood monocytes: activated cells are preferentially influenced by the neuropeptide. Scand. J. Immunol. 31: 529.
16. Kimball, E.S., F.J. Persico, and J.L. Vaught. 1988. Substance P, neurokininA, and neurokinin B induce generation of IL-1-like activity in P388D1 cells. J. Immunol. 141: 3564
17. Wagner, F., R. Fink, R. Hart, and H. Dancygier. 1987. Substance P enhances interferon production by human peripheral blood mononuclear cells. Regul. Pept. 19: 355.
18. Rameshwar, P., D. Ganea, and P. Gascon. 1993. In vitro stimulatory effect of substance P on hematopoiesis. Blood 81: 391.
19. Dexter, T.M., T.D. Allen, and L.G. Lajtha. 1977. Conditions controlling the proliferation of haemopoietic stem cells in vitro. J. Cell Physiol. 91: 335.
20. Perkins, S., and R.A. Fleischman. 1990. Hematopoietic microenviroment: origin, lineage, and transplantability of the stromal cells in long-term bone marrow cultures from chimeric mice. Blood 75, 620.
21. Pascual, D.W., and K. L. Bost. 1990. Substance P production by macrophage cell lines: a possible autocrine function for this neuropeptide. Immunology 71: 52.
22. Nilsson, J., M.A. von Euler, and C.J. Dalsgaard. 1985. Stimulation of connective tissue cell growth by Substance P and Substance K. Nature 315: 61.
23. Nakanishi, S. 1991. Mammalian tachykinin receptors. Annu. Rev. Neuroscience 14: 123.
24. Namen, A.E., A.E. Schmierer, C.J. March, R.W. Overell, L.S. Park, D.L. Urdal, and D.Y. Mochizuki. 1988. B cell precursor growth-promoting activity: purification and characterization of a growth factor active on lymphocyte precursors. J. Exp. Med. 167: 988.
25. Namen, A.E, S. Lupton, K. Hjerrild, J. Wignall, D.Y. Mochizuki, A. Schmierer, B. Mosley, C.J. March, D. Urdahl, S. Gillis, D. Cosman, and R.G Goodwin. 1988. Stimulation of B cell progenitors by cloned murine IL-7. Nature 333: 571.

26. Morrissey, P.J., R.G. Goodwin, R.P. Nordan, D. Anderson, K.H. Grabstein, D. Cosman, J. Sims, S. Lupton, B. Acres, S.G. Reed, D. Mochizuki, J. Eisenman, P.J. Conlon, and A.E. Namen. 1989. Recommmbinant IL-7, pre-B cell growth factor, has co-stimulatory activity on purified mature T cells. J. Exp. Med. 169: 707.

27. Grabstein, K.H., A.E. Namen, K. Shanebeck, R.F. Voice, S.G. Reed, and M.B. Widmer. 1990. Regulation of T cell proliferation by IL-7. J. Immunol. 144: 3015.

28. Conlon, P.J., P.J. Morrissey, R.P.Nordan, K.H. Grabstein, K.S. Prickett, S.G. Reed, R. Goodwin, D. Cosman, and A.E. Namen. 1989. Murine thymocytes proliferate in direct response to Interleukin-7. Blood 74: 1368.

29. Quesenberrry, P.J., S. Rudnick, and D.C. Dale. 1991. Hematopoietic growth factors. Blood suppl (Hematology 1991, Education program: ASH meetings) p. 80.

30. Zsebo. K.M., J. Wypych, I.K. McNiece, H.S. Lu, K.A. Smith, S.B. Karkare, R.K. Sachdev, V.N. Yuschenkoff, N.C. Birkett, L.R. Williams, V.N. Satyagal, T. Weifong, R.A. Bosselman, E.A. Mendiaz, and K.E. Langley. 1990 Identification, purification, and biological characterization of hematopoietic stem cell factor from buffalo rat liver-conditioned medium. Cell 63: 213.

31. Witte, O.N. 1990. Steel locus defines new multipotent grwoth factor. Cell 63: 5.

32. Huang, E., K. Nocka, D.R. Beier, T-Y. Chu, J. Buck, H-W. Lahm, D. Wellner, P. Leder, and P. Besmer. 1990. The hematopoietic growth factor KL is encoded by the *Sl* locus and is the ligand for *c-kit* tyrosine kinase receptor. Cell 63: 225.

33. Flanagan, J.G., and P. Leder. 1990. The kit ligand: a cell surface molecule altered in steel mutant fibroblasts. Cell 63: 185.

34. Allen, T.D., T.M. Dexter, and P.J Simmons. 1990. Marrow biology and stem cells. in: *Colony-Stimulating Factors: Molecular and Cell Biology.* (Dexter, T.M., Garland, J.M. and Testa, N.G. ed.) p.1, Marcel Dekker, Inc., New York.

35. Parnet, P., Payan, D.G., Kerdelhue, B., and Mitsuhashi, M. 1990. Neuroendocrine interaction on lymphocytes: testosterone-induced modulation of the lymphocyte Substance P receptor. J. Neuroimmunology. 28: 185.

36. Gordon, M.Y., G.P. Riley, S.M. Watt, and M.F. Greaves. 1987. Compartmentalization of a haematopoietic growth factor (GM-CSF) by glycosaminoglycans in the bone marrow microenvironment. Nature 326: 403.

37. Maggio, J.E. 1988. Tachykinins. Annu. Rev. Neurosci. 11:13.

38. Regoli, D.G. G. Drapeau, S. Dion, and P. D'Orleans-Juste. 1989. Receptors for substance P and related neurokinins. Pharmacology. 38: 1.

8

MALARIA TRANSMISSION-BLOCKING IMMUNITY

Identification of Epitopes and Evaluation of Immunogenicity

Nirbhay Kumar, Isabelle Ploton, Gary Koski, Cheryl Ann-Lobo, and Carmen Contreras

Department of Molecular Microbiology and Immunology
Johns Hopkins School of Public Health
615 N. Wolfe Street, Baltimore, Maryland 21205

INTRODUCTION

The eradication of malaria was conceived in the mid-1950s as an activity of limited duration. With the recognition in the late 1960s that eradication was not feasible, attention turned to control and newer approaches to combat this major cause of human misery. According to published estimates, malaria parasites annually infect 250 to 300 million people, cause 100-120 million clinical cases, and kill 1 to 2 million people. Including recent estimates of malaria from sub-Saharan Africa, these numbers could be as high as 500-900 million infections annually. Insecticide resistance in the mosquitoes, drug resistance in the parasites are, among many others (poverty, ignorance, social disruptions and often inadequate national resources, etc.), major causes for resurgence of malaria worldwide (Oaks et al. 1991). Over the last 20 years, major scientific advances have been made in the understanding of the immuno-biology of host-parasite interactions. The major goal of malaria research at the molecular and immunological fronts has been the identification and characterization of vaccine candidates.

The possibility of developing a vaccine against malaria is supported by the following facts:

- People living in areas of intense malaria transmission acquire partially protective immunity resulting in reduced prevalence, incidence, morbidity, and mortality.
- Purified immunoglobulins from adults living in endemic areas have, in passive transfer studies shown a protective role for antibodies.
- Immunization with radiation-attenuated *Plasmodium falciparum* sporozoites completely protects volunteers against experimental challenge.
- Immunization with purified sexual stage parasites (gametes) results in complete blocking of transmission of parasites from vaccinated hosts to mosquitoes.

Immunobiology of Proteins and Peptides VIII
Edited by M. Z. Atassi and G. S. Bixler, Jr., Plenum Press, New York, 1995

65

However, acquisition of immunity against malaria is a slow process and does not lead to sterilizing immunity. In addition, the acquired immunity is short lived, requiring continuous antigenic challenge. Antigenic diversity and antigenic variation probably account for the slow and short lived nature of acquired immunity. The nature of acquired immune mechanisms and the antigenic targets of natural protective immunity remain poorly understood. Four types of malaria vaccines (Miller et al., 1986, Mitchell, 1989) are currently being developed, a pre-erythrocytic vaccine targeting sporozoites during invasion and further development in the hepatocytes, a blood-stage vaccine to limit the replication of erythrocytic stage parasites, an anti-disease vaccine to prevent the immunopathology associated with malaria, and a transmission blocking vaccine that targets sexual reproduction of parasites developing within the mosquito midgut. The goal of studies in our laboratory is the development of a malaria transmission blocking vaccine, thus the discussion in this paper will be limited to biology and immunology of sexual stages of malaria parasite.

THE RATIONALE FOR A MALARIA TRANSMISSION-BLOCKING VACCINE

Antigens expressed in the sexual stages (male and female gametocytes) are immunogenic during natural infection and thus the immune host has reduced capacity to transmit malaria to others. This form of immunity directed against epitopes in the antigens expressed in the sexual stages is termed malaria transmission-blocking immunity (TBI) (Carter et al., 1986, Kaslow, 1993, Alano and Carter, 1990, reviews). This form of immunity against the sexual and mosquito midgut stages of the parasite blocks only the exit of the parasites from a host into the vector population, thus would be an attractive public health strategy to control or limit transmission of malaria.

A vaccine against the sexual and mosquito midgut stages of human parasite could have several important roles:

- It can reduce or even eliminate (in certain circumstances) transmission of malaria in an endemic region.
- It can be used as a component of a cocktail vaccine to block the spread of mutants not affected by immunity induced by vaccines against other stages.
- It can be used in combination with chemotherapeutic approaches to prevent the spread of drug -resistant parasites.
- It can also serve as a "quarantine vaccine" to prevent reintroduction of malaria in the areas from which malaria has been eradicated but where the transmission potential remains.

While an anti-asexual stage vaccine could also have an impact on reduction of transmission, the most effective way to immunize against the release of undesirable parasite genes (e.g. drug resistant, asexual vaccine escape mutants etc.) into the parasite population at large would be to directly target the parasite stage responsible for spread through the vector population. One can even make a compelling argument that an anti-asexual stage vaccine should preferably be deployed only when combined with transmission blocking vaccine components. It is realized that an anti-asexual stage (which infect humans) vaccine would directly protect the host against parasitism and the disease. However, if a transmission blocking vaccine were to be deployed on a community wide basis, all members of the population, including those not immunized, would experience the same reduction in inoculations of malaria. The commonly expressed view that the recipients of a transmission blocking vaccine would not benefit is thus false.

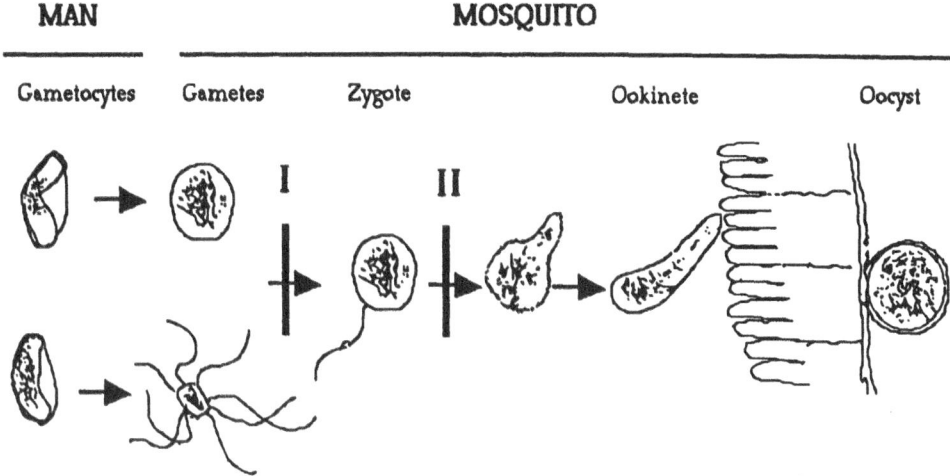

Figure 1. Malaria transmission-blocking immunity. I: Block fertilization. II: Block sygote development.

THE BIOLOGICAL AND IMMUNOLOGICAL BASIS OF MALARIA TBI

Sexual stages and mosquito midgut stages of the parasites have been shown to induce TBI in several animal malaria models. The target stages of this immunity include the newly fertilized zygote and the mature ookinete. The major effector mechanism is mediated by antibodies recognizing antigens in these stages of the parasite. The antibodies react with parasite antigens in a complement-dependent as well as complement-independent manner and block the development of parasites in the midgut of the mosquitoes. In addition to antibodies, non-specific factors such as cytokines (tumor necrosis factor-alpha and interferon-gamma) have also been shown to reduce infectivity of gametocytes when ingested by the mosquitoes (Carter et al., 1986, Naotunne et al., 1991). Two major stages of the parasite development affected by transmission blocking antibodies are shown in the figure 1.

TARGET ANTIGENS OF P. FALCIPARUM TBI AND MAPPING OF EPITOPES

Several antigens have been identified as the candidate antigens for the development of transmission blocking vaccines (Table 1) (Kaslow, 1993, review). These antigens can be broadly categorized as antigens expressed either before fertilization or after gametogenesis and fertilization of gametes in the mosquito midgut. Production of mAbs and their evaluation in transmission blocking assays (membrane feeding) allowed identification of all these putative target antigens. Prior to publication of our studies (Wizel and Kumar, 1991), a prominent feature in almost all the target antigens was the reduction-sensitive (conformational) nature of the epitopes recognized by *P. falciparum* transmission blocking mAbs. In the studies from our laboratory we identified four hybridomas, from a total of approximately 800, producing transmission blocking mAbs recognizing a reduction-insensitive epitope in the 27 kDa protein also shared with two other antigens of 230 kDa and 48/45 kDa (Wizel and Kumar, 1991). We have since mapped the target epitope to a 16 amino acid region in the 27 kDa protein (Ploton et al., 1995). We have also compared the sequence of the gene

Table 1. Target antigens of malaria transmission-blocking immunity

Antigen (kDa)	Stage	Location	Target of TBI	Cloned	Epitope
230	Gametocyte/ gamete	Surface	Yes	Yes	Conformational
48/45	Gametocyte/gamete	Surface	Yes	Yes	Conformational
27	Gametocyte	Cytoplasmic	Yes	Yes	Linear
25	Zygote /Ookinete	Surface	Yes	Yes	Conformational
16	Gametocyte/sporozoite	Vaccuole /surface	No	Yes	
11.1	Gametocyte	Granule	±	Yes	
28	Zygote /Ookinete	Surface	±	Yes	

encoding the 27 kDa protein in eight different isolates of *P. falciparum* and found it to be totally conserved except for a single base change in the 7G8 isolate resulting in a conservative amino acid substitution outside the region of the C^3-epitope (Ploton and Kumar, unpublished). Because the target epitope recognized by the four transmission blocking mAbs produced in our laboratory is Continuous, Cross-reacting and Conserved, we have designated it as the C^3-epitope. In ongoing studies, we are mapping T and B epitopes to assist in the construction of either a recombinant or a synthetic immunogen to be used as a subunit transmission blocking vaccine.

In various studies on mapping B and T epitopes, our general approach has been to use full length and overlapping truncated fragments of the 27 kDa protein expressed in *E. coli*. Various recombinant products were then used as immunogens for the induction of transmission blocking antibodies in mice. DNA encoding these fragments was obtained by polymerase chain reaction (PCR) and cloned in the vectors pMG1 (expression with or without NS-1, the non-structural protein-1 of influenza virus A, as the fusion polypeptide at the amino terminus of the cloned insert) and pRSET (expression of cloned inserts as fusion proteins containing 6 histidine residues which in turn allowed a one step purification of all the fragments using a Nickel-Agarose matrix) (Fig. 2).

Recombinant fragments were initially used in a western immuno-blot experiment with various mAbs. These mAbs reacted with the r-27 kDa protein and amino-terminal fragments F1 (amino acid residues 1-102) and F1a (residues 1-56) (Ploton et al., 1995). Synthetic peptides (20-mer with 10 amino acid overlap) were then used in an ELISA to precisely map the B epitope. Three out of four transmission blocking mAbs (6B6, 16C3 and 19F1) recognized peptide 1 (residues 1-20) and peptide 2 (residues 11-29) suggesting that the epitope must lie in the overlapping region. Indeed peptide 5 (residues 10-25) (C^3-epitope) was recognized strongly by these mAbs (Ploton et al., 1995). These results also showed that a fourth mAb (11G12) reacted only with peptide 2 and peptide 5. Various analysis led us to conclude that these transmission blocking mAbs are directed against two overlapping linear epitopes contained within the region of amino acid residues 10-25. Peptides 1, 2 and 5 when conjugated with KLH were also strongly immunogenic in mice and produced antibodies capable of recognizing the native 27 kDa protein in gametocytes of *P. falciparum* (Ploton et al., 1995).

IMMUNOGENICITY OF RECOMBINANT PROTEINS IN MICE

Purified r-27 kDa protein was used for immunization of five different inbred strains of mice (Balb/c, $H-2^d$; C57Bl/6, $H-2^b$; C3H, $H-2^k$; SJL, $H-2^s$; RFM, $H-2^f$) and CD-1 outbred mice (intraperitoneally, 50 micrograms / mouse in CFA for primary and in IFA for booster

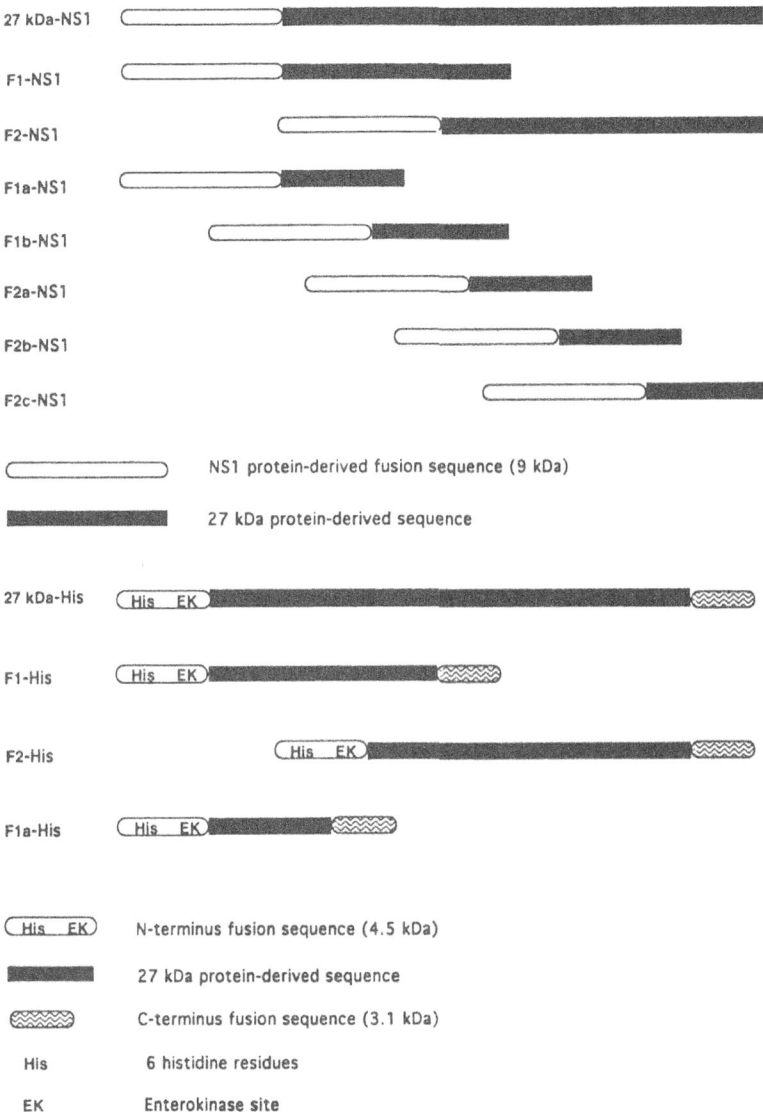

Figure 2. Recombinant expression of 27kDa and overlapping fragments in plasmids pMG1 and pRSET. **A.** pMG1 expression system; **B.** pRSET expression system.

immunizations). All the mice responded to immunization with a boostable immune response. Table 2 shows ELISA titers of the pooled sera (five mice) against the r-27 kDa protein and P-5 representing the C^3-epitope (Koski, 1994). Immunizations in alum with or without QS-21 have also revealed comparable titers in Balb/c mice (Ploton et al., unpublished).

Pooled sera from immunogenicity studies (above) were tested for analysis of the isotypes of C^3-peptide-specific antibodies by ELISA. Significant differences were found in the pattern of isotypes from strain to strain for two different recombinant constructs of immunogens (Fig. 2) (Ploton and Kumar, unpublished). These sera have also been tested for transmission blocking activity by membrane feeding. The sera from CD-1 and RFM were

Table 2. ELISA end-point titers of anti-27kDa and anti-P5
antibodies in inbred strains of mice immunized and
boosted twice with recombinant 27 kDa.

Strain	r-27kDa	P-5
Balb/c	100,000	6,400
C57Bl/6	50,000	6,400
C3H	200,000	6,400
SJL	50,000	25,000
RFM	400,000	400,000

the strongest blockers with still significant blocking shown by the sera from SJL mice and
only moderate blocking by other sera (Kumar et al. unpublished). Clearly further studies are
needed to find correlation (if any) between the antibody titers and transmission blocking or
between the antibody isotypes and transmission blocking.

CLONING 27 kDA PROTEIN-SPECIFIC MURINE T CELLS AND T EPITOPE MAPPING

Spleen cells from immunized mice (Balb/c and C57Bl/6 strains) were used to clone
27 kDa protein-specific T cells by *in vitro* rounds of stimulation and rest. Enriched T cells
were fused with BW5147 mouse thymoma to obtain T cell hybridomas. Positive hybridomas
were cloned by limiting dilution. Five T cell clones (CD4[+]) selected (three from Balb/c and
two from C57Bl/6) resulted in identification of five T cell epitopes dispersed throughout the
molecule. Interestingly, a single 20-mer peptide was recognized by T cells from both the
strains of mice. The strategy for mapping of T epitopes initially employed crude bacterial
lysates containing recombinant fragments of the 27 kDa protein (step1) as antigens during
proliferation. Based on regions found to be responsible for proliferation, synthetic peptides
were used for fine epitope mapping (step 2) (Koski, 1994). We are now investigating T cell
priming by these peptide epitopes and their ability to prime for *in vivo* help in the production
of antibodies against the C[3] epitope.

IMMUNOLOGICAL CONSIDERATION FOR SUBUNIT VACCINE DEVELOPMENT

In addition to recognition of an antigen by blocking antibodies, the antigen itself must
be able to induce similar blocking antibodies if it is to serve as a vaccine candidate. This
may however, be of particular concern for antigens that require formation of proper
conformation. On the other hand, if the target epitopes are linear in nature, recombinant
products expressed in *E. coli* or even a synthetic immunogen could potentially elicit blocking
antibody responses. The C[3]-epitope described above thus offers an attractive candidate target
for TBI.

There are, however, other biochemical and immunological issues that must be
addressed so as to rationally design an effective vaccine.

- *Is the antigen a target of immune response during natural infection?* As mentioned
  above, antigens in various life cycle stages of malaria parasites are targets of
  immune responses during natural infection. Many antigen candidates for TBI have

also been shown to be targets of such natural immune responses (Carter et al., 1986, Kaslow, 1993 for reviews). Several studies have investigated these aspects and the 230 kDa, the 48/45 kDa doublet as well as the 27 kDa protein have been found to be targets of an immune response during infection. While the response against the 230 and the 48/45 kDa proteins was restricted to only a small percentage of the population, the 27 kDa protein as well as the peptide representing the C^3-epitope have been found to be recognized by more than 70% of the people living in the endemic areas (Graves et al., 1988, Riley et al., 1994, Quakyi et al., 1989, Ong et al., 1990, Carter et al., 1989, Good et al., 1988, Sjoberg, et al., 1992, Contreras and Kumar, unpublished). In contrast, the 25 kDa protein which is not present in the gametocyte stage of the parasite, as expected, is not a target of natural immune response (Quakyi et al., 1989). Restricted recognition of antigens could be due to diversity or polymorphism of epitopes, and genetic restriction of the immune response (Good et al., 1988). Recent studies on analysis of immune responses to malaria antigens in people have also indicated non-genetic factors influencing the development of immunity (Riley et al., 1994, Sjoberg et al., 1992).

- *Are the target antigens and epitopes conserved in parasites from different geographical locations?* If the antigens and epitopes are conserved, it would greatly facilitate the development of a vaccine. On the other hand, if some of the critical epitopes are polymorphic, this compounded by restricted immune response would complicate the process of vaccine development. The 27 kDa protein proposed as the vaccine candidate neither suffers from restricted immune response nor is it polymorphic.

- *Opportunity for natural boosting (anamnestic response)?* It is a well known fact that partially protective immunity in people does play a critical role in suppressing clinical morbidity and mortality. One goal for any vaccine development thus would be to induce better immunity than the immunity resulting from natural exposure to antigens during infection. Helper T cells play an important role in helping antibody production, isotype switching and more importantly generation of immunological memory. Priming of memory T cells is thus critical for inducing a long lasting immunity. Such memory T cells stimulated by recall antigens in the gametocytes during natural infection could every time boost the immune response and thus maintain an effective level of antibodies capable of reducing transmission. It is therefore essential to understand human immune responses and identify epitopes responsible for priming of human T cells. We have recently produced several 27 kDa-specific human T cell clones and studies are in progress to map epitopes recognized by these T cells.

- *Validation of a prototype vaccine.* This is a particularly important issue. Evaluation of *P. falciparum* transmission blocking antibodies can only be done using the membrane feeding assay (feeding gametocytes to mosquitoes and following parasite development in the presence and absence of test antibodies). It is tedious and often highly variable. Moreover it remains an *in vitro* assay. In collaboration with Drs. T. V. Rajan and L .D. Shultz we have developed a NOD-SCID (non-obese diabetic -SCID) mouse model for maintenance of *P. falciparum* and transmission to mosquitoes (Moore et al., 1995). This model could be adapted for *in vivo* evaluation of transmission blocking antibodies after immunization with passive transfer of antibodies.

Thus an ideal malaria transmission blocking vaccine when delivered appropriately (adjuvants?, live vectors?) must therefore be able to induce production of blocking antibodies, must be able to prime memory T cells in a majority of individuals and preferably not be based on polymorphic antigens and epitopes. Studies currently in progress are now addressing some of these questions to allow a rational development of a malaria transmission blocking vaccine.

ACKNOWLEDGMENTS

These studies were supported by research grants from the National Institutes of Health (AI24704), USAID (HRN-6001-A-00-2035-00), WHO, and John D. and Catherine T. MacArthur Foundation.

REFERENCES

Alano, P. and Carter, R., 1990, Sexual differentiation and malaria parasites. Ann.Rev. Microbiol. 44:429-449.

Carter, R., Kumar, N., Quakyi, I., Good, M.F., Mendis, K., Graves, P.M. and Miller, L., 1986, Immunity to sexual stages of malaria parasites. Progress in Allergy. 41:193-214.

Carter, R., Graves, P.M., Quakyi, I.A. and Good, M.F., 1989, Restricted or absent immune responses in human populations to *Plasmodium falciparum* gamete antigens that are targets of malaria transmission blocking antibodies. J.Exp.Med. 169:135-147.

Good, M.F., Miller, L.H., Kumar, S., Quakyi, I.A., Keister, D., Adams, J.H., Moss, B., Berzofsky, J.A., Carter, R., 1988, Limited immunological recognition of critical malaria vaccine candidate antigens. Science 242:574-577.

Graves, P.M., Carter, R., Burkot, T., Quakyi, I.A. and Kumar, N., 1988, Antibodies to *Plasmodium falciparum* gamete surface antigens in Papua New Guinea sera. Parasite Immunol. 10: 209-218.

Kaslow, D.C., 1993, Transmission-blocking immunity against malaria and other vector-borne diseases. Curr. Op.Immunol. 5: 557-565.

Koski, G., 1994, Mapping of T cell epitopes in the 27 kDa protein target of *Plasmodium falciparum* transmission-blocking immunity. Ph.D. thesis, Johns Hopkins University.

Miller, L.H., Howard, R.J., Carter, R., Good, M.F., Nussenzweig, V. and Nussenzweig, R., 1986, Research toward malaria vaccines. Science 234: 1349-1356.

Mitchell, G.H., 1989, An update on candidate malaria vaccines. Parasitology. 98:S29-S47.

Moore, J.M., Kumar, N., Shultz, L.D. and Rajan, T.V., 1995, A scid mouse model for maintenance of *Plasmodium falciparum* and transmission to mosquitoes. J.Exp.Med. (submitted after revision).

Naotunne, T.D.S., Karunaweera, N.D., Giudice, G.D., Kularatne, M.U., Grau, G.E., Carter, R.and Mendis, K.N., 1991, Cytokines kill malaria parasites during infection crisis: extracellular complementary factors are essential. J.Exp.Med. 173: 523-529.

Oak, S.C., Mitchell, V.S., Pearson, G.W., Carpenter, C.J.. 1991, Malaria: obstacles and Opportunities, National Academy Press.

Ong, C.S., Zhang, K.Y., Graves, P.M., Dow, C., Looker, M., Rogers, N.C., Chiodini, P.L., Targett, G.A., 1990, The primary antibody response to malaria patients to *Plasmodium falciparum* sexual stage antigens which are potential transmission blocking vaccine candidates. Parasite Immunol. 12:447-456.

Ploton, I.N., Wizel, B., Viscidi, R. and Kumar, N., 1995, Mapping of two overlapping linear epitopes in Pfg27 recognized by *Plasmodium falciparum* transmission-blocking monoclonal antibodies. Vaccine (in press).

Quakyi, I.A., Otoo, L.N., Pombo, D., Sugars, L.Y., Menon, A., DeGroot, A.S. Johnson, A., Alling, D., Miller, L.H. and Good, M.F., 1989, Differential non-responsiveness in humans of candidate *Plasmodium falciparum* vaccine antigens. Am.J.Trop.Med.Hyg. 41:125-134.

Riley, E.M., Nennett, S., Jepson, A., Hassan-King, M., Whittle, H., Olerup, O. and Carter, R., 1994, Human antibody response to Pf230, a sexual stage-specific surface antigen of *Plasmodium falciparum*: non-responsiveness is a stable phenotype but does not appear to be genetically regulated. Parasite Immunol. 16:55-62.

Sjoberg, K., LepersJ-P. Raharimalala, L., Larsson, L., Olerup, O., Marbiah, N.T., Troy-Blomberg, M., Perlman, P., 1992, Genetic regulation of human anti-malaria antibodies in twins. Proc.Natl.Acd.Sci. 89:2101-2104.

Wizel, B. and Kumar, N., 1991, Identification of a continuous and cross-reacting epitope for *Plasmodium falciparum* transmission-blocking immunity. Proc.Natl.Acad.Sci. 88:9533-9537.

9

EXPERIMENTAL FELINE LYME BORRELIOSIS AS A MODEL FOR TESTING *BORRELIA BURGDORFERI* VACCINES

Michael D. Gibson, M. T. Omran, and Colin R. Young

Department of Veterinary Anatomy and Public Health
College of Veterinary Medicine
Texas A&M University
College Station, Texas 77843-4458

INTRODUCTION

Lyme disease is a multisymptom disease resulting in dermatologic, rheumatologic, neurologic and cardiac abnormalities which develop in stages with various clinical pictures (1) following infection by a newly discovered spirochete, *Borrelia burgdorferi* (2).

Since its first diagnosis in Lyme, Connecticut, USA, Lyme disease has been identified in all continents throughout the world (3). It is found over most of Europe (4) Asia (5,6) and some parts of Africa (7,8). The clinical picture of Lyme disease is similar in some respects to other spirochetal diseases such as syphilis, relapsing fever and leptospirosis (9). The early stage is characterized by distinctive Erythema Chronicum Migrans, annular skin lesions, regional adenopathy and these may be accompanied by flu like symptoms and signs. Many cases develop the second or disseminated stage; characterized by multiple secondary erythema migrans, lymphocytoma, migratory musculoskeletal pain, acute arthritis, early neurologic manifestations (lymphocytic meningitis, cranial neuritis, radiculoneuritis, encephalitis and myelitis), generalized adenopathy, splenomegaly and cardiac abnormalities (carditis, conduction defects and arrhythmia). During the third or late stage severe malaise and fatigue, Acrodermatitis chronica atrophicans, chronic arthritis, late neurologic manifestations (chronic progressive encephalomyelitis, focal encephalitis, mild encephalopathy, distal axonopathy and neuropathy in patients with acrodermatitis), chronic arthritis and chronic fatigue are manifested.

Borrelia burgdorferi belongs to the order *Spirochaetales* that forms two families of gram negative helical bacteria with common morphologic features. *Borrelia* species are the only pathogenic spirochetes transmitted to vertebrates by blood sucking arthropods (9,10). *Borrelia burgdorferi* like other *Borrelia* stains well with Giemsa (11) and can be demonstrated in the tissue using silver impregnation methods (12,13,14).

Borrelia spirochetes are believed to be primarily transmitted by ticks. Transmission of the spirochete from the vector to the host occurs either by way of saliva or regurgitation

Immunobiology of Proteins and Peptides VIII
Edited by M. Z. Atassi and G. S. Bixler, Jr., Plenum Press, New York, 1995

73

(15). The tick feeds as an obligate blood and tissue fluid feeding ectoparasite, on terrestrial vertebrates including birds (16), during each of its three successive postembryonic stages or instars i.e. larva, nymph and adult (17).

Different species of ticks are involved in the transmission of Lyme disease. *Ixodes Scapularis, Dermacentor variabilis* and *Amblyomma americanum* are the principal vectors of transmission in the USA (18,19). Outside the USA *Borrelia* spirochetes have been demonstrated in other ticks and blood sucking insects such as *I. persulcatus* in Asia (5,22). Furthermore, some blood sucking arthropods such as mosquitoes, horseflies (18,23) and cat fleas *Ctenocephalides felis* (24) have been reported to be infected with *Borrelia burgdorferi*, although their involvement in the transmission of Lyme disease is still understudy. Oral infection cannot be ruled out, since mice and ducks have been experimentally infected by the oral route (23).

The list of wild animals which can act as a reservoir for *Borrelia burgdorferi* differs according to geographical location. The agent has been recovered from rodents, rabbits and medium sized mammals, wolves, coyotes, foxes, deer as well as large animals such as buffalos and bears (25). Migratory birds may play an important role both as a local reservoir of infection and long-distance dispersal agents for *Borrelia burgdorferi* infected ticks (26). This potential reservoir of infection is underscored by the fact that over 95 species of migratory birds have been shown to be infected with *Borrelia burgdorferi*.

Disease signs have also been demonstrated in domestic animals. Arthritis sometimes occurs in naturally infected horses and cows. Infection in horses led to swelling of legs, weight loss, dermatitis, conjunctivitis, uveitis, nasal discharge and cough, while mastitis, spontaneous abortion, myocarditis, interstitial nephritis, glomerulonephritis and pneumonitis have been demonstrated in infected cows (27,28,29,30,31).

Many different animal models of Lyme borreliosis have been investigated, however an animal model demonstrating the complete clinical picture of human Lyme borreliosis has not been established. For example naturally infected white-footed mice (32) and experimentally infected rabbits (33,34) develop skin lesions and spirochetemia, while experimentally infected laboratory rats (35,36) and irradiated hamsters (37,38) develop arthritis.

As Lyme disease is zoonotic it is prudent to study closely the animals, particularly pets, in the proximity of man which may act as a reservoir for the disease. Dogs can naturally be infected with *Borrelia burgdorferi* and show signs of Lyme disease (39); likewise with cats (40,41). Other blood sucking insects such as fleas (24) may act as vectors for transmitting Lyme borreliosis. This study was initiated to determine whether cats could be i) infected with different isolates of *borrelia burgdorferi*, ii) will different isolates cause any pathological changes in cats and iii) does the clinical picture in cats resemble human Lyme borreliosis?

MATERIALS AND METHODS

Borrelia burgdorferi

Three strains of *Borrelia burgdorferi* were used in this study; B31 - a reference strain isolated from *Ixodes scapularis* from Shelter Island, New York (42); TX1579 - isolated from *A. americanum* or Lone Star Tick from Uvalde County in Texas (24); TX532 - isolated from a pool of 5 cat fleas from Fort Bend County in Texas (24).

Cats (Phase I)

Twenty uninfected normal healthy cats (obtained from the College of Veterinary Medicine, Auburn University, Auburn, Alabama) were divided into four groups each containing 5 cats. Group I cats were used as non-infected control cats. Groups 2,3 and 4 were used as test cats. Groups 2,3, and 4 each was injected with 10^6 live *Borrelia burgdorferi* of a different isolate intradermally in a single site in the sacral region. Ten days post injection each group of cats was split into two subgroups. Uninfected cat fleas or Lone Star ticks were placed onto either of these subgroups for a period of 7 days. At this time the ticks and cat fleas were removed and examined by IFA for the presence of *Borrelia burgdorferi*. Thereafter, the cats were bled every other week, and monthly one cat per group was sacrificed; post mortem and histopathological studies were carried out on all tissues and organs.

Cats (Phase II)

In an attempt to simulate natural Lyme borreliosis infection in the domestic feline, fifteen uninfected normal healthy specific pathogen free (SPF) outbred cats were used, with an average weight 2.93 kg, SD $\pm$ .28 kg and 17-19 months of age and these cats were divided into groups of four each. Group 1, three cats were used as non-infected negative control cats. Groups 2, 3, and 4 (each containing 4 cats) were used as test cats. Group 2 cats were exposed to experimentally infected cat fleas *C. felis* with strain TX532, while group 3 cats were exposed to experimentally infected cats fleas *C. felis* with strain TX1579, for a period of one week. Group 4 cats were divided into 2 subgroups; Group A which consisted of two cats, one was inoculated with 10^6 live *B. burgdorferi* (TX532) intradermally in a single site in the sacral region while the other was inoculated intravenously with 10^6 live *B. burgdorferi* (TX532). Group B which consists of two cats, one was inoculated with 10^6 live *B. burgdorferi* (TX1579) intradermally in a single site in the sacral region while the other was inoculated intravenously with 10^6 live *B. burgdorferi* (TX1579). The two cats were monitored daily; every two weeks they were bled. Bimonthly, one cat per group was euthanized and necropsied and the histopathological studies were carried out on all other organs and tissues.

Ticks and Fleas

The *Amblyomma americanum* or Lone Star tick and the cat fleas *Ctenocephalides felis* were obtained from the Department of Entomology, Texas A&M University, College Station, Texas.

Blood Smears Staining

Hemacolor® stain (EM Diagnostic Systems, Inc., New Jersey, USA) was used. It is a stain set for rapid manual staining of blood smears with color pattern similar to that of Wright Stain and Wright-Giemsa stains. The blood cells in the blood smear stain in the following manner: The erythrocytes cytoplasm stains pink; polymorphonuclear neutrophils stain with a purple nucleus, purplish-grey cytoplasm with lilac-violet pink granules; eosinophils stain with a purple nucleus, purplish-grey cytoplasm with red/orange granules; basophils stain with a dark-blue nucleus and dark-purple cytoplasm; monocytes stain with violet nucleus and blue-gray cytoplasm; lymphocytes stain with dark-blue nucleus and blue-grey cytoplasm.

Immunofluorescent Antibody (IFA)

All arthropods were evaluated for spirochetal infection as follows: arthropods were washed by immersion in 70% ethanol for 1 min, 30% ethanol for 1 min, and 3% hydrogen peroxide for 30 seconds, then transferred to and rinsed twice in sterile PBS. The midgut and all the internal organs from each tick were compressed on a glass slide and identification of spirochetes was assessed by IFA for reactivity to monoclonal antibodies H9724, H5332, H68, and H4825 (kindly donated by Dr. Alan Barbour, University of Texas Health Science Center, San Antonio, TX) as described elsewhere (43). Specimen slides were prepared for IFA examination with fluorescein isothiocyanate conjugaged goat anti-mouse Ig (Cappel, Organon Teknika Corp., West Chester PA). For positive controls, slides prepared with the B31 strain of *Borrelia burgdorferi* were included in each run of specimens.

Enzyme-Linked Immunosorbent Assay (ELISA)

Antigenic preparations were prepared by sonicating the spirochete suspension in an ice bath for 30 second bursts with a Virsonic 300 (Virtis Scientific, Gardner, New York) at 40% of the maximal intensity setting.

Assays were performed with modifications, according to Envall and Pearlmann (44). Flat bottom 96 well immunoassay plates (Costar, Cambridge, Mass, USA) were coated overnight with 100 μl of *B. burgdorferi* (20 μl protein/ml) in PBS buffer (pH 7.4) at 4°C. The plates were drained then washed once with 0.01 M PBS containing 0.05% Tween 20. 300 μl of Biotin conjugated mouse anti-cat (Sigma Laboratories, St. Louis, Missouri, USA) was added and the plates were incubated for 2 hrs at room temperature. Following this the plates were washed three times with 0.01 M PBS containing 0.05% Tween 20. The assay was developed by the addition of 100 μl PNNP substrate (Zymed, San Francisco, California, USA), and absorbance read at 405 nm using Biotek Instruments, Winsooki, Vermont, USA).

PHASE I

Results

In 13 out of 15 infected cats the white blood corpuscle differential count indicated cycles of reduction in the neutrophil count accompanied by an increase in the lymphocyte and eosinophil counts compared to the control, cat inoculated with *B. burgdorferi* isolate B31, cat inoculated with *B. burgdorferi* isolate TX532, and cat inoculated with *B. burgdorferi* isolate TX1579. During the course of infection of these cats we noted the appearance of an 'atypical cell' in the blood films. Following staining with Hemacolor® this 'atypical cell' had the following characteristics; 6-7μ in diameter, a single compact round dark blue nucleus 1.5-3 μ in diameter, blue-grey cytoplasm. A summary of hematologic findings is shown in table 1.

PHASE I

Discussion

Earlier studies on cats have shown that cats seroconvert and develop antibodies against *Borrelia burgdorferi*. In some instances infected cats present with a history of fever, fatigue, anorexia, and lameness while in another study no clinical signs nor gross or histological abnormalities were found (45,46).

The 'atypical cells', found in our study differ from the enlarged cells (occasionally similar to Reed-Sternberg cells of Hodgkin's disease) or the atypical enlarged plasmacytoid

Table 1. A summary of hematologic findings

Hematologic finding[1,2]	B31	TX532	TX1579	Control
Neutropenia	4/5	1/5	3/5	2/5
Neutrophilia	3/5	0/5	2/5	0/5
Lymphocytopenia	2/5	0/5	2/5	0/5
Lymphocytosis	3/5	1/5	3/5	2/5
Monocytosis	1/5	4/5	2/5	1/5
Eosinophilia	2/5	3/5	1/5	2/5
Atypical cells	4/5	5/5	4/5	1/5
Cyclic change	3/5	5/5	4/5	0/5
($\uparrow$lymphocytes, $\uparrow$ eosinophils, $\downarrow$ neutrophils, $\uparrow$antibody)				

[1]Data based upon changes and percentage distribution of WBC in test animals compared with mean of control animals ($\bar{x} \pm 2\sigma$) and standard values.
[2]Nº of infected cats/Nº of cats in group.

cells (sometimes binucleate) reported for certain cases of human Lyme borreliosis (47). The size and color pattern of these cells suggest that they may be lymphocytes.

CNS changes in the cat in the form of perivascular lymphocytic infiltration and mild spongiosis is comparable to findings in humans.

The moderate to severe cytoplasmic swelling and fatty degeneration of the hepatocytes and pericholangitis accompanied with lymphocytic and plasma cell infiltration was similar to findings in human Lyme borreliosis (48).

The synovial membrane hypertrophy with multiple aggregates of lymphocytes and plasma coupled with other signs are suggestive of Lyme arthritis.

Table 2. A summary of the histopathologic findings

	Histopathologic findings	B31	TX532	TX1579	Ctrl
Cerebrum & thalamus	Spongiform encephalopathy and/or perivascular reaction	3/5	0/5	2/5	2/5
Cerebellum & medulla oblongata	Spongiform encephalopathy and/or perivascular reaction	3/5	1/5	0/5	3/5
Liver	Hydropic degeneration, lymphocytic infiltration, reticuloendothelial hyperplasia, periportal proliferation	5/5	5/5	5/5	1/5
Stifle joint & tarsus	Synovial hyperplasia and/or non-suppurative meningitis	1/5	1/5	0/5	0/5
Spleen	Lymphoid hyperplasia and/or plasmocytosis	5/5	5/5	3/5	1/5
Kidneys	Degeneration, lipidosis, and/or lymphocytic infiltration	4/5	2/5	2/5	2/5
Mesenteric lymphnodes	Lymphoid hyperplasia	5/5	5/5	5/5	2/5
Colon	Lymphoid hyperplasia and/or mucous and muscular layer hyperplasia	3/5	3/5	3/5	0/5
Rectum	Lymphoid hyperplasia and/or mucous and muscular layer hyperplasia	3/5	3/5	3/5	0/5
Lungs	Alveolar emphysema, pneumonitis, and/or lymphocytic infiltration	2/5	2/5	2/5	1/5

The mild alveolar emphysema, mild multifocal interstitial pneumonitis, plasmacy-tosis, and lymphoid hyperplasia observed in the lungs of our infected cats is supportive of spirochetemia. A summary of the histopathological findings is shown in table 2.

PHASE II

Results

Gross clinical observations at approximately 6 months post-exposure to the experi-mentally infected fleas revealed that two cats demonstrated skin lesions at the site of the arthropod containment chamber. These skin lesions were erythematous for 6-7 days.

One test cat demonstrated marked change in behavior; exhibiting pronounced irrita-bility, mild photophobia and aggressive disposition. The behavioral change occurred ap-proximately 7 months after exposure to the infected fleas and did not change to the point of euthanasia.

Another cat demonstrated stiffness and lameness of the left hind limb 5 months post-inoculation. The lameness subsided after 1 week.

Serology

The antibody titer to *B. burgdorferi* of the test cats was measured. If there was a four fold increase in the titer, the cat was declared seroconverted.

TX532 (*C. felis*)

Two cats which were exposed to experimentally infected fleas seroconverted by week 3 after exposure (titer 1:2048). One cat seroconverted by week 6 reaching a titer of 1:2048. Another cat reached a titer of 1:512 by week 3 post-inoculation. One cat which was inoculated I.D. seroconverted by week 3 with a titer of 1:2048, while the cat injected I.V. demonstrated a titer of 1:512 by week 3 post-inoculation.

1579 isolate (*A. americanum*)

Two cats and seroconverted by week 6 after exposure to the experimentally infected fleas and reached a titer of 1:2048. Another cat exhibited no increase in the titer above 1:512, while the cat which was injected I.V. with *B. burgdorferi* seroconverted by week 3. The cat which was injected I.D. also seroconverted by week 3 revealing a titer of 1:2048.

Histopathological Data

Control Cats

The liver of two of the three control cats exhibited very minimal biliary duct proliferation. One cat demonstrated a slight increase in the muscularis mucosa of the colon.

532 Cats

Two cats which were exposed to infected fleas and one cat injected I.V. exhibited mild synovial hyperplasia of the stifle joint. The colon, ileum and rectum demonstrated various degrees of thickness with enlarged mesenteric lymph nodes. The mesenteric lymph nodes revealed hyperplasia.

1579 Cats

Three out of four cats exposed to experimentally infected fleas and the cat injected I.D. exhibited mild to moderate biliary duct proliferation, periportal hepatocyte swelling and periportal cholangitis. Prominent Peyer's patches were found in the ileum of all of the test cats with the mesenteric lymph nodes exhibiting hyperplasia and cortical depletion.

The muscle wall of the colon was mildly thickened in one cat, while markedly thickened in two cats. Also, the rectum in these cats exhibited more folds and an increase in lymphoid follicles. The spleen exhibited lymphoid hyperplasia which is characteristic with chronic infection.

PHASE II

Discussion

The establishment of patterns corresponding to human Lyme borreliosis has been achieved in this domestic feline model. Although spirochete recovery eluded our efforts, correlation has been demonstrated in the test cats applicable to the three recognized and defined stages of human Lyme borreliosis.

Skin lesions found at the site of the cat flea containment chamber are representative and similar to the lesions seen in human Lyme borreliosis. The onset of these lesions appearing at approximately 6 months in the cats, as opposed to human clinical presentation at 3 days to 3 weeks, may be dose related and/or a manifestation of the particular borrelia isolate.

The two cats which exhibited lameness were experimentally infected with the TX1579 isolate and though the lameness abated, it is similar in nature to tertiary stage human Lyme borreliosis.

In comparing I.D. to I.V. routes of inoculation, the titers were elevated in the I.D. route versus the I.V. route and this may be explained by the belief that in the I.V. route the bacteria are cleared faster than in the I.D. route where the contact with the macrophages is slower.

CONCLUSIONS

Phase I

In conclusion our study demonstrated that cats were susceptible to infection with different isolates of *Borrelia burgdorferi*. They exhibited minimal clinical signs, however the histopathological changes conformed with that reported in human Lyme borreliosis, therefore cats may act as reservoirs for *Borrelia burgdorferi*.

Phase II

Cat fleas, *Ctenocephalides felis* have been suspect in the transmission of Lyme borreliosis. However to date no confirmatory literature exists delineating natural borreliosis infection through normal vector to host routes with the cat flea. We conclude that natural infection is a possibility due to the multitude of clinical and histopathological observations in the experimental cat model.

SUMMARY

The feline model investigated establishes that domestic cats may act as an animal model for evaluating the pathogenesis of Lyme borreliosis. Specifically this feline model demonstrates: First, that animals seroconvert following either needle injection of, or arthropod delivery of, *Borrelia burgdorferi*. Clinical findings obtained are consistent with those observed in human Lyme disease; histopathological observations are also consistent with those observed in human Lyme disease. Therefore, cats may also be used as a representative animal model for measuring immune protection against Lyme borreliosis. Specifically we are exploring the protective capacity of *Borrelia burgdorferi* antigenic compounds in cats, namely OspA, OspB, OspC, heat shock proteins, flagellar antigens and various protective immunological combinations.

ACKNOWLEDGMENTS

- Paravax, Inc.
- College of Veterinary Medicine, Texas A&M University
- Lyme Disease Foundation
- Ciba Geigy Corp.

REFERENCES

1. Steere, A.C. 1989. Lyme disease. *N Engl J Med*. 321:586-596.
2. Steere, A.C., Grodzicki, R.L., and Kornblatt, A.N. 1983. The spirochetal etiology of Lyme disease. *N Engl J Med*. 308:733.
3. Schmid, G.P. 1985. The global distribution of Lyme disease. *Rev Infect Dis* 7:41-50.
4. Dekonenko, E.J., Steere, A.C., and Berardi, V.P. 1988. Lyme borreliosis in the Soviet Union: a cooperative US-USSR report. *J Infect Dis*. 748-753.
5. Ai, C.X., Wen, Y.X., and Zhang, Y.G. 1988. Clinical manifestations and epidemiological characteristics of Lyme disease in Hailin county, Heilongjiang Province, China. *Ann NY Acad Sci*. 539:302-313.
6. Isogai, E. Isogai, H., and Sato, N. 1990. Antibodies to *Borrelia burgdorferi* in dogs in Hokkaido. *Microbiol Immunol*. 34:1005-1012.
7. Fivaz, B.H. and Petney, T.N. 1989. Lyme disease—a new disease in southern Africa? *J S Afr Vet Assoc*. 60:155-158.
8. Haberberger, R.L. Jr., Constantine, N.T., Shwan, T.G. et al. 1989. Lyme disease agent in Egypt. *Trans R Soc Trop Med* 83:556.
9. Schmid, G.P. 1989. Epidemiology and clinical similarities of human spirochetal diseases. *Rev Infect Dis*. Supp 6:S1460-S1469.
10. Barbour, A.G. and Hayes, S.F. 1986. Biology of *Borrelia* species. *Microbiol Rev*. 50:381-400.
11. Burgdorfer, W. 1984. Discovery of the Lyme disease spirochete and its relation to tick vectors. *Yale J. Biol. Med*. 57:515-520.
12. Duray, P.H., Kusnitz, A. and Rayan, J. 1985. Demonstration of the Lyme disease spirochete by modified Dieterle stain methods. *Lab. Med*. 16:685-687.
13. de Koning, J., Bosma, R.B. and Hoogkamp-Korstanje, A.A. 1987. Demonstration of spirochetes in patients with Lyme disease in a modified silver stain. *J. Med. Microbiol*. 23:261-267.
14. Swisher, B.L. Modified Steiner procedure for microwave staining of spirochaetes and nonfilamentous bacteria. *J. Histotech*. 10:241-243.

15. Burgdorfer, W. 1989. Vector/host relationships of the Lyme disease spirochete *Borrelia burgdorferi*. *Rheum. Dis. Clin. North Am.* 15:748-753.

16. Anderson, J.F. 1989. Vector/host relationships of the Lyme disease spirochete *Borrelia burgdorferi*. *Rheum. Dis. Clin. North Am.* 15:748-753.

17. Bosler, E.M. 1992. Tick Vectors and Hosts. In: *Lyme Disease*. P.K. Coyle, ed. Mosby Year Book, St. Louis, p. 18-26.

18. Magnarelli, L.A. and Anderson, J.F. 1988. Ticks and biting insects infected with the etiologic agent of Lyme disease, *Borrelia burgdorferi*. *J. Clin. Microb.* 26:1482-1486.

19. Schulze, T.L., Bowen, G.S., and Bosler, E.M. 1984. *Amblyomma americanum*: A potential vector of Lyme disease in New Jersey. *Science*. 224:601-603.

20. Piesman, J. and Stinsky, R.J. 1988. Ability of *Ixodes pacificus*, *Dermacentor variabilis* and *Amblyomma americanum* (Acri: Ixodidae) to acquire, maintain and transmit Lyme disease spirochetes, *Borrelia burgdorferi*. *J. Med. Entomol.* 25:336.

21. Rousselle, C., Floret, D., Cochat, P., et al. 1989. Encephalite aigue a *Borrelia burgdorferi* (maladie de Lyme) ches un enfant algerien. *Pediatrie* 44:265-269.

22. Kawabata, M. 1987. Lyme disease in Japan and its possible incriminated tick vector, *Ixodes persulcatus*. *J. Infect. Dis.* 156:854.

23. Post, J.E. 1990. Lyme disease in large animals. *NJ Med.* 87:575-577.

24. Tetlow, G.J., Fournier, P.V. and Rawlings, J.A. 1991. Isolation of *Borrelia burgdorferi* from arthropods collected in Texas. *Am. J. Trop. Med. Hyg.* 44:469-474.

25. Cohen, D., Bosler, E.M., Bernard, W., et al. 1988. Epidemiologic studies of Lyme disease in horses and their public health significance. *Ann. NY Acad. Sci.* 539:244-257.

26. Weisbrod, A.R. and Johnson, R.C. 1989. Lyme disease and migrating birds in Staint Croix River Valley. *Appl. Environ. Microbiol.* 55:1921-1924.

27. Burgess, E.C., Gendron-Fitzpatrick, A. and Wright, W.O. 1987. Arthritis and systemic disease caused by *Borrelia burgdorferi* infection in a cow. *J.A.V.M.A.* 191:1468-1470.

28. Burgess, E.C. and Mattison, M. 1987. Encephalitis associated with *Borrelia burgdorferi* infection in a horse. *J.A.V.M.A.* 191:1457-1458.

29. Burgess, E.C. 1988. *Borrelia burgdorferi* infection in horses and cows. *Ann. NY Acad. Sci.* 539:234-243.

30. Bosler, E.M., Cohen, D.P., and Schulze, T.L. 1988. Host responses to *Borrelia burgdorferi* in dogs and horses. *Ann. NY Acad. Sci.* 539:244-257.

31. Post, J.E., Shaw, E.E. and Wright, S.D. 1988. Suspected borreliosis in cattle. *Ann. NY Acad. Sci.* 539:488.

32. Magnarelli, L.A., Anderson, J.E. and Chappell, W.A. 1984. Geographic distribution of humans, raccoons, and white-footed mice with antibodies to Lyme disease spirochetes in Connecticut. *Yale J. Biol. Med.* 57:619-626.

33. Burgdorfer, W. 1984. The New Zealand white rabbit: An experimental host for infecting ticks with Lyme disease spirochetes. *Yale J. Biol. Med.* 57:609-612.

34. Kornblatt, A.N., Steere, A.C. and Brownstein, D.G. 1984. Experimental Lyme disease in rabbits: Spriochetes found in erythema migrans and blood. *Infect. Immun.* 46:220-223.

35. Barthold, S.W., Moody, K.D., and Terwillinger, G.A. 1988. An animal model for Lyme arthritis. *Ann. NY Acad. Sci.* 539:254-273.

36 Barthold, S.W., Moody, K.D., and Terwillinger. 1988. Experimental Lyme arthritis in rats infected with *Borrelia burgdorferi*. *J. Infect. Dis.* 157:842-846.

37 Schmitz, J.L., Schell, R.F. and Hejka, A. 1988. Introduction of Lyme arthritis in LSH hamsters. *Infect Immunol.* 56:2236-2342.

38 Hejka, A., Schmitz, J.L., England, D.M. 1989. Histopathology of Lyme arthritis in LSH hamsters. *Am. J. Path.* 134:1113-1123.

39 Lissman, B.A. 1990. Lyme disease in small animals. *NJ Med* 87:573-574.

40 Magnarelli, L.A., Anderson, J.F., Levine, H.R. and Levy, S.A. 1990. Tick parasitism and antibodies to *Borrelia burgdorferi* in cats. *J.A.V.M.A.* 197:63-66.

41 Burgess, E.C. 1992. Experimentally induced infection in cats with *Borrelia burgdorferi*. *Am. J. Vet. Res.* 53:1507-1511.

42 Burgdorfer, W., Barbour, A.G., Hayes, S.F., et al. 1982. Lyme disease - a tick-borne spirochetosis? *Science*. 216:1317-1319.

43 Barbour, A.G., Tessier, S.L. and Todd, W.J. 1983. Lyme disease spirochetes and ixodid tick spirochetes share a common surface antigenic determinant defined by monoclonal antibody. *Infect. Immun.* 41:795-804.

44 Envall, E. and Perlmann, P. 1972. Enzyme-linked immunosorbent assay, ELISA. II. Quantitation of specific antibodies by enzyme-labeled anti-immunoglobulin in antigen coated tubes. *J. Immunol.* 109:129-135.

45 Burgess, E.D. 1992. Experimentally induced infection of cats with *Borrelia burgdorferi*. *Am. J. Vet. Res.* 53:1507-1511.

46 Magnarelli, L.A., Meegan, J.M., Anderson, J.F., et al. 1984. Comparison of indirect fluorescent antibody test with an enzyme-linked immunosorbent assay for serological studies of Lyme disease. *J. Clin. Micro.* 20:181-184.

47 Duray, P.H. 1992. Target organs of *Borrelia burgdorferi* infections: Functional responses and histology. In: *Lyme disease molecular and immunologic approaches*. Scutzer, S.E., ed. Lold Spring Harbor Laboratory Press, New York, p.11-30.

48 Goellner, M.H., Agger, W.A, and Duray, P.H. 1988. Hepatitis due to recurrent Lyme disease. *Ann. Intern. Med.* 108:707.

10

LIPOSOMAL VACCINES

Shawn Green, Anne Fortier, Jan Dijkstra, John Madsen, Glenn Swartz,
Leo Einck, Ed Gubish, and Carol Nacy

EntreMed, Inc.
9650 Medical Center Drive
Rockville. Maryland 20850

In the past decade, medical research made rapid advances in technologies associated with vaccine development, in particular molecular engineering and peptide synthesis. These new technologies opened the door to development of vaccines effective against newly emerging pathogens and even proteins or antigens that cause diseases of noninfectious origin. However, realization of the potential of subunit or peptide vaccines has been hampered by fundamental issues inherent in the presentation of these antigens to an immune system. In general, such vaccines fail to stimulate strong immune responses because they present antigens that (a) are not oriented in space or in the context of other surface proteins and molecules in the same way they are on the pathogen itself, (b) don't replicate or stimulate inflammatory mediators that call in immune cells, and (c) don't persist, the most critical issue for compromising tissue integrity and getting the immune response interested. Our laboratory addressed these issues by orienting the specific antigen in a lipid bilayer as it would appear naturally, and by providing an adjacent adjuvant in that same bilayer to stimulate immune reactivity to the entire complex of antigens. We describe in this paper a platform vaccine technology that combines a liposome for insertion of different antigen classes by different means with lipid A to enhance antigen immunogenicity. These vaccine formulations specifically induce both humoral (antibody) and cellular (cytotoxic cell) immunity and selectively stimulate immunologic memory in those cases where an antigen-specific challenge response is desired.

Vaccines under development in our laboratories include those that induce immunologic control of several classical infectious threats, but also target antigens involved in neoplastic, metabolic, and angiogenic disorders, such as diabetic retinopathy, hypercholesteremia, and multidrug-resistant cancers. We review here the basis of the vaccine technology and the rationale behind several representative vaccines for different antigen types.

LIPOSOMES AS DELIVERY VEHICLES FOR ANTIGENS

Liposomes are lipid vesicles that form spontaneously when their substituent phospholipids coalesce in an aqueous environment into spherical membranes referred to as lipid bilayers, or lamellae (Figure 1).

Immunobiology of Proteins and Peptides VIII
Edited by M. Z. Atassi and G. S. Bixler, Jr., Plenum Press, New York, 1995

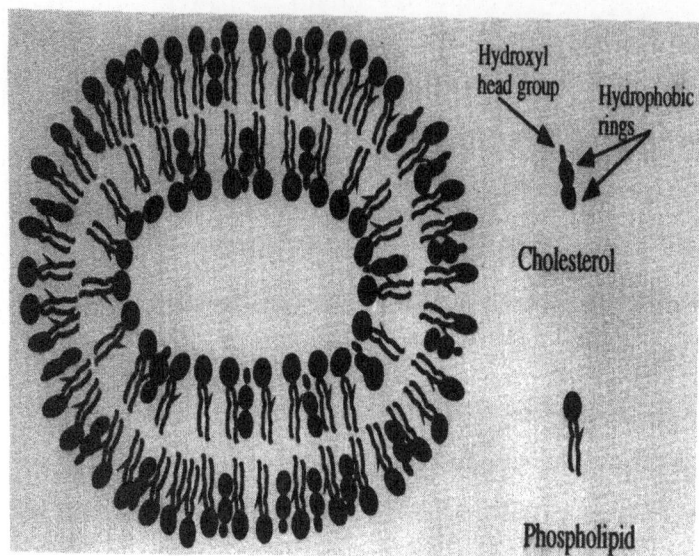

Figure 1. Representation of an stable unilamellar liposome with cholesterol in the lipid bilayer.

Depending on the method of preparation, liposomes form as unilamellar vesicles or multilamellar vesicles with a varying number of spherical concentric lamellae. These lipid particles are phagocytosed by macrophages and phagocytic cells of the liver, spleen, and bone marrow, the major antigen-presenting cells for initiating immune responses. Since liposomes are amenable to manipulation by changing their lipid constituents, one can rationally design a suitable delivery system to enhance immunogenicity of particular classes of antigens, including proteins, peptides, carbohydrates, and lipids.

Liposomes can decrease antigen or adjuvant toxicity by sequestration of a toxic aqueous solution within the aqueous lumen of liposomes (which also decreases dilution of encapsulated hydrophilic drugs or antigens), or by shielding the in vivo environment from toxic hydrophobic amphipathic compounds by burying those substances in the liposomal bilayers, as has been done with lipid A (1,2). Liposomes can be made more resistant to disintegration and biological degradation by inclusion of cholesterol, which lends stability to the preparations (3,4). Since mammalian biological membranes also contain phospholipids and cholesterol, immunogenicity is then regulated by the incorporated antigens and adjuvants, rather than the liposomal carrier. Saturated carrier lipids, such as dimyristoylphosphatidylcholine (DMPC) and dimyristoylphosphatidylglycerol (DMPG), increase the useful life of liposomes and make the vesicles more amenable to quality assurance and control. Monophosphoryl lipid A (MPLA) is used as an adjuvant in all the vaccines under development in our laboratories. It is as potent a B cell adjuvant as LPS, and induces recruitment of antigen processing cells, production of cytokines, and stimulation of T cell regulated immunity (5), but is 100-10,000 fold less toxic in vivo in several animal models (6).

We use different methods tailored to the unique requirements of the antigen to incorporate it into liposomes. Antigens are either attached to the outside of the bilayer by conjugation or through the use of palmitic acid tails, intercalated into the hydrophobic interior of the bilayer (hydrophobic antigens), or encapsulated in the aqueous space(s) within the liposome (hydrophilic antigens).

Since the first report on immunoadjuvant activity of liposomes in 1974 (7), many studies show that liposomes enhance immune responses to various antigens (8-10). The natural target of liposomes is the macrophage, which initiates immunological reactions by processing and presenting antigens to T lymphocytes. Slow release of the antigen with the gradual degradation of the lamellae enhances the immunogenicity of the preparation and increases exposure of the host to the antigen, either intra- or extracellularly. Liposome encapsulated antigens elicit a T cell-dependent immune response, including production of both IgM and IgG antibodies, whereas liposome-associated haptens (small portions of molecules) elicit a T cell-independent IgM response only. Incorporation of an adjuvant into the liposome membrane enhances immunogenicity of liposome antigens and can switch T cell-independent antigens to T-cell dependency. Thus, the vaccine can be tailored to produce long-lived immunity and immunological memory for certain indications, but only short-term recognition and clearance (IgM antibodies) for indications where greater control over the degree of host response to the antigen is desirable.

ANTI-SEPSIS VACCINE: DUAL ROLE OF ADJUVANT AS ANTIGEN

Septic shock is the catastrophic culmination of events which begin with the inappropriate introduction of a pathogen into sterile tissue and progress to a full-scale battle between the imperative of the pathogen (self duplication) and the agenda of the host (tissue integrity). To identify strategies for control of sepsis, one must recognize the syndrome as a complex and progressive process. Sepsis should be viewed as a management issue, with a pertinent strategy for each stage of disease applied as appropriate to reduce the risk of progression.

Since the entire sequence of activities begins with the replication of a bacterium or fungus in tissue, one possibility to reduce risk might be to activate the immune response to classes of septic shock pathogens in advance of invasive medical procedures. To this end, we developed and are refining a vaccine against gram negative bacteria, organisms which account for progression to shock in about 45% of all reported sepsis-related deaths (11). The vaccine targets a common component in all gram negative bacteria, the lipid A portion of the bacterial outer membrane lipopolysaccharide (LPS). LPS, a complex molecule which is shed by gram negative bacteria as they grow and replicate, is a potent stimulant of immune reactions and most certainly contributes to the lethal consequences of septic shock by its manifold activities on immune cells. A potent result of the interaction of LPS and macrophages is the release of tumor necrosis factor alpha (TNF-α), a cytokine that can mimic in its entirety the symptoms of septic shock (12). Both LPS and TNF-a are targets for other septic shock interventions (12,13). While gram negative bacteria of different classes have characteristic and different polysaccharides associated with their specific LPS, the lipid anchor (lipid A) of each of these organisms is quite similar (Figure 2).

The anti-sepsis vaccine packages lipid A in lipid vesicles composed of phosphatidylcholine, phosphatidylglycerol, and cholesterol so that the inherent toxic properties of lipid A are masked, yet it can still act as an adjuvant to stimulate immunologic reactions, and it can still act as the antigen to which immune cells respond. Incorporation of detoxified monophosphoryl lipid A (MPLA) into liposomes detoxifies it even further (over ten-thousand fold), and MPLA included in the liposome formulation of a malaria vaccine was not pyrogenic, even at the highest doses, in a recent clinical trial (14).

Vaccination of mice with this preparation results in the generation of lipid A-specific antibodies (IgM and IgG) and protection against two gram negative pathogens, *Francisella tularensis* and *Escherichia coli* (Table 1). These data suggest that it is possible to intervene in the usual interactions of a gram negative pathogen and its mammalian host by stimulating the immunologic recognition of lipid A prior to infection. Vaccination with lipid A in

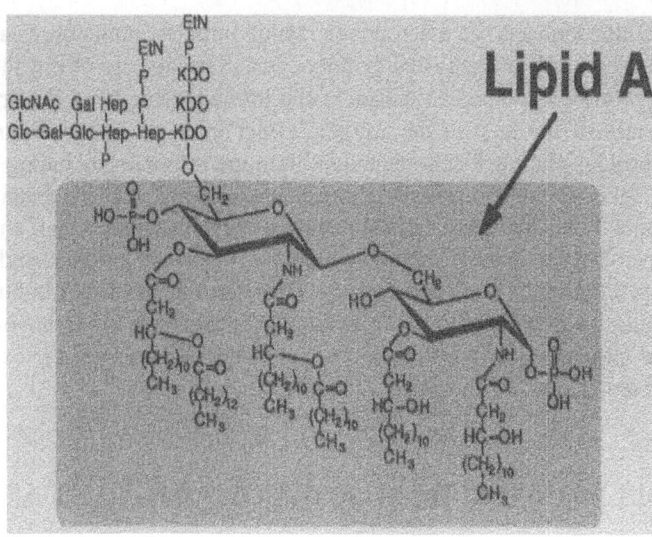

Figure 2. General structure of lipid A of the lipopolysaccharide (LPS) of gram negative Enterobacteriaceae.

liposomes facilitated survival of animals challenged with ten to a thousand-fold more bacteria in the infectious inoculum.

Mice were rendered neutropenic by the administration of cyclophosphamide 1 and 4 days prior to infectious challenge (15). Since the vaccine targets lipid A as the antigen, and since lipid A is present in all LPS, we asked whether antibodies to lipid A could neutralize the toxic effects of LPS in vivo. Mice were vaccinated and then treated with Actinomycin D to increase sensitivity to LPS in an endotoxemia model of gram negative sepsis. Vaccinated animals were challenged with several unrelated gram negative LPS (Table 2). In addition to providing protection against an infectious challenge, the lipid A vaccine also increased the amount of LPS that the animals could tolerate.

Mice were rendered sensitive to lps by the simultaneous administration of actinomycin d with the lps challenge (16). The lipid A vaccine may be an interesting adjunctive therapy to all the current management strategies, particularly under controlled conditions where septic shock is a risk, i.e., elective surgery, chemotherapy, or long-term debilitating illness. The vaccine reduces the progressive lethal effects of several gram negative pathogens and also facilitates neutralization of LPS effects in vivo. As part of a management strategy for this catastrophic illness, it could be administered several weeks in advance of a surgical or therapeutic procedure and coupled with antibiotic prophylaxis and passive adjunctive therapies. Such a comprehensive management strategy for sepsis should reduce the incidence of septic shock due to gram negative bacteria.

Table 1. Protection of BALB/c mice against gram negative bacteria with a lipid A vaccine

Mice challenged with	Mice treated with	Antibody titer	Log_{10} LD_{50}	Log protection
F. tularensis	Control liposomes	<1:40	1	—
	Lipid A liposomes	>1:6000	4	3
E. coli[a]	Control liposomes	<1:40	3	—
	Lipid A liposomes	>1:6000	4+	1+

Table 2. Protection of BALB/c mice against endotoxemia with a lipid A vaccine

Mice challenged with	Mice treated with	Antibody titer	LPS LD_{50} (ng)	Fold protection
S. minnesota LPS	Control liposomes	<1:40	30	—
	Lipid A liposomes	>1:6000	600	20
E. coli LPS[a]	Control liposomes	<1:40	50	—
	Lipid A liposomes	>1:6000	500	10

ANTI-CHOLESTEROL VACCINE: THE LIPID ANTIGENS

Cholesterol is antigenic. In fact, naturally occurring antibodies reactive with cholesterol are found in serum of virtually all species tested (17,18). Immunological intervention as an alternative to conventional approaches for lowering elevated serum cholesterol is actually an old concept. Bailey showed in 1964 that immunizing rabbits with albumin-conjugated cholesterol esters reduces the development of dietary-induced aortic atherosclerosis (19,20). Immunization of rabbits with LDL also inhibits development of atherosclerosis in rabbits fed a high cholesterol diet (21). These early studies suggest that production of anticholesterol antibodies and immunologic clearance of opsonized LDL may indeed be a useful strategy to lower serum cholesterol levels.

Dr. Carl Alving tested the liposomal anticholesterol vaccine at the Walter Reed Army Institute of Research (22). Six outbred unvaccinated rabbits served as controls; forty rabbits were vaccinated. All rabbits were fed a high-fat (0.5% cholesterol) diet for several months. Vaccinated rabbits developed antibodies that reacted with cholesterol. At termination of the experiment, aortas were removed for analysis of atherosclerotic disease by assessment of intensity of staining with Sudan IV (Table 1). The aorta was divided into four regions: 1-the ascending aorta (and arch), 2-the descending thoracic aorta, 3-middle descending aorta, and 4-the abdominal aorta above the iliac bifurcation. Significant reduction in plaque formation occurred in regions 2 and 4, where large sections were actually free of atheromas in the vaccinated rabbits. Sections 1 and 3 (where hemodynamic sheer forces of blood are the greatest), also had fewer and less dense atheromas in the vaccinated rabbits compared to controls. In addition, rabbits fed a 0.5% cholesterol diet had serum cholesterol levels in the thousands of mg/dl, well over 4 times that seen in human dietary hypercholesterolemia. On average, vaccination reduced total serum cholesterol levels by >20% (22).

Vaccination of mice with the same anticholesterol vaccine induced a 1000-fold increase in anti-cholesterol activity in serum of immunized compared to non-immunized

Table 3. Atherosclerosis in individual sections of rabbit aorta[a]

	Percent sudanophilic area (mean)		
Region of interest	Nonimmunized rabbits (N = 6)	Immunized rabits (N = 40)	p value
Entire vessel	32	20	0.13
Region 1	44	38	0.63
Region 2	32	12	0.03*
Region 3	32	22	0.20
Region 4	12	5	0.05*

[a]Adapted from Alving, et al., ref 22.
*Significant at.

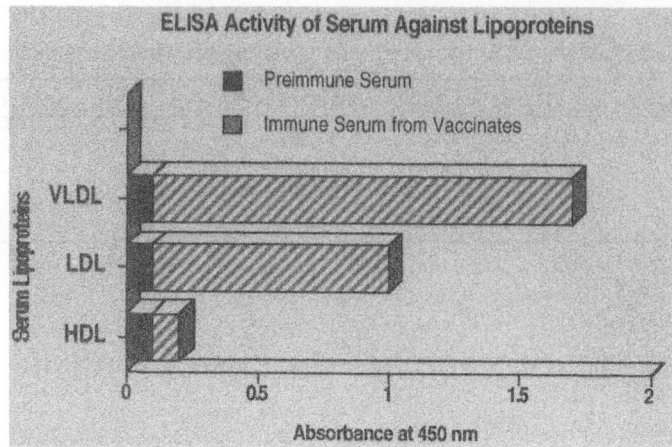

Figure 3. ELISA activity of serum against lipoproteins. Reactivity of normal and immune mouse serum in an ELISA for serum lipoproteins.

mice. Serum IgM antibody titers were more than 1/25,000, whereas titers for IgG1 and IgG2a were less than 1/25 (23). Antibody switching (and presumable T cell memory) was not stimulated by the vaccine. For optimal clearance of serum cholesterol through vaccination, then, periodic restimulation with booster inoculations would be required.

Vaccine-induced anticholesterol antibodies preferentially bound to VLDL and LDL, and far less to HDL (24, Figure 3). The aqueous interface of all lipoproteins is occupied by phospholipids, apoproteins, and unesterified cholesterol, whereas the cholesterol esters and triglycerides are largely hidden within the lumen of the lipoprotein. All three lipoproteins (HDL, LDL, VLDL) contain roughly similar amounts of unesterified cholesterol. The preferential binding of anti-cholesterol serum to VLDL suggests that antibody accessibility to the polar head of cholesterol may be critical. Of the three fractions evaluated, HDL contains 4 times as much protein to unesterified cholesterol as LDL and VLDL: the unesterified cholesterol in the surface bilayer of HDL may be masked by an excess of surface apoprotein. Cholesterol in cell membranes is also not accessible to the anticholesterol antibodies, either because of concentration or steric hindrance by cell-surface proteins.

Cholesterol regulates the physical properties of mammalian cell membranes and serves as a precursor for steroid hormone synthesis. Failure to metabolize excess circulating cholesterol can result in vascular diseases such as atherosclerosis. Current therapeutic strategies use drugs to reduce dietary cholesterol and interfere with cholesterol metabolism or intestinal absorption by upregulating LDL-R in the liver, further facilitating clearance of serum cholesterol. Our approach, in contrast, is to augment the non-LDL-R scavenger pathway through the generation of antibodies to cholesterol. This strategy may facilitate clearance of LDL directly through its ingestion by cells with receptors for antibody, as well as up regulate the LDL-R to enhance clearance of nonopsonized LDL.

ANTI-P-GLYCOPROTEIN VACCINE: THE PALMITOPEPTIDE ANTIGENS

The advent of molecular cloning ushered in a whole new era in vaccine development: a way to tailor the vaccine to specific molecules and actually refine the vaccine to target only the 'business' end of the particular protein by making peptides that mimic either the binding site or the active portion of the molecule. While this breakthrough invigorated vaccinologists world-wide, practitioners of the art realized that the refined approach to synthetic immunity induction has its own problems. Not the least among these is the rapid clearance of small proteins and peptides, with an inefficient recruitment of inflammatory and immune cells. These inherent problems have been recognized for over a decade, and a variety of techniques to improve immunogenicity of subunit vaccines were attempted, including engineering T cell epitopes into proteins, linking peptides to carrier molecules, and adding adjuvants. Our laboratory addressed these issues by orienting the specific antigen in a lipid bilayer as it would appear naturally, and by providing an adjacent adjuvant in that same bilayer to stimulate immune reactivity to the entire complex of antigens. The vaccine against multi-drug resistant (MDR) cancer is an example of a peptide vaccine under development in our labs.

In a manner analogous to the development of bacterial resistance to chronic antibiotic treatment, rapidly growing cancer cells in an environment of chronic chemotherapeutics become desensitized to their anticancer effects. Indeed, these cancers are frequently resistant to more than one type of drug. Within the past decade, a molecule termed P-glycoprotein was identified as a major cause of MDR; cells which develop drug resistance express of this glycoprotein molecule on their surfaces. P-glycoprotein functions as a pump to rapidly expel cancer chemotherapeutic drugs from inside the cell before they are able to exert their therapeutic actions (25).

Dr. Claude Nicolau (CBR Harvard Medical School) developed a P-glycoprotein vaccine. The essential strategy involved the cloning of various epitopes of the P-glycoprotein molecule. The human MDR1 cDNA encodes a 1280-amino acid protein with 12 predicted transmembrane domains. A model of the multidrug transporter, based on the deduced amino acid sequence of the human protein, is shown in Figure 4.

Critical extracellular epitopes of the P-glycoprotein molecule were cloned and sequenced and synthetic peptides made (26). The two ends of the peptides were attached to palmitic acid for insertion into the liposomes (Figure 5).

Liposomes containing these external epitopes of P-glycoprotein were immunogenic when administered to mice. High titer antibodies against the P-glycoprotein antigen were obtained within three weeks, and the sera from the immunized mice restored doxorubicin sensitivity to a strain of MDR cells (Table 4):

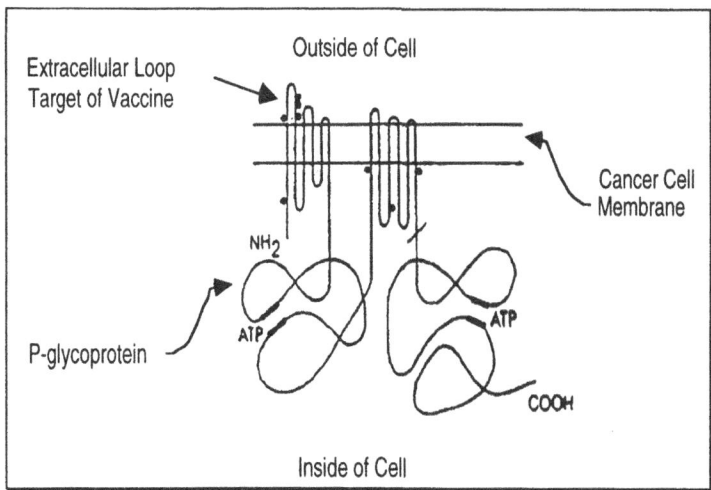

Figure 4. Schematic representation of the putative structure of the MDR protein and its orientation in the cell membrane.

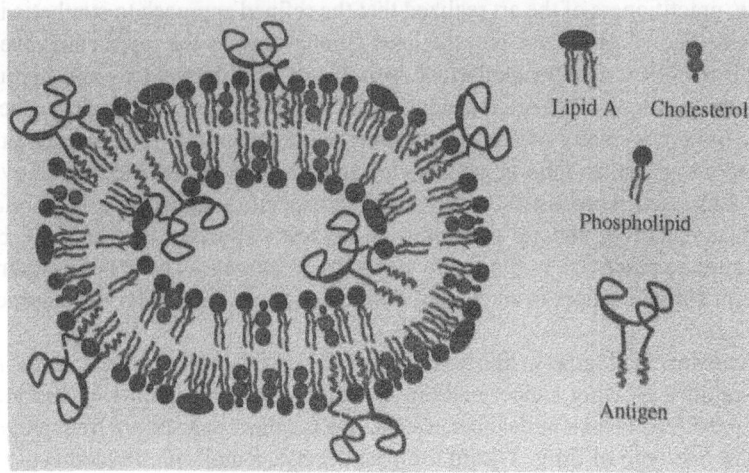

Figure 5. Simplified representation of the P-glycoprotein vaccine formulation

At present, there is no effective method to prevent or treat multidrug resistant tumors. Strategies for disabling P-glycoprotein, preventing the expression of the P-glycoprotein gene, or inactivating tumor glutathione have all been the target of extensive research. A method of immunologically disabling the transport function of the P glycoprotein may be an alternative approach to control of MDR

SUMMARY

Liposomes have been used therapeuticaly to deliver drugs to certain anatomical sites. The use of liposomes to deliver antigens, although not a new concept, has received less attention. At least two vaccines of nearly identical liposome base composition to our vaccines have been tested in humans. A malaria vaccine study showed that the liposomal preparation is quite safe: reaction profiles of volunteers receiving the vaccine demonstrated little reactivity and virtually no pyrogenicity (14). The concentration of MPLA in the vaccine was substantially higher (nearly 50,000 times) than the pyrogenic dose of free lipid A. The same vaccine, but different antigen (gp120, an HIV protein), was tested in volunteers and had the same lack of toxicity (27). In both studies, antibodies and cytotoxic cells specific for the respective antigens were produced. We have several subunit vaccines under development for infectious diseases (gram negative sepsis, fungal infections, protozoan infections), metabolic disorders (hypercholesterolemia, diabetic retinopathy, macular degeneration), and

Table 4. Reversal of drug resistance in vitro by antibody to P-glycoprotein[a]

Cells	% survival	LD_{50}
Normal L1210	26	4×10^{-8} M
MDR L1210	62	4×10^{-5} M
MDR L1210 + anti-liposome control	61	4×10^{-5} M
MDR L1210 + anti-p-glycoprotein	40	4×10^{-7} M

[a]Adapted from Tosi, et al., ref 26.

neoplastic diseases (multi-drug resistant cancer, primary and metastatic tumors, and angiogenic hyperproliferative disorders). In each case, one or more antigens were identified that might be useful in immunologic control of biologic proliferation (i.e., pathogen or tumor growth, rise in serum cholesterol, growth of blood vessels). We anticipate that at least one of these vaccines will be ready for testing in humans in the next calendar year.

ACKNOWLEDGMENTS

The authors would like to thank Mr. Paul Troxell for providing support in preparing this manuscript.

REFERENCES

1. Ramsey, R.B., Hamner, M.B., Alving, B.M., Finlayson, J.S., Alving, C.R., and Evatt, B.L. (1980) Effects of Lipid A and liposomes containing lipid A on platelet and fibrinogen production in rabbits. *Blood* 56: 307.
2. Richardson, E.C., Banerji, B., Seid Jr, R.C., Levin, J., and Alving, C.R. (1983) Interaction of lipid A and liposome-associated lipid A with *Limulus polyphemus* amoebocytes. *Infect. Immun.* 39: 1385.
3. Weinstein, J.N., and Leserman, L.D. (1984) Liposome as drug carriers in cancer chemotherapy. *Pharmacol. Ther.* 24: 207.
4. Lippert, J.L., and Peticolas, W.L. (1971) Laser Raman investigation of the effect of cholesterol on conformational changes in dipalmitoyl lecithin multilayers. *Proc. Natl. Acad. Sci. USA* 68: 1572.
5. Johnson, A.G., Tomai, M., Solem, L., Beck, L., and Ribi, E. (1987) Bacterial endotoxin and host immune responses. *Rev. Infect. Dis.* 9: S512.
6. Takayama, K., Qureshi, N., Ribi, E., and Cantrell, J.L. (1984) Separation and characterization of toxic and nontoxic forms of lipid A. *Rev. Infect. Dis.* 6: 439.
7. Allison, A. C., and Gregoriadis, G. (1974) Liposomes as immunological adjuvants. *Nature* 252: 252.
8. Gregoriadis, G. (1990) Immunological adjuvants: a role for liposomes. *Immunol. Today* 11: 89.
9. van Rooijen, N. (1990) Liposomes as carrier and immunoadjuvant of vaccine antigens. *In*: A. Mizrahi (Ed.) *Bacterial Vaccines.*, p. 255, NY: Alan R. Liss, Inc.
10. Alving, C. R. (1992) Immunologic aspects of liposomes: presentation and processing of liposomal protein and phospholipid antigens. *Biochim. Biophys. Acta.* 1113: 307.
11. Martin, M.A., Wenzel, R.P., Gorelick, K.J. (1991) Gram negative bacterial sepsis in hospitals in the United States: natural history in the 1980s. *Prog. Clin. Biol. Res.* 367: 111.
12. Tracey, K.S., Beutler, B., Lowery, S.F., et al. (1986) Shock and tissue injury induced by recombinant human cachectin. *Science* 234: 470.
13. Ziegler, E.J., Fisher, C.J., Sprung, C.L., et al. (1991) Treatment of gram negative bacteremia and septic shock with HA-1A human monoclonal antibody against endotoxin. A randomized, double-blind, placebo-controlled trial. *N. Engl. J. Med.* 324: 429.
14. Fries, L.F., et al. (1992) Liposomal malaria vaccine in humans: A safe and potent adjuvant strategy. *Proc. Natl. Acad. Sci.* 89: 358.
15. Golenbock, D.T., Leggett, J.E., Rasmussen, P., Graig, W.A., Raetz, C.R.H., and Proctor, R.A. (1988) Lipid X protects mice against lethal *Escherichia coli* infection. *Infect. Immunity* 56: 779.
16. Djkstra, J., Mellors, J.W., and Ryan, J.L. (1989) Altered in vivo activity of liposome-incorporated lipopolysaccharide and lipid A. *Infect. Immunity* 57: 3357.
17. Alving, C.R., Swartz, G.M., Jr., and Wassef, N.M. (1989) Naturally occuring autoantibodies to cholesterol in humans. *Biochem. Soc. Transactions* 17: 637.
18. Swartz, G.M., et. al. (1988) Antibodies to cholesterol. *Proc. Natl. Acad. Sci. USA* 85: 1902.
19. Bailey, J.M., Bright, R., and Tomar, R. (1964) Immunization with a synthetic cholesterol-ester antigen and induced atherosclerosis in rabbits. *Nature* 201: 407.
20. Bailey, J.M., and Butler, J. (1967) Synthetic cholesterol-ester antigens in experimental atherosclerosis. *In*: N.R. Di Luzio and R. Paoletti (Eds.) *The Reticuloendothelial System and Atherosclerosis.*, p.433-441, New York: Plenum.
21. Gero, S., et al. (1959) Inhibition of cholesterol atherosclerosis by immunization with lipoprotein. *Lancet* 1: 11.

22. Alving, C.A., et al. Anticholesterol vaccine: protection of rabbits against diet-induced atherosclerosis. (submitted for publication).

23. Aniagolu, J., Swartz, G.M., Dijkstra, J., Madsen, J.W., Raney, J.J., Green, S.J. Analysis of anticholesterol antibodies using hydrophobic membranes (submitted for publication).

24. Swartz, G.M., Aniagolu, J., Dijkstra, J., Raney, J.J., Madsen, J.W., Green, S.J. Binding of vaccine-induced anticholesterol antibodies to serum lipoproteins (submitted for publication).

25. Gupta S., Kim, C.H., Tsuruo, T., and Gollapudi, S. (1992) Preferential expression and activity of multidrug resistance gene 1 product (P-glycoprotein), a functionally active efflux pump, in human CD8 + T cells: a role in cytotoxic effector function. *J. of Clin. Immunol.* 12: 451.

26. Tosi, P-F., Radu, D., Johnson, J. Hayes, T.K., and Nicolau, C. In vitro reversal of multidrug resistance phenotype by autoantibodies against P-glycoprotein (submitted for publication)

27. Cassatt, D.R., White, W.I., Madsen, J.M., Birke, S.J., Woods, R.M., Wassef, N.M., Alving, C.R., and Koenig, S. (1995) Induction of antibody and cytotoxic T lymphocyte responses to a liposome-associated HIV peptide. *Vaccine (in press).*

11

PROTECTION STRATEGIES AGAINST BOTULINUM TOXIN

J. L. Middlebrook

Toxicology Division
U. S. Army Medical Research Institute of Infectious Diseases
Frederick, Maryland 21702-5011

INTRODUCTION

The clostridial neurotoxins are the most toxic substances known to science. The neurotoxin (tetanus toxin) produced from *Clostridium tetani* is encountered by humans as a result of wounds, and remains a serious public health problem in developing countries around the world. Humans are usually exposed to the neurotoxins (botulinum toxins) produced by *Clostridium botulinum* through food poisoning, although there are rare incidents of wound botulism and colonizing infections in neonates known as infant botulism (Tacket and Rogawski, 1989).

There is currently available a botulinum toxoid approved for investigational human use that confers protection from five of the seven toxin serotypes. This toxoid was developed in the 1960s with the technology available at that time. It is composed of partially purified culture supernatants that are treated exhaustively with formaldehyde. The material is only 10% neurotoxin by weight, or in other words, it contains 90% irrelevant material. There are other shortcomings, including (i) a significant incidence and degree of reactogenicity; (ii) multiple, and well-spaced vaccinations are required to achieve a protective titer; (iii) titers fall off quickly unless regular boosters are given; (iv) there are limited supplies. However, the most serious shortcoming of this toxoid is its cost; production is extremely expensive. Ironically, although preparation of crude toxin for offensive purposes is a low-tech, inexpensive process, the production of toxin for vaccines requires dedicated, high-containment laboratory space, immunized and trained staff and relatively long times for processing and safety testing. The cost problem cannot be overcome so long as native and highly lethal toxin remains a part of the vaccine production process.

Although the lethal effects of clostridial neurotoxins have been known for centuries, vaccines were not developed until the 1920s for tetanus toxin and the 1960s for botulinum toxins. There have been no further advances in vaccines for these toxins, principally for two reasons. First, little was known about the molecular details of how the toxins brought about their highly lethal effects. Second, virtually no data were available on the primary structures of the toxins. During the past 10 years, some important information has emerged concerning

Immunobiology of Proteins and Peptides VIII
Edited by M. Z. Atassi and G. S. Bixler, Jr., Plenum Press, New York, 1995

93

the structure-activity relationships and molecular mechanisms of action of the clostridial neurotoxins (Simpson, 1989). It is now clear that these toxins act via the same general mechanisms as do many other bacterial toxins. Namely, one region of the molecule recognizes and binds to a structure (receptor) on the target cell (neuron), a second region promotes, or, in combination with the receptor, assists transport of the toxin into the cell, and a third region acts intracellularly as an enzyme and inactivates a critically required cellular function.

Recently, the second impediment to new vaccine development has been overcome. With the techniques of molecular cloning, the full amino acid sequences for tetanus (Fairweather and Lyness, 1986) and botulinum toxin serotype A (Thompson *et al.*, 1990) have been deduced. These data, along with structure-activity studies, permit the design and synthesis of nontoxic fragments or peptides of the botulinum toxins for testing as new generation vaccines. Since it is known that antibodies to the receptor recognition regions of other bacterial toxins are very effective at neutralization, this domain of the clostridial neurotoxins is a good vaccine candidate. In the case of tetanus toxin, receptor binding function is carried by the 50 kDa carboxyl-terminal portion of the molecule, in a large fragment produced by papain cleavage (defined as fragment C by Helting and Zwisler, 1977). This fragment of tetanus toxin proved to be a protective immunogen when produced in several heterologous expression systems (Clare *et al.*, 1991). Since a similar large fragment of the botulinum toxins has yet to be identified via protein chemistry, we used the sequence homology of tetanus and the botulinum toxins to select a fragment of approximately the same size and position and investigated its potential as a botulinum toxin vaccine.

METHODS AND RESULTS

We began this work by attempting to use the natural gene sequence for botulinum toxin serotype A fragment C (new terminology H_C) as a structural gene for product expression. This was accomplished using the clone pCBA3 (Thompson *et al.*, 1990) and polymerase chain reaction technology. Oligonucleotide primers encoding *Bam* HI and *Hind* III restriction sites were used to amplify the region of interest (amino acids 861-1296), then the product was purified by gel electrophoresis. After cleavage by restriction endonucleases, the gene and recipient vector were again gel-purified, then ligated. The H_C-encoding gene was cloned into two possible expression vectors. One was a fusion protein vector (pMAL-p) sold by New England Biolabs, while the other was pKK233 (Promega) which should produce H_C alone. After transformation of *Escherichia coli* cells, the cultures were induced, and both cells and supernatants examined for H_C production. Fragment H_C was not reproducibly detected in experiments using the pKK233 construct. Experiments using the pMAL-p construct did yield material that could be detected by immunologic techniques. Table 1 shows the results from one experiment where enzyme-linked immunosorbent assay (ELISA) was used.

ELISA-reactive material was not found in the culture supernatants (data not shown). Cells were harvested, broken and centrifuged to produce a supernatant and pellet fraction. Transformed cells carrying the vector without insert did not yield ELISA-reactive material in either fraction (Table 1). Cells transformed with pMAL carrying the H_C-encoding insert produced ELISA-reactive product distributed in both the pellet and supernatant fractions (Table 1).

While attempting to purify the maltose-binding protein-H_C product, we elected to test the crude cell preparation (pellet and supernatant combined) for its potential to elicit immunity to botulinum toxin. The results were encouraging and are shown in Table 2.

Mice vaccinated only once with crude material from cells transformed with vector-carrying H_C insert were protected two weeks later from intraperitoneal challenge with a

Table 1. ELISA analysis of H_C-maltose-binding protein fusion
product expression in *E. coli* (OD_{405})

Sample dilution	pMAL-p control		H_C-pMAL		
	Supernant	Pellet	Supernatant	Pellet	Blank
102400	0.21	0.26	0.47	0.44	0.24
25600	0.27	0.26	0.87	0.77	0.24
6400	0.23	0.33	1.10	1.19	0.22
1600	0.24	0.35	1.29	1.20	0.17
400	0.32	0.37	1.23	1.37	0.14
100	0.08	0.34	0.69	1.52	0.13

relatively low dose (3 LD_{50}) of toxin . Mice vaccinated with control cell material all died. Two vaccinations (at 0 and 2 weeks) protected mice from up to 1200 LD_{50} (Table 2) at 4 weeks. Meanwhile, our purification efforts were unsuccessful, due primarily to the insoluble nature of the product. Therefore, we decided to construct an entirely synthetic gene encoding the H_C sequence and to incorporate features that we believed would markedly enhance its expression.

Studies with heterologous gene expression in *E. coli* identified two unfavorable structural factors linked with low levels of product. One is the incidence of codons that are rarely used by *E. coli* in the heterologous gene. The second is a high fraction of A-T in a heterologous gene. Examination of the natural gene for H_C revealed that both factors could present a problem. A large number of codons frequently used by Clostridia are rarely employed by *E. coli* and are found in the H_C gene. The A-T content of the natural H_C gene is also high, approaching 80% (Figure 1A).

To avoid these problems, a synthetic gene was designed by first reverse-translating the amino acid sequence of serotype A H_C using the single codon most preferred by *E. coli* for any given residue. This sequence was altered to introduce restriction sites at four approximately equally-spaced intervals along the gene, while conserving fidelity of the encoded protein (Figure 2).

Finally, some of the triplets were changed to use a redundant *E. coli*-preferred codon that had one less A or T. This design yielded a gene with a reduced A-T content as shown by the plot in Figure 1B. To construct this artificial gene, nucleotide oligomers 60-114 bases long were synthesized and the appropriate 5' positions chemically phosphorylated. The oligomers were designed to be complementary and to overlap on their ends by 6-7 bases. Four of five subunits comprising the gene were constructed by mixing the appropriate oligomers, allowing them to anneal and ligating the assembly. The resulting subunit was cloned into a vector and amplified. The fifth subunit was constructed by cloning and assembling three large oligonucleotides of 120-130 bases in shuttle vectors. After excision

Table 2. Vaccination of mice with natural gene-encoded H_C-maltose binding
protein fusion product (# deaths/total)

Challenge dose (#LD_{50})	One vaccination		Two vaccinations	
	Control vector	Vector with insert	Control vector	Vector with insert
3	3/3	0/5	6/12	0/11
30				1/3
300				1/3
1200				1/3

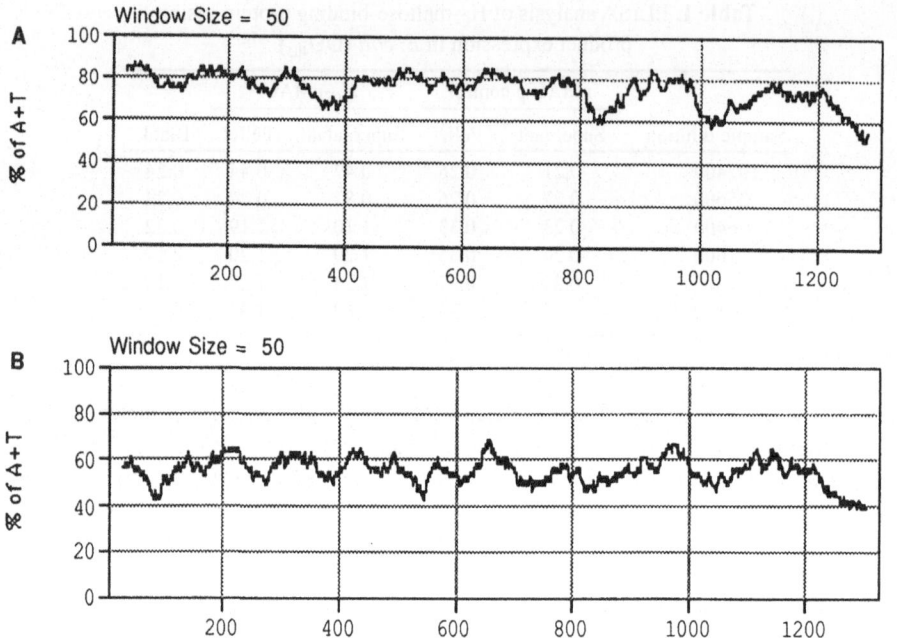

Figure 1. Base composition plot for the natural and synthetic genes for botulinum toxin serotype A H$_C$. With the nucleotide sequence for the indicated gene, the Macintosh program MacVector was used to calculate and display the fraction of A-T throughout the genes. **A**: Natural gene. **B**: Synthetic gene.

using the appropriate endonucleases, the five subunits were cloned into an assembly vector carrying a multiple cloning site designed to yield the complete gene.

The synthetic gene was sequenced to assure its fidelity. Seven errors were identified, all representing one or more deletions. These errors were corrected by cloning the gene into the pALTER vector (Promega), synthesizing the appropriate repair oligomers and carrying out *in vitro* mutagenesis. Resequencing confirmed that the repaired version of the gene was absolutely correct.

Expression of the serotype-A synthetic H$_C$ gene was attempted by cutting it out of the pALTER vector and cloning it into pKK233. Both the parent vector and that with the H$_C$ insert were used to transform *E. coli* cells and look for the production of immunoreactive material. Initially, the cells were centrifuged, broken by sonication, and a whole-cell lysate subjected to SDS polyacrylamide gel electrophoresis, followed by Western-blot analysis. A strong immunostaining band was observed at a position corresponding to 50 kDa. This band was not observed in identical samples from the cells transformed with pKK233 that were not carrying the insert.

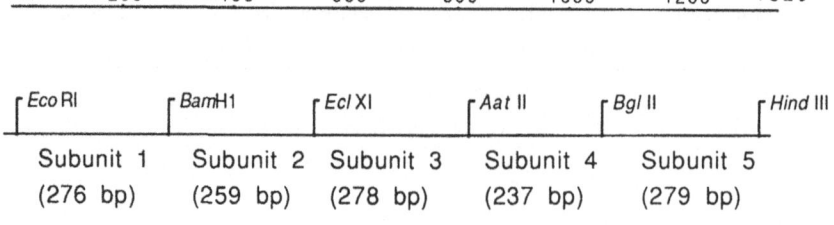

Figure 2. Size and subunit map of synthetic gene for botulinum toxin serotype A H$_C$.

Table 3. Protection of ICR mice vaccinated with H_C derived from
the synthetic gene (# deaths/total)

Calculated challenge dose (#LD_{50})	Control (vector without insert)	Experimental (vector with insert)
4	2/3	0/3
10		0/3
30	3/4	0/3
100		0/3
300		0/3
1000		0/3
3000		0/1

Animals received vaccinations of crude lysate at 0, 2 and 4 weeks.
Challenges were ip with serotype A at 5 weeks.

Before attempting to purify this presumptive H_C for characterization studies, we thought it prudent to determine if the material was immunologically functional. We therefore immunized mice with crude, whole-cell lysates from control and insert-carrying transformed cells. Since we had no estimate of the level of immunity which would be conferred, we began by challenging vaccinated animals with relatively low doses of toxin, increasing by steps of one-half a logarithm up to 1000 LD_{50} (Table 3) of botulinum toxin serotype A; each animal in the groups of 3 survived. Not anticipating the level of protection achieved, we had one remaining immunized, unchallenged animal; it survived 3000 LD_{50} (Table 3).

Assuming that the challenge doses of 3000 LD_{50} and below (< 3 ng) were insufficient to act as a booster within a few days, we subjected the same animals to another challenge series and observed no deaths up to 1,000,000 LD_{50} (data not shown).

In order to have a more direct comparison of the efficacy of H_C and the current toxoid, purified *E. coli*-produced H_C was prepared and adsorbed to alum. Mice were then vaccinated with that material and a second group with the toxoid, also adsorbed to alum. Mice were then challenged with various doses of toxin with the results depicted in Table 4. Clearly, vaccination with this toxin fragment confers a high degree of protection from botulinum neurotoxin.

We have also carried out limited studies into alternative expression systems. In one approach, the H_C gene was slightly altered to change restriction sites and cloned into a baculovirus expression vector. As produced in baculovirus, serotype A H_C is soluble and will elicit a level of immunity in mice equal to that produced by the crude *E. coli* product. In our first construct, the level of protein expression was low. We have since completed a collabo-

Table 4. Protection of ICR mice vaccinated with H_C or botulinum
toxoid (# deaths/total)

Challenge dose (#LD_{50})	H_C (1μg/vaccination[1])	Toxoid (1μg/vaccination[1])
10^2	6/6	6/6
10^3	6/6	6/6
10^4	6/6	6/6
10^5	6/6	6/6
10^6	4/6	3/6

[1]Mice were vaccinated at 0, 2 and 4 weeks. Challenge with toxin was
on week 5. Unvaccinated control animals did not survive the toxin,
0/12 for a challenge dose of 10^3 LD_{50}.

rative study designed to boost expression to a level acceptable for large-scale vaccine production, if the baculovirus system were chosen to serve that role.

CONCLUSIONS

Our goal was to produce a new, less expensive human vaccine for botulinum toxin. Thus far, our data indicate that our synthetic gene-encoded H_C is a vaccine candidate which is completely nontoxic and, therefore, very safe, both from production and vaccination standpoints. The data also demonstrate vaccination with genetically-engineered H_C can produce levels of immunity in mice similar to those attained with the current botulinum toxoid. Because of the marked increase in safety of the H_C polypeptide as a vaccine and the projected decrease in costs per dose, we are now moving to complete the necessary animal testing preparatory for application to the FDA for human testing.

TS: M. Clayton, J. Clayton, D. Brown. E. Brown and H. LaPenotiere conducted various aspects of the research described in this paper.

REFERENCES

1. Clare, J.J., Rayment, F.B., Ballantine, S. P., Sreekrishna, K. and Romanos. M.A. (1991) *Bio/Technology* 9, 455-460.
2. Fairweather, N.F. and Lyness, V.A. (1986) *Nucleic Acid Research* 14, 7809.
3. Helting, T.B. and Zwisler, O. (1977) *J. Biol. Chem.* 252, 187-193.
4. Simpson, L.L. (1989) *Botulinum Neurotoxin and Tetanus Toxin* (Academic Press, New York, pp. 153-178.
5. Tacket, C.O. and M.A. Rogawski. 1989. Botulinism p. 351-378. *In* L. L. Simpson (ed.) Botulinum Neurotoxin and Tetanus Toxin. New York: Academic Press, Inc.
6. Thompson, D.E., Brehm, J.K., Swinfield, T-J., Shone, C.S., Atkinson, T., Melling, J. and Minton, N.P. (1990) Eur, *J. Biochem.* 189, 73.

12

COLLAGEN ARTHRITIS IN T CELL RECEPTOR CONGENIC MICE

A Unique Approach to Study the Role of T Cell Receptor Genotypes in Autoimmune Arthritis

Gerald H. Nabozny and Chella S. David

Department of Immunology
Mayo Medical School
Rochester, Minnesota 55905

INTRODUCTION

Identifying the genes responsible for susceptibility to human rheumatoid arthritis (RA) has been an area of intense investigation. Evidence showing that genetic factors play a role in RA susceptibility is certainly irrefutable (1) and the association between particular HLA class II alleles of the major histocompatibility complex (MHC) and RA is widely accepted (2). However, genetic analysis of RA patients and multiplex families indicates that MHC genes alone account for approximately 50% of RA heritability (3). Thus, additional non-MHC genes appear to play an important role in RA. Given the intimate association between the T cell receptor (TCR), antigenic peptide and MHC, a logical candidate which may also influence RA are the V alpha (VA) and V beta (VB) genes of the TCR. Analysis of the TCR repertoire from lymphocytes infiltrating RA synovium suggest, in some cases, biased VB gene utilization (4). Also, population studies indicate that certain VB polymorphisms may be more common in RA patients. Most recently, analysis of multiplex RA families suggested that a gene within or near the TCRB locus may confer RA susceptibility (5). Unfortunately, these findings, albeit compelling, cannot address the functional consequences of possessing such genetic markers. Therefore, understanding the mechanisms underlying a particular genetic association in RA may lead to the development of a highly specific therapy to treat this disease.

To facillitate the study of RA, researchers have turned to animal models. An experimental model which shares many of the immunologic, genetic and histopathologic features of RA is the mouse model of collagen induced arthritis (CIA). Moreover, the utilization of specific breeding strategies to develop congenic and recombinant mouse strains provides an added dimension to study the immunogenetics of autoimmune arthritis. In this review, we will briefly discuss some of the recent findings obtained studying CIA susceptble mouse strains congenic for the known TcrVB genotypes. The use of TCR congenic mice is

Immunobiology of Proteins and Peptides VIII
Edited by M. Z. Atassi and G. S. Bixler, Jr., Plenum Press, New York, 1995

99

a novel approach to explore the role of TCR genes in autoimmune disease. The results summarized below sheds light on the influence of these genotypes mediating arthritogenic responses and may provide clues to the role of human TCR geneotypes in RA.

T CELL RECEPTOR VB GENOTYPES OF THE MOUSE

Advances in molecular biology have led to an accelerated and clearer understanding of the organization of the TCR region in both mice and humans. In the mouse, the TcrVB genes are well characterized and some intriguing characteristics have been uncovered. Of interest, is the observation that some mouse strains bear large deletions within the TcrVB genome (6) (Table 1). For example, five mouse strains have been identified which lack the TcrVB 5, 8, 9, 11, 12 and 13 genes. This deletion has been designated as the TcrVBa haplotype and accounts for approximately 50% of the TcrVB genome. Also, Haqqi et al. (7) reported that the RIIIs/J mouse bears a massive deletion of approximately 70% of the TcrVB genome. This unique deletion was designated TcrVBc and comprised the genes deleted in the TcrVBa haplotype along with the TcrVB 6, 15, 17 and 19 genes.

That TcrVB haplotypes are important in CIA was illustrated by Banerjee et al. (8). Analysis of offspring between a cross of two arthritis resistant strains, SWR (H-2^q,TcrVBa) and B10 (H-2^b, TcrVBb) revealed a strong association between CIA susceptibility, possession of the arthritis susceptible H-2^q haplotype and at least one copy of the wild type TcrVBb genome. This observation strongly suggested that TCR haplotypes can play a major role in dictating an arthritogenic response. However, extensions of this observation are limited given the potential contribution of background genes. Therefore developing lines of CIA susceptible mice which differ only at the TcrVB locus would provide an excellent system to clearly address the influence of TcrVB genotypes in arthritis.

COLLAGEN ARTHRITIS IN TcrVB CONGENIC MICE: DISTINCT ARTHRITIS PHENOTYPES DUE TO A PARTICULAR TcrVB GENOTYPE

To directly address the influence of TcrVB gene deletions in CIA, we took advantage of two arthritis-susceptible MHC haplotypes, H-2^q and H-2^r, for our studies. These two haplotypes display distinct patterns of arthritis susceptibility when immunized with type II collagen (CII) derived from various species (9). Also, analysis of T cell responses against purified cyanogen bromide (CB) fragments of CII has revealed that dominant T cell epitopes recognized by H-2^q and H-2^r strains reside on different regions of the CII molecule (10,11). Therefore, generation of TcrVB congenic mice on these two H-2 backgrounds establishes an outstanding system to study the influence of a TCR genotype on two distinct arthritogenic responses.

To dissect the role of TcrVB genotypes on H-2^q restricted arthritogenic responses, the TcrVBa and TcrVBc haplotypes were introduced into the CIA susceptible B10.Q

Table 1. T cell receptor VB genotypes of the mouse

TcrVB	VB genes deleted
TcrVB-b	None
TcrVB-a	5, 8, 9, 11, 12, 13
TcrVB-c	5, 6, 8, 9, 11, 12, 13, 15, 17, 19

Table 2. Characteristics of CIA in B10.Q-VB congenic mice[a]

Strain	Frequency of animals developing CIA (%)	Disease characteristics
B10.Q	> 70	severe arthritis
B10.Q-VBa	> 70	severe arthritis
B10.Q-VBc	< 20	mild arthritis

[a]Data summarized from Nabozny et al. (12).

background. Following immunization with bovine CII, we observed that B10.Q-VBa mice showed no difference in arthritis susceptibility or disease severity versus B10.Q animals. However, B10.Q-VBc mice were highly resistant to CIA (Table 2). The differential suscep-tibility between B10.Q-VBa and B10.Q-VBc mice to CIA was direct demonstration that TcrVB genes absent in the TcrVBc haplotype but present in TcrVBa (VB 6, 15, 17 and 19) were responsible for the induction and/or perpetuation of CIA. Support for this interpretation comes from our recent demonstration that CIA onset could be significantly delayed in B10.Q-VBa mice following the in vivo depletion of VB6$^+$ T cells (12). Thus, in the B10.Q background, it appears that the TcrVBc genotype can render animals resistant to CIA.

To examine the role of truncated TcrVB genomes on H-2^r restricted arthritogenic responses, the TcrVBa and VBc genotypes were introduced into the B10.RIII background. Unlike B10.Q-VBa mice, which remained CIA susceptible, B10.RIII-VBa animals presented a delayed onset of arthritis when compared to B10.RIII animals. Thus, it appeared that TcrVB genes absent in the VBa deletion participate in the H-2^r restricted arthritogenic response. Introduction of the TcrVBc genotype led to a further alteration of CIA development; approximately 50% of B10.RIII-VBc mice developed CIA. Also, arthritis onset was delayed and the severity of CIA significantly reduced (13) (Table 3). This observation is in contrast to B10.Q-VBc mice which are highly resistant to CIA. Therefore, in the B10.RIII back-ground, our findings indicated that a multitude of TcrVB genes, some of which are absent in the TcrVBa and VBc genotypes participate in the arthritogenic response.

On balance, the use of TcrVB congenic mice to study CIA demonstrates that Tcr genotypes indeed influence autoimmune responses. Also, these results clearly show that possession of a particular TcrVB genotype leads to a distinct disease phenotype. This heterogeneity in disease outcome is similar to the disease heterogeneity observed in RA patients and may explain, in part, the underlying mechanisms responsible for such a phenomenon.

FUNCTIONAL ANALYSIS OF TcrVB CONGENIC MICE

Given the altered development of CIA in B10.RIII-VB congenic mice, this line provides an excellent system to dissect CII specific T cell responses in vitro. A thorough

Table 3. Characteristics of CIA in B10.RIII-VB congenic mice[a]

Strain	Frequency of animals developing CIA (%)	Disease characteristics
B10.RIII	> 85	severe arthritis
B10.RIII-VBa	> 85	delayed onset severe arthritis
B10.RIII-VBc	~ 50	delayed onset moderate to mild arthritis

[a]Data summarized from Nabozny et al. (13).

Table 4. T cell responsiveness against the CB 10 fragment
of bovine CII in B10.RIII-VB congenic mice[a]

Strain	CB10 proliferation (stimulation index)
B10.RIII	3.7
B10.RIII-VBa	2.4
B10.RIII-VBc	1.7

[a]Mice were immunized with 200 μg bovine CII in
CFA. Fourteen days later, the draining lymph
node cells were removed and challenged in vitro
with 25 μg/ml bovine CII CB 10.

understanding of these responses may shed light on the nature of the arthritogenic response
in vivo. Previously, we showed that the predominant T cell response against CII in B10.RIII
mice is directed at the CB 10 fragment of the molecule (11). Thus, elucidating the nature of
CB 10 specific T cell responses in B10.RIII-VB congenic mice may be reflective of CIA
development in vivo. Indeed, Table 4 shows that a hierarchy exists in the magnitude of T
cell proliferation against CB 10 in the B10.RIII-VB congenic strains.

The association between arthritis development and CB 10 responsiveness opens the
door towards understanding the functional consequences of possessing a particular TcrVB
genotype. Further in vitro studies may provide a clearer picture regarding TCR utilization
and epitope recognition in mice bearing a truncated TcrVB genome. Such information may
advance our understanding regarding the flexibility of the autoimmune repertoire.

FUTURE PROSPECTS

The data summarized above points to the strength of utilizing TcrVB congenic mice
to study the impact of particular TcrVB genotypes in autoimmune arthritis. To enhance and
complement this approach, researchers may also take advantage of TcrVA congenic animals.
In humans, there is experimental evidence suggesting extensive allelic polymorphism within
the TcrVA and VB loci (14). Analysis of V region polymorphisms show that they consist of
single amino acid substitutions involving predicted framework positions (15) or regions
involved in MHC binding (16). Thus, it is logical to suggest that such polymorphisms can
influence the shaping of the TCR repertoire which, in turn, may play a role in autoimmune
disease. In mice, sequence polymorphisms in the coding regions of mouse TcrVA genes has
been shown (17). Restriction fragment length polymorphism analysis of inbred mouse strains
revealed that TcrVA polymorphisms can be designated into 5 distinct TcrVA haplotypes (17).
These haplotypes are genuinely inherited since no recombinations have been observed in
over 1,000 backcross mice (18,19). The functional consequences of TcrVA haplotypes on
immune responses remains poorly defined. However, recent studies utilizing TcrVA congenic
mice has shown that TcrVA haplotypes can influence the frequencey of certain TcrVA bearing
T cells in the periphery (20). Thus, germline sequence VA polymorphisms do play a role in
shaping the TCR repertoire. Together with our findings concerning TcrVB haplotype
influences in CIA, one would predict that the combination of particular VA and VB genotypes
may add a new dimension to studies examining the influence of TcrV genes in autoimmune
responses. Also, the evidence that TCR haplotypes may play a role in RA susceptibility
strengthens the usefullness of TCR congenic animals for the study of human autoimmune
diseases. Hopefully, further study of TCR congenic mice will provide valuble clues regarding

the functional influence of TCR haplotypes in RA. Such information may lead to improved approaches directed towards the design of TCR based therapies for autoimmune disease.

ACKNOWLEDGMENTS

This work was supported by NIH grant AR30752, and a grant from the Minnesota chapter of the Arthritis Foundation. Dr. G.H. Nabozny is supported by a postdoctoral fellowship from the Arthritis Foundation.

REFERENCES

1. Wordsworth, B.P. and Bell, J.I., 1992, The immunogenetics of rheumatoid arthritis. *Springer Semin. Immunopathol.* 14:59-78.
2. Stastny, P., 1978, Association of the B cell alloantigen DRw4 with rheumatoid arthritis. *N. Eng. J. Med.* 298:869-871.
3. Rigby, A.S., Silman, A.J., Voelm, L., Gregory, J.C., Ollier, W.E.R., Khan, M.A., Nepom, G.T. and Thomson, G., 1991, Investigating the HLA component in rheumatoid arthritis: an additive (dominant) mode of inheritance is rejected, a recessive mode is preferred. *Genet. Epidemiol.* 8:153-175.
4. Jenkins, R.N., Nikaein, A., Zimmermann, A., Meek, K. and Lipsky, P., 1993. T cell receptor VB gene bias in rheumatoid arthritis. *J. Clin. Invest.* 92:2688-2701.
5. McDermott, M., Kastner, D.L., Holloman, J.D., Schmidt-Wolf, g., Lundberg, A.S., Sinha, A.A. et al., 1995, The role of T cell receptor beta chain genes in susceptibility to rheumatoid arthritis. *Arthritis and Rheum.* 38:91-95.
6. Behlke, M.A., Chow, H.S., Huppi, K. and Loh, D.Y., 1986, Murine T cell receptor mutants with deletions of beta-chain variable regions genes. *Proc. Nat. Acad. Sci.* 475:361-367.
7. Haqqi, T.M., Banerjee, S., Anderson, G.A. and David, C.S., 1989, RIIIs/J (H-2^r)-an inbred mouse strain with a massive deletion of T cell receptor VB genes. *J. Exp. Med.* 169:1903-1911.
8. Banerjee, S., Haqqi, T.M., Luthra, H.S., Stuart, J.M. and David, C.S., 1988, Possible role of V beta T cell receptor genes in susceptibility to collagen induced arthritis in mice. *J. Exp. Med.* 167:832-844.
9. Wooley, P.H., Luthra, H.S., Griffiths, M.M., Stuart, J.M., Huse, A., and David, C.S., 1985, Type II collagen induced arthritis in mice IV. Variations in immunogenetic regulation provide evidence for multiple arthritogenic epitopes on the collagen molecule. *J. Immunol.* 135:2443-2451.
10. Brand, D.D., Myers, L.K., Terato, K., Whittington, K.B., Stuart, J.M., Kang, A.H. and Rosloniec, E.F., 1994, Characterization of the T cell determinants in the induction of autoimmune arthritis by bovine alpha 1(II)-CB11 in H-2q mice. *J. Immunol.* 152:3088-3097.
11. Nabozny, G.H., Griffiths, M.M., Harper, D.S., Luthra, H.S. and David, C.S., 1995, Identification of a cynogen bromide fragment of porcine type II collagen capable of modulating collagen arthritis in B10.RIII (H-2^r) mice. *Autoimmun.* (in press).
12. Nabozny, G.H., Bull, M.J., Hanson, J., Griffiths, M.M., Luthra, H.S. and David, C.S., 1994, Collagen-induced arthritis in T cell receptor V beta congenic B10.Q mice. *J. Exp. Med.* 180:517-524.
13. Nabozny, G.H., Hanson, J., Griffiths, M.M., Luthra, H.S. and David, C.S., 1995, Altered development of collagen induced arthritis in T cell receptor V beta congenic B10.RIII mice. *Autoimmun.* (in press).
14. Cornelis, F., Pile, K., Loveridge, J., Moss, P., Harding, R., Julier, C. and Bell, J., 1993, Systematic study of human alpha beta T cell receptor V segments shows allelic variations resulting in a large number of distinct T cell receptor haplotypes. *Eur. J. Immunol.* 23:1277-1283.
15. Reyburn, H., Cornelis, F., Russell, V., Harding, R., Moss, P. and Bell, J., 1993, Allelic polymorphism of human T-cell receptor V alpha gene segments. *Immunogen.* 38:287-291.
16. Maksymowych, W.P., Gabriel C.A., Luyrink, L., Melin-Aldana, H., Elma, M., Giannini, E.H., Lovell, D.J., Van Kerckhove C., Leiden, J., Choi, E., et al. 1992. Polymorphism in a T-cell receptor variable gene is associated with susceptibility to a juvenile rheumatoid arthritis subset. *Immunogen.* 35:257-262.
17. Jouvin-Marche, E., Morgado, M.G., Trede, N., Marche, P.N., Couez, D., Hue, I., Gris, C., Malissen, M. and Cazenave, P.A., 1989, Complexity, polymorphism, and recombination of mouse T-cell receptor alpha gene families. *Immunogen.* 30:99-104.
18. Rubin, B., Wegener, A.M., Liabeuf, N. and Jorgensen, A.W., 1990, Biological transfer of the CBA Tcra locus into C57BL/6 mice. *Immunogen.* 31:207-210.

19. Gleditsch, L.,Snodgrass, R. and Bogen, B., 1990, No recombinations between Tcra-V and Tcra-C gene segments in 669 backcross mice. *Immunogen.* 32:297-303.
20. Gleditsch, L. and Bogen, B., 1992, A TcrA congenic mouse: V alpha epitope expression is influenced by both Tcra haplotypes and background genes. *Immunogen.* 35:153-160.

13

THE BLOOD-BRAIN BARRIER IN VIRUS-INDUCED DEMYELINATION

C. J. R. Welsh, B. V. Sapatino, A. Petrescu, and J. Piedrahita

Department of Veterinary Anatomy and Public Health
College of Veterinary Medicine
Texas A&M University
College Station, Texas 77843-4458

INTRODUCTION

The central nervous system (CNS) has been considered to be an immunologically privileged site (1). This is partly due to the paucity of professional antigen presenting cells within the CNS and the lack of lymphatic drainage (2). In addition, there is very little expression of major histocompatibility antigens (MHC) within the CNS. Finally, the blood-brain barrier (BBB) normally protects the CNS from immunological damage. The BBB is formed by astrocytes and specialized cerebrovascular endothelial cells (CVE) which line the lumen of the blood vessels (3). The endothelial cells have relatively few pinocytic vesicles and no constitutive MHC Class II expression unlike endothelial cells that supply other organs (4).

The immunologically privileged status of the brain has recently been questioned by Werkle's group who demonstrated that activated T cells continually pass through the BBB in surveillance mode (5). In inflammatory conditions of the CNS such as multiple sclerosis (MS) and its experimental analogues: experimental allergic encephalomyelitis (EAE) and Theiler's virus-induced demyelination (TVID), T cells enter the CNS and cause tissue damage. In order to gain access to the CNS, these pathogenic T cells must transmigrate through the BBB. Therefore, interactions between T cells and CVE are the one of the initiating events in the pathogenesis of all these conditions.

The etiology of MS is unknown although epidemiological studies have implicated an infective agent as a probable initiating factor (6). Furthermore, there are a number of viral infections that lead to demyelination in animals (7). Therefore, in order to understand the pathogenesis of MS it would seem appropriate to study an animal model of virus-induced demyelination. One of the best animal models of MS is the demyelination induced by Theiler's virus. Theiler's murine encephalomyelitis virus (TMEV) is a Picornavirus which causes an asymptomatic gastrointestinal infection and occasionally paralysis (8). There are two main strains of Theiler's virus which are classified according to their neurovirulent characteristics. The virulent GDVII strains (GDVII, FA) of TMEV cause a fatal encephalitis

Immunobiology of Proteins and Peptides VIII
Edited by M. Z. Atassi and G. S. Bixler, Jr., Plenum Press, New York, 1995

following intracranial infection (9). The persistent TO strains (BeAn, DA, WW, Yale) cause, in susceptible strains of mice, a primary demyelinating disease which is similar to MS (10).

In the early phase of Theiler's virus infection, the virus is thought to enter the CNS by axonal transport (11), replication in CVE (12, 13) or by infection of macrophages (14, 15). Once the virus has gained access to the CNS, it infects neurons lytically and then establishes a persistent infection of astrocytes, oligodendrocytes and cells of the macrophage/microglial lineage (16).

The immunological events that occur following TMEV infection have been studied by a number of investigators. CD 8 and CD4 T cells play an important role in early viral clearance, but in later disease these T cell subsets have been implicated in the demyelinating process (17). The CD4 T cells assist B cells to produce antibody which are probably the most important mediators of Picornavirus clearance (18). In addition, CD4 T cells secrete IFN-γ which has been shown to inhibit the replication of Theiler's virus in vitro (19) and to have a protective role in vivo (20). CD8 T cells clearly are important in viral clearance as demonstrated by in vivo depletion experiments (21) and studies with transgenic mice (22, 23). CD8 depleted mice fail to clear virus from the CNS and developed more severe demyelinating disease than the immunocompetent controls (21). β-$_2$ microglobulin knock out mice were constructed on a TMEV-resistant background and these mice were shown to lack functional cytotoxic T cells (22,23). Histological evidence of demyelination developed in the knock out mice following intracranial infection with Theiler's virus (22,23). These investigations clearly implicate a role for cytotoxic CD8 T cells in viral clearance. Indeed cytotoxic T lymphocyte (CTL) activity has been detected in TMEV infected SJL/J mice by two groups (24, 25).

The later demyelinating disease induced by TMEV is partly mediated by direct viral lysis of oligodendrocytes (26) but also by immune mechanisms including virus specific delayed type hypersensitivity (DTH) T cells (27), autoimmunity (28) and cytotoxic T cell reactivity (29). The relative contributions made by each of these mechanisms to the demyelinating process, remains to be elucidated.

It is evident that T cells play a dual role in TVID. In the early disease, both CD4 and CD 8 T cells mediate viral clearance and in the late disease these T cells take on a pathogenic role and cause tissue destruction. In order to perform either of these functions, the T cells must enter the CNS by transmigration across the BBB. Thus, understanding the interactions between T cells and CVE will provide important insights into the pathogenesis of TVID. To investigate the role of the BBB in TVID, it was imperative to derive completely pure populations of CVE from strains of mice that are resistant and susceptible to TVID (30). Having established cloned CVE cell lines from SJL/J, CBA (susceptible to TVID) and BALB/c (resistant to TVID) mice, the effect of TMEV infection on these cells was determined. Three main questions were addressed:- (1) Is susceptibility to TVID related to the ability of the virus to replicate in CVE ? (2) Can CVE be persistently infected with TMEV? (3) Does TMEV infection alter the expression of MHC Class I and II on CVE?

MATERIALS AND METHODS

Cerebrovascular Endothelial Cell Isolation and Culture

CVE were obtained from the brains of 2-4 week old SJL, CBA or BALB/c mice (Harlan Sprague Dawley Inc., Indianapolis, IN) by a previously published method (30). Briefly, the brains were minced in phosphate buffered saline (PBS) containing 1% bovine serum albumin (BSA) and then centrifuged (250xg). The pellet was homogenized in a Wheaton B homogenizer and passed through a series of Nitex nylon filters, 149 μm, 74 μm and a 20 μm. The filtrate was washed once in PBS/BSA solution at 450xg. Then the filters and the filtrate were digested with solution of 0.1% collagenase/dispase (Boehringer Mannheim, Indianapolis IN), at 37^0C for three hours, with occasional shaking. The digest was then centrifuged at 150xg and the resulting pellets were washed once at 150xg and were resuspended in 40 ml of Iscove's modification of Dulbecco's modified Eagle medium (IMDM) supplemented with 10% inactivated fetal bovine serum (FBS), 20 μg/ml endothelial cell growth supplement (Collaborative Cell Research, Bedford, MA) and 50·to 100 μg/ml porcine heparin (Elkins-Sinn Inc., Cherry Hill, NJ.). The plastic ware was previously coated with 1mg/ml collagen type IV (Sigma's type VI Sigma Chemical Co, St Louis, MO) and human serum fibronectin (10 μg/ml, Boehringer Mannheim).

When the endothelial cells reached confluence they were passaged conventionally using trypsin/EDTA. Completely pure cultures of cloned SJL, CBA and BALB/c CVE were obtained by the limiting dilution technique (30).

Characterization of Cerebrovascular Endothelial Cells

CVE were characterized by uptake of acetylated low density lipoprotein (Biomedical Technologicals Inc., Stoughton, MA.) and by immunostaining with monoclonal antibody to Factor VIII-related antigen (abV Immune Response, Inc., Derry, NH) and antibodies to angiotensin converting enzyme (a gift from Dr Auerbach, University of Wisconsin) (30).

Viruses

Theiler's virus strains BeAn and GDVII were kindly supplied by Dr H.L. Lipton, Department of Neurology, University of Colorado Health Sciences Center, Denver, CO, and the DA strain by Dr. Raymond Roos, Department of Neurology, University of Chicago, IL. The viruses were propagated in BHK-21 cells. The stock virus was titered, aliquoted and stored at -70° C (31).

Acute Infection of CVE and BHK-21 Cells

BHK-21 cells were grown in DMEM (Gibco BRL, Grand Island, NY) with 5% FBS (Irvine Scientific, Irvine, CA.). CVE cells were grown in IMDM (Gibco) supplemented with 10% FBS for growth and 1% FBS for post inoculation maintenance of cell cultures. The titration of viruses was performed both in BHK-21 cells and CVE cells by cytopathic effect (CPE). CVE cells were seeded in 96-well plates at a concentration of 10^5 cells per well. When the cells reached confluency, each well was washed once with serum-free IMDM and then inoculated in quadruplicate with 0.05 ml virus suspension per well, in tenfold dilutions (in IMDM). After 45 min of viral adsorption, the cells were washed in serum-free IMDM and incubated with 1% FBS-containing IMDM for CVE and DMEM with 1% FBS for BHK-21 cells. The CPE was assessed daily and the TCID-$_{50}$ was calculated by the Reed and Muench formula.

Persistent Infection of CVE

Cloned CVE from SJL/J, CBA, and BALB/c mice were grown to a monolayer and infected with the BeAn strain of TMEV at an m.o.i of 1 (32). After the adsorption period of 45 mins, the inoculum was removed, the cells washed in serum-free medium and then incubated in IMDM containing 1% FBS and 100 units/ml of either IFN–γ or TNF-α (Genzyme, Cambridge, MA), or in the case of virus control, the cells were cultured with 1% FBS/IMDM, without cytokines.

Four days post-infection, the surviving cells were rescued by collection and growth in 10% FBS containing medium. The medium was changed daily for up to ten days, during which time the rescued cells grew to confluence. The cells were then passaged every three to four days, which is consistent with the culturing of non-infected CVE.

Detection of Theiler's Viral Polyprotein 1 (VP1) Antigen in PI-CVE

Cells were collected by trypsinization from various PI-CVE cells lines and washed once in PBS containing 1% BSA and 0.1% sodium azide (PBS-BSA). All the following steps were performed at 4°C. The cells were fixed in acetone/ethanol (1:1 v/v) for 20 minutes, washed by addition of PBS-BSA and then centrifuged.

The PI-CVE were then incubated with either DA mAB1 or DA mAB2 monoclonal antibodies (33) directed against VP1 of DA (a gift from Dr R. Roos, Department of Neurology, University of Chicago Medical Center, Chicago, IL) followed by FITC-conjugated goat anti-mouse IgG antibody (Cappel, Durham, NC) in IMDM containing 1% FBS. Finally the cells were washed in PBS-BSA and the percentage of positive cells was estimated by analysis of 10^4 cells in a flow cytometer with the argon laser set at 488nm (EPICS V, Coulter Diagnostics, Hialeah, FL, U.S.A.). The windows were gated to exclude dead cells. The background of fluorescent staining with second antibody alone was subtracted from each value obtained.

Detection of MHC Class I and Class II on CVE

The expression of MHC Class I and II was determined on acutely infected CVE, persistently infected CVE and uninfected CVE. The expression of MHC on acutely infected CVE was determined at 24, 48 and 72 hours post infection and the persistently infected CVE at passage 6 and 12. Uninfected cultures of CVE were treated with 100units/ml recombinant mouse IFN-γ (Genzyme Corporation, Cambridge, MA) for 1-3 days. In all cases, the cells were then trypsinized, washed in PBS containing 1%FBS+ and 0.1% sodium azide (FACS-PBS) and incubated with the appropriate monoclonal antibody directed against MHC Class I or II. Following further washing with PBS-FACS, the cells were incubated with a fluorescein-labeled sheep anti-mouse IgG antibody (Amersham, Arlington Heights, IL), for 45 mins and then washed and resuspended in PBS for flow cytometric analysis as described above.

The specificity of the monoclonal antibodies was tested by their ability to bind macrophages derived from mice with the appropriate haplotype and their lack of reactivity with CVE derived from mice with different haplotypes.

Reverse Transcriptase-T7 RNA Polymerase-Dependant Amplification of mRNA (RT-TRDA) for IL1-β

RNA from PI-CVE and non-infected BALB/c and SJL/J CVE was extracted and analyzed by RT-TRDA (34, 35). Briefly, the mRNA was reverse transcribed with a poly dT primer containing the T7 RNA polymerase promoter in it's 5' end and ds cDNA copies prepared with the use of Klenow and T4 DNA polymerase. The resulting population was then amplified by T7 RNA polymerase in the presence of radiolabeled CTP and hybridized to a slot blot containing probes for the genes shown in table 4 and the control genes, histone and GAPDH (Table 4). The slot blots were prepared by adding 250ng of purified insert to each slot. The data were normalized to GAPDH and histone was used as the negative control.

Table 1. A comparison of the replication of Theiler's virus in BHK-21 cells and CVE derived from SJL/J, CBA and BALB/c mice

| Theiler's virus strain | Expt# | BHK-21 | Cells | | |
| | | | CVE | | |
			BALB/c	CBA	SJL/J
GDVII	I	6.26*	5.5	6	5.5
	II	7.0	3.4	3.4	3.4
DA	I	4.63	5.0	5.26	5.0
	II	5.87	2.69	3.87	3.87
BeAn	I	5.66	5.25	5.26	5.0
	II	4.69	3.0	3.69	3.4

*The results are expressed as LOG_{10} $TCID_{50}$/ml.
Experiment I:- cells infected with 10^6 p.f.u of virus.
Experiment II:- cells infected with 10^4 p.f.u of virus.

The data were analyzed separately for each strain by two-way analysis of variance (ANOVA) to determine which genes were affected by persistent viral infection.

RESULTS

The Replication of Theiler's Virus in CVE

All strains of Theiler's virus infected CVE derived from all three strains of mice (Table 1). BeAn and DA virus replicated to similar titers in all CVE and to slightly higher titers in BHK-21 cells. The virulent GDVII strain of TMEV replicated to similar titers in CVE as the persistent strains but to higher titers in BHK-21 cells. This was particularly noticeable at low virus inoculum (experiment II).

Virus or IFN-γ Induced Expression of MHC Class I on CVE

Two replicate experiments were performed in order to examine the upregulation of Class I on CVE following either infection with TMEV or treatment with IFN-γ. The mean results of these experiments are shown in Figure 1. The BeAn strain of TMEV induced detectable levels of MHC Class I on SJL/J and CBA CVE by 48 hours post infection (Figure 1). After 72 hours infection, 54% of the SJL/J CVE and 46% of the CBA CVE were expressing MHC Class I. The virus did not induce detectable levels of Class I on BALB/c CVE. IFN-γ induced high levels of MHC Class I on both SJL/J and CBA CVE by 24 hours but Class I was only detectable at 48 hours in the case of BALB/c CVE (Figure 1).

Virus or IFN-γ Induced Expression of MHC Class II on CVE

TMEV did not induce the expression of MHC Class II on any of the CVE tested (Figure 2a, b & c). IFN-γ did induce the expression of MHC Class II on SJL/J and CBA CVE (Figure 2a & b) but not BALB/c CVE (Figure 2c).

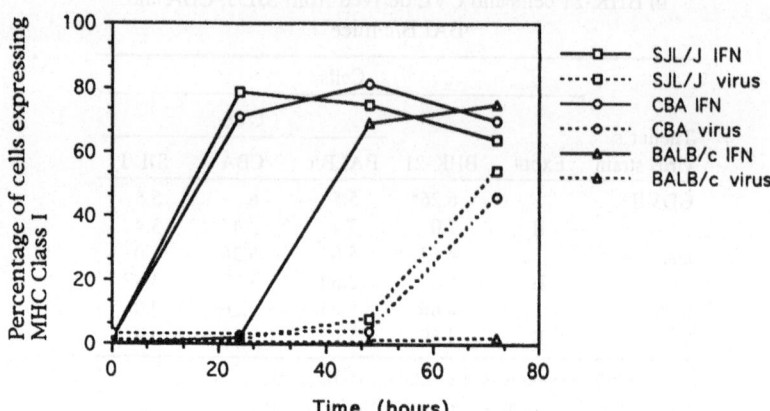

Figure 1. Virus or IFN-g induced MHC Class I expression on CVE derived from SJL/J, CBA and BALB/c mice. The percentage of SJL/J, CBA and BALB/c CVE cells expressing MHC Class I following treatment with either IFN-g (100 units/ml) or Theiler's virus. MHC Class I was measured by flow cytometry at time 0, 24, 48 and 72 hours post treatment.

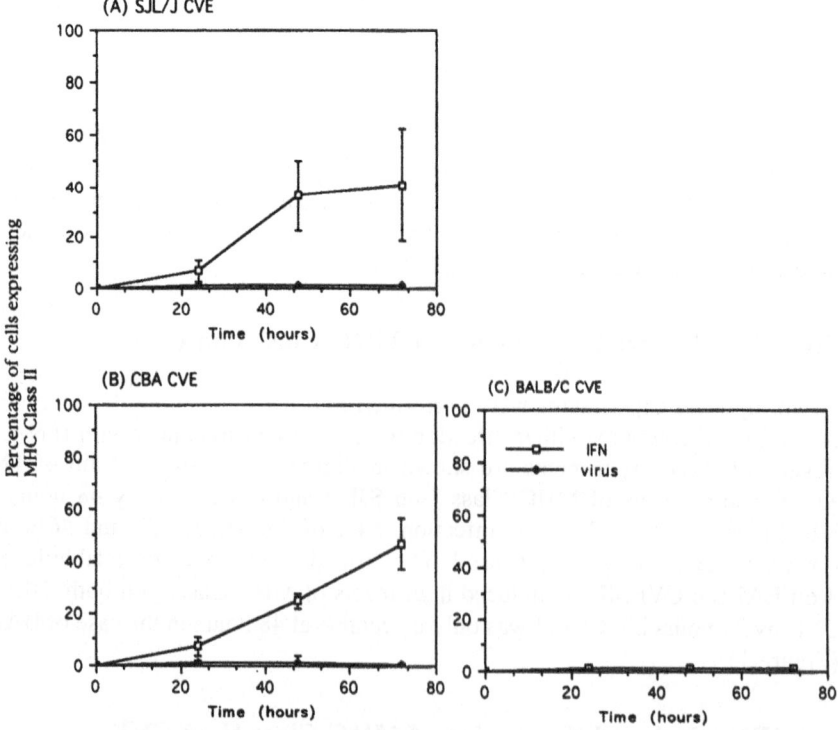

Figure 2. Virus or IFN-g induced MHC Class II expression on CVE derived from SJL/J, CBA and BALB/c mice. The percentage of SJL/J, CBA and BALB/c CVE cells expressing MHC Class II following treatment with either IFN-g (100 units/ml) or Theiler's virus. MHC Class II was measured by flow cytometry at time 0, 24, 48 and 72 hours post treatment.

Table 2. Percentage positive PI-CVE stained with monoclonal antibodies to TMEV as analyzed by flow cytometry

		% Positive	
Strain of mouse	Monoclonal antibody#	Experiment I	Experiment II
BALB/c	DA mAB 1	2	—
	DA mAB 2	20	38
CBA	DA mAB 1	18	—
	DA mAB 2	31	36
SJL/J	DA mAB 1	23	—
	DA mAB 2	30	39

Percentage of PI-CVE Expressing Viral Antigen

Between 20-39% of the PI-CVE expressed VP-1 antigen as detected by monoclonal antibody DAmAB 2 and between 2-23% expressed the DA mAB1 epitope (Table 2). Antigen was only detected on fixed cells but not on unfixed cells (data not shown). There was no obvious difference between the percentages of BeAn-VP-1 positive cells from BALB/c, CBA or SJL/J PI-CVE.

MHC Class I & II Expression on PI-CVE

A total of nine BeAn PI-CVE cell lines were established, 3 SJL/J, 3 CBA and 3 BALB/c. The nine PI-CVE lines were examined for expression of MHC Class I and II at passage 6 and 12. MHC Class I was only detected on one of the BALB/c PI-CVE at passage 6 (data not shown) but when retested at passage 12 this line did not express detectable levels of Class I (Figure 3). SJL/J and CBA PI-CVE expressed high levels of Class I at passage 6 and 12 (Figure 3). Significant levels of Class II were not detected on any of the PI-CVE (data not shown).

Upregulation of Expression of IL-1β mRNA in PI-CVE

The message for IL-1β was significantly ($p < 0.05$) upregulated from 0.65 to 3.4 in BALB/c PI-CVE and from 1.4 to 5.4 in the case of SJL/J PI-CVE, compared to non-infected BALB/c and SJL/J CVE, respectively (Table 3). Significant increases in gene expression were also noted in the case of IL-3 and IL-5 in both PI-CVE and non-infected cells. A decrease in expression of GM-CSF was noted in the BALB/c PI-CVE compared to non-infected BALB/c CVE.

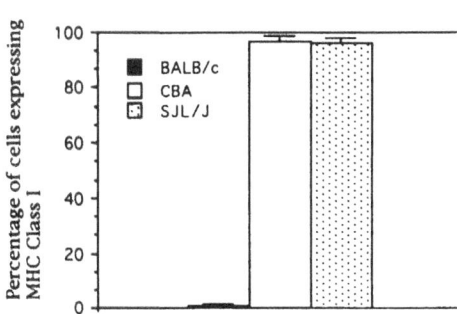

Figure 3. The expression of MHC Class I on SJL/J, CBA and BALB/c PI-CVE. The percentage of persistently infected SJL/J, CBA and BALB/c CVE cells expressing MHC Class I. MHC Class I was measured by flow cytometry at passage 12 after the persistent infection was established.

Table 3. RT-TRDA results from SJL/J and BALB/c PI-CVE compared
to uninfected CVE

Gene	BALB/c CVE	BALB/c PI-CVE	SJL/J CVE	SJL/J PI-CVE
GAPD	1 ± 0	1 ± 0	1 ± 0	1 ± 0
HIST	$0.31 \pm .03$	0.47 ± 0.11	0.52 ± 0.07	0.52 ± 0.06
IL-1β	$0.65^@ \pm 0.28$	$3.4^@ \pm 0.7$	$1.4^@ \pm 0.53$	$5.4^@ \pm 1.03$
IL-3	$4.5^* \pm 1.8$	$5.4^* \pm 1.1$	$6.58^* \pm 1.62$	$5.77^* \pm 0.91$
IL-5	$4.7^* \pm 0.18$	$5.8^* \pm 1.3$	$2.82^* \pm 1.63$	$5.5^* \pm 0.31$
IL-6	0.62 ± 0.29	0.97 ± 0.19	1.23 ± 0.3	1.4 ± 0.16
CSF-1	0.46 ± 0.06	0.58 ± 0.18	0.69 ± 0.2	0.88 ± 0.27
CSF-1R	0.36 ± 0.04	0.50 ± 0.17	0.55 ± 0.08	0.45 ± 0.09
GM-CSF	1.3 ± 0.08	$0.67^# \pm 0.2$	0.53 ± 0.1	0.69 ± 0.1
SCF-1	0.4 ± 0.06	0.46 ± 0.11	0.39 ± 0.04	0.45 ± 0.09
TGFβ2	0.46 ± 0.01	0.57 ± 0.11	0.56 ± 0.1	0.68 ± 0.05
TGFβ3	1.24 ± 0.1	0.98 ± 0.04	1.01 ± 0.15	0.85 ± 0.06
RARA	0.37 ± 0.06	1.2 ± 0.27	1.38 ± 0.19	1.42 ± 0.26
RARB	0.19 ± 0.04	0.39 ± 0.11	0.43 ± 0.09	0.48 ± 0.07
RARG	0.18 ± 0.04	0.36 ± 0.08	$>0.33 \pm 0.05$	0.34 ± 0.07
β2M	0.32 ± 0.05	0.47 ± 0.13	0.41 ± 0.11	0.55 ± 0.03
β-TUB	$2.8 \pm .68^*$	$3.2 \pm 2.2^*$	$1.1 \pm .29^*$	$2.15 \pm 1.08^*$

R = receptor, GAPD = glyceraldehyde 3-phosphate dehydrogenase, Hist = histone, IL = interleukin, CSF = colony stimulating factor, GM-CSF = granulocyte/macrophage colony stimulating factor, SCF = stem cell factor, TGF = transforming growth factor, RARA retinoic acid receptor, β2M = beta 2 microglobulin, β-TUB = Beta-tubulin.
Values are gene/GAPD. ± = standard deviation.
*Significant changes in gene expression between infected CVE and non-infected.
@Statistically significant difference between infected and non infected CVE p< 0.05.
#Gene expression down regulated.

DISCUSSION

The present study describes the in vitro replication of Theiler's virus in various CVE cell lines, the establishment of persistently infected CVE, and the correlation between susceptibility to demyelination and the ability of Theiler's virus to induce the expression of MHC Class I on acutely and persistently infected CVE. The relevance of these observations will be discussed in relation to the pathogenesis of TVID and MS.

Theiler's virus replicates in cerebrovascular endothelial cells derived from all three strains of mice. Therefore there is no correlation between susceptibility to TVID and the ability of Theiler's virus to replicate in CVE. The virulent GDVII strain of Theiler's virus replicated to higher titers in BHK-21 cells and CVE than the persistent strains of Theiler's virus. The higher efficiency of GDVII replication is related to its neurovirulence.

In the natural infection with Theiler's virus, CVE may represent an important route of viral entrance into the CNS. Many of the viruses that infect CVE have also been implicated in the etiology of multiple sclerosis or shown to cause demyelination in animals. These viruses include HSV-1 (7), mouse hepatitis virus (MHV-4 JHM) (36), mumps (37) and measles (38).

Both acute and persistent Theiler's virus infection of CVE resulted in the upregulation of MHC Class I expression on CVE derived from susceptible but not resistant strains of mice. Viral infection and the subsequent expression of MHC Class I on CVE may allow CVE to become targets for cytotoxic T cells. The T cell-mediated lysis of virus-infected CVE may be one of the initiating events in the pathogenesis of Theiler's virus-induced demyelination. The fact that BALB/c CVE did not express MHC Class I following infection with Theiler's virus may be related to the resistance of this strain of mouse to TVID. In addition to their defect in virus-induced MHC Class I expression, BALB/c CVE also showed a delayed upregulation of MHC Class I following treatment with IFN-γ when compared to CVE derived from susceptible strains of mice. Thus, there appears to be a correlation between susceptibility to demyelination and the ability of IFN-γ to induce early high levels of MHC Class I on CVE.

Genetic studies have shown that one of the genes involved in susceptibility to TVID maps to the MHC Class I gene (39) which implicates cytotoxic T cells in susceptibility and/or resistance to TVID. Cytotoxic T cells have been shown to play a role in both the clearance of Theiler's virus during early infection and in the later demyelinating process. In susceptible strains of mice, CD8 T cell depletion prior to infection with Theiler's virus, results in ineffective viral clearance from the CNS. The CD 8 depleted animals subsequently developed more severe demyelination than their immunocompetent counterparts (21). When CD8 T cell depletion was performed after the early clearance of virus, the procedure alleviated the demyelinating process (29). Studies with β2-microglobulin deficient mice (22,23) have also implicated cytotoxic T cells in viral clearance in resistant strains of mice. Transgenic mice have been constructed with the resistant MHC Class I genotype (H-2D^b) on a susceptible background FVB (H-2^q) (40). Following intracranial infection with Theiler's virus, these mice clear the virus from the CNS and so persistent viral infection is not established. The mice do not subsequently develop demyelination. In addition to these findings, cytotoxic T lymphocyte activity has been described in Theiler's virus-infected mice. (24,25). The target cells in these studies were fibroblasts and glioma cell lines (24) and transfected L-cells (25). However, Kurtz et al were unable to detect CTL against acutely-virus infected CVE in TMEV infected SJL/J mice (12). The production of persistently infected CVE described in the present study, may prove to be a more appropriate target cell than the acutely infected CVE.

The role of CD8 T cells in the late demyelinating phase of the disease has not been fully elucidated. The β2-microglobulin deficient mice, infected with Theiler's virus, developed signs of demyelination which suggests that cytotoxic T cells are not the main mediators of demyelination (22, 23). These mice developed high delayed type hypersensitivity (DTH) responses to the virus which are only detected in susceptible strains of mice (22). Since virus-specific CD4 DTH T cells have been implicated in the demyelinating phase of the disease (27, 31, 41), it is possible that CD8 T cells may play a regulatory role suppressing the pathogenic CD4 DTH T cells (21, 22). However, CD8 depletion of Theiler's virus-infected SJL/J mice has been reported to inhibit the demyelinating process (29). Thus, while CTLs are apparently involved in viral clearance, their role in the demyelinating process remains an obscure.

Assuming that the in vitro results described here can be extrapolated to the in vivo situation, the following sequence of events may be hypothesized to occur. Theiler's virus may gain access to the CNS via replication in CVE. TMEV infection of CVE in susceptible mice would result in early expression of MHC Class I. TMEV-infected CVE would then be available as targets for CTLs. Interestingly, cytotoxic T cells against CVE, have been described in MS patients suffering exacerbation of their disease (42). Thus CTL activity against CVE may be important in the pathogenesis of both MS and TVID, by breaking down the BBB and allowing the influx of other immune cells into the CNS.

In the presence of activated T cells or natural killer cells, local production of IFN-γ would elicit both MHC Class I and II expression in CVE's derived from susceptible mice (19, 43). The CVE could then become antigen presenting cells and allow increased migration of T cells into the CNS. Virus-specific CD4 T cells would then be able to recruit macrophages into the CNS. Macrophages mediate demyelination by bystander mechanisms such as the release of TNF-α which damages oligodendrocytes directly (44).

Infection of resistant BALB/c mice would not result in the increased expression of Class I on CVE and so they are unlikely to be targets for CTL's. This may prevent the entrance of CTLs into the CNS although virus-infected CVE may be targets for natural killer (NK) cells which have also been implicated as mediators of viral clearance (45). IFN-γ does not upregulate the expression of MHC Class II on BALB/c CVE and this may result in the decreased T cell traffic into the CNS (43).

The fact that CVE can be persistently infected with Theiler's virus in vitro may have some interesting in vivo consequences. Persistent infection of CVE may provide a reservoir of infectious virus which could re-infect the CNS. This would result in the characteristic perivascular infiltration of immune cells seen in TVID. Recurrent episodes of demyelination may follow after the release of TMEV from persistently infected CVE (46).

Increased expression of IL-1β gene was observed in both persistently infected SJL/J and BALB/c CVE. IL-1 has been shown to upregulate the expression of adhesion molecules on CVE and also to lead to antigen presentation (47). Persistently infected SJL/J CVE treated with IFN-γ have been shown to express MHC Class II which would also enhance their antigen presenting capabilities (32).

In conclusion, Theiler's virus infection of CVE may prove to be an important route for viral entry into the CNS. Infection of CVE in SJL/J and CBA mice results in the upregulation of MHC Class I which may render these cells susceptible to CTLs. Local production of cytokines in the vicinity of the BBB will also upregulate the expression of adhesion molecules, MHC Class I and Class II on CVE which will result in increased cell traffic into the CNS. Persistently infected CVE may prove to be important reservoirs of virus and also targets for CTLs. Further in vivo experimentation is required in order to fully elucidate the role of CVE in the pathogenesis of TVID.

ACKNOWLEDGMENTS

Supported by NIH/NINDS grant R01 NS 29133, and Smokeless Tobacco Research Council Project Grant 0370-02 and a grant from the College of Veterinary Medicine, Texas A&M University. We would also like to acknowledge Betty Rosenbaum and Roger Smith of the Department of Veterinary Pathobiology, Texas A&M University, for the flow cytometry experiments.

REFERENCES

1. Barker, C.F .and R. E. Billingham. 1977. Immunologically privileged sites. Adv. Immunol. 25: 1.
2. Cserr, H.F., and P.M. Knopf. 1992. Cervical lymphatics, the blood-brain barrier and the immunoreactivity of the brain: a new view. Immunol.Today 13: 507.
3. Janzer, R.C., and M.C. Raff. 1987. Astrocytes induce blood-brain barrier properties in endothelial cells. Nature 325: 253.
4. Male, D., and G. Pryce. 1988. Kinetics of MHC gene expression and mRNA synthesis in brain endothelium. Immunol. 63: 37.
5. Werkle, H., C. Linington, H. Lassman, and R. Meyermann. 1986. Cellular immune reactivity within the CNS. Trends. Neurosci. 9: 271.
6. Acheson, E.D. 1977. Epidemiology of multiple sclerosis. Brit. Med. Bull. 33: 9.
7. Allen, I. and B.J. Brankin. 1993. Pathogenesis of multiple sclerosis- the immune diathesis and the role of viruses. Neuropathol. and Exp. Neurol. 52: 95.
8. Theiler, M. 1934. Spontaneous encephalomyelitis of mice- a new virus. Science 80: 122.

9. Theiler, M. and S. Gard. 1940. Encephalomyelitis of mice. I. Characteristics and pathogenesis. J. Exp. Med. 72: 79.

10. Lipton, H.L. 1975. Theiler's virus infection in mice: an unusual biphasic disease process leading to demyelination. Inf. & Immun. 11: 1147.

11. Rodriguez, M., J.L. Leibowitz, H.C. Powell, and P.W. Lampert. 1983. Neonatal infection with the Daniel's strain of Theiler's murine encephalomyelitis virus. Lab. Invest. 49: 672.

12. Kurtz, C.I.B., R.M. McCarron, M. Spatz, and R.S. Fujinami. R.S. 1994. Characterization of a murine central nervous system-derived cell line: infectability and presentation of a viral antigen. J. Neuroimmunol. 51: 35.

13. Zurbriggen, A. and R.S. Fujinami. 1988. Theiler's virus infection in nude mice: viral RNA in vascular endothelial cells. J. Virol. 62: 3589.

14. Clatch, R.J., S.D. Miller, R. Metzner, M.C. Dal Canto, and H.L. Lipton. 1990. Monocytes/ macrophages isolated from the mouse central nervous system contain infectious Theiler's murine encephalomyelitis virus (TMEV). Virology 176: 244.

15. Levy, M., C. Aubert, and M. Brahic. 1992. Theiler's virus replication in brain macrophage cultured in vitro. J. Virol. 66: 3188.

16. Aubert, C., M. Chamorro, and M. Brahic. 1987. Identification of Theiler's virus infected cells in the central nervous system of the mouse during demyelinating disease. Micro. Pathol. 3: 319.

17. Welsh, C.J.R., W.F. Blakemore, P. Tonks, P. Borrow, and A.A. Nash. 1989. Theiler's murine encephalomyelitis virus infection in mice: a persistent viral infection of the central nervous system which induces demyelination. Chapter in "Immune responses, Virus Infection and Disease (pp125-147). Published by Oxford University Press. Editor Nigel Dimmock.

18. Borrow, P., C.J.R. Welsh, and A.A. Nash. 1993. Study of the mechanisms by which CD4+ T cells contribute to protection in Theiler's murine encephalomyelitis. Immunol. 80: 502.

19. Welsh, C.J.R., B.V. Sapatino, B. Rosenbaum, and R. Smith. Theiler's virus and cerebrovascular endothelial cells. I Characersitics of cerebrovascular endothelial cells acutely infected with Theiler's virus. Manuscript submitted to J Neuroimmunology.

20. Kohanawa, M., A. Nakane, and T. Minagawa. 1993. Endogenous gamma interferon produced in central nervous system by systemic infection with Theiler's virus in mice. J. Neuroimmunol. 48: 205.

21. Borrow, P., P. Tonks, C.J.R. Welsh, C.J.R. and A.A. Nash. 1992. The role of CD8+ T cells in the acute and chronic phases of Theiler's virus-induced disease in mice. J. Gen. Virol. 73: 1861.

22. Pullen, L.C., S.D. Miller, M.C. Dal Canto, and B.S. Kim. 1993. Class I-deficient resistant mice intracerebrally inoculated with Theiler's virus show an increased T cell response to viral antigens and susceptibility to demyelination. Eur. J. Immunol. 23: 2287.

23. Fiette, L., C. Aubert, M. Brahic, and C.P. Rossi. 1993. Theiler's virus infection of β2-microglobulin deficient mice. J. Virol. 67: 589.

24. Lindsley, M.D., R. Thiemann R., and M. Rodriguez. 1991. Cytotoxic T cells isolated from the central nervous system of mice infected with Theiler's virus. J. Virol. 65: 6612.

25. Rossi, P.C., A. McAllister, L. Fiette, and M. Brahic. 1991. Theiler's virus infection induces a specific cytotoxic T lymphocyte response. Cell. Immunol. 138: 341.

26. Roos, R.P. and R. Wollmann. 1984. DA strain of Theiler's murine encephalomyelitis virus induces demyelination in nude mice. Ann. Neurol. 15: 494.

27. Clatch, R.J., H.L. Lipton, and S.D. Miller. 1987. Class II-restricted T cell responses in Theiler's murine encephalomyelitis virus (TMEV)-induced demyelinating disease. Part 2. Survey of host immune responses and central nervous system virus titers in inbred mouse strains. Micro. Pathol. 3: 327.

28. Welsh, C.J.R., P. Borrow, P. Tonks, D. Dean and A.A. Nash. 1995. Autoimmunity in Theiler's virus-induced demyelination. Manuscript in preparation.

29. Rodriguez, M. and S. Sriram. 1988. Successful therapy of Theiler's virus-induced demyelination (DA strain) with monoclonal anti-Lyt2 antibody. J. Immunol. 140: 2950.

30. Sapatino, B., C.J.R. Welsh, C.A. Smith, B. Bebo, and D.S. Linthicum. 1993. Cloned mouse cerebrovascular endothelial cells that maintain their differentiation markers for Factor VIII, low density lipoprotein and angiotensin-converting enzyme. In Vitro Cell. Dev. Biol.29A: 923.

31. Welsh, C.J.R., P. Tonks, A.A. Nash, and W.F. Blakemore. 1987. The effect of L3T4 T cell depletion on the pathogenesis of Theiler's murine encephalomyelitis virus infection in CBA mice. J. Gen. Virol. 68: 1659.

32. Sapatino, B.V., A. Petrescu, B. Rosenbaum, J. Peidrahita, R. Smith, and C.J.R. Welsh. Theiler's virus and cerebrovascular endothelial cells. II Characersitics of cerebrovascular endothelial cells persistently infected with Theiler's virus. Manuscript submitted to J Neuroimmunology.

33. Nitayaphan, S., M.M. Toth, and R.P. Roos. 1985. Neutralizing monoclonal antibodies to Theiler's murine encephalomyelitis viruses. J. Virol. 53: 651.

34. Eberwine, J., H. Yeh, K. Miyashiro, Y. Cao, S. Nair, R. Finnell, M. Zettel, and P. Coleman. 1992. Analysis of gene expression in single live neurons. Proc. Natl. Acad. Sci. USA 89: 3010.

35. Tecott, L.H., J.D. Barchas, and J.H. Eberwine. 1988. In situ transcription: specific synthesis of complimentary DNA in fixed tissue sections. Science 240: 1661.

36. Joseph, J., J.L. Grun, F.D. Lublin, and R.L. Knobler. 1993. Interleukin-6 induction in vitro in mouse brain endothelial cell and astrocytes by exposure to mouse hepatitis virus (MHV-4, JHM). J. Neuroimmunol. 42: 47.

37. Friedman, H.M. 1989. Infection of endothelial cells by common human viruses. Rev. Infect. Dis. 11: 5700.

38. Kirk, J., A.-L. Zhou, S. McQuaid, S.L. Cosby, and I.V. Allen. 1991. Cerebral endothelial cell infection by measles virus in subacute sclerosing panencephalitis: ultrastructural and in situ hybridization evidence. Neuropath. and Appli. Neurobiol. 17: 289.

39. Bureau, J.F., X. Montagutelli, S. Lefebvre, J.L. Guenet, M. Pla, and M. Brahic. 1992. The interaction of two groups of murine genes determines the persistence of Theiler's virus in the central nervous system. J. Virol. 66: 4698.

40. Azoulay, A., M. Brahic, and J.F. Bureau. 1994. FVB mice transgenic for the H-2D^b gene become resistant to persistent infection by Theiler's virus. J. Virol. 68: 4409.

41. Gerety, S., J., R.J. Clatch, H.L. Lipton, R.G. Goswami, M.K. Rundell, and S.D. Miller. 1991. Class II-restricted T cell responses in Theiler's murine encephalomyelitis virus-induced demyelinating disease. IV Identification of an immunodominant T cell determinant, on the N-terminal end of the VP-2 capsid protein in susceptible SJL/J mice. J. Immunol. 146: 4322.

42. Tsukada, N., M. Matsuda, K. Miyagi, and M. Yanagisawa. 1993. Cytotoxicity of T cells for cerebral endothelium in multiple sclerosis. J. Neuro. Sci. 117: 140.

43. Welsh, C.J.R., B. Sapatino, B. Rosenbaum, R. Smith, and D.S. Linthicum. 1993. Correlation between susceptibility to demyelination and interferon gamma induction of major histocompatibility complex class II antigens on murine cerebrovascular endothelial cells. J. Neuroimmunol. 48: 91.

44. Selmaj, K.W. and C.S. Raine. 1987. Tumor necrosis factor mediates myelin and oligodendrocyte damage in vitro. Ann. Neurol. 23: 339.

45. Paya C.V., A.K. Patick, P.J. Leibson, and M. Rodriguez. 1989. Role of natural killer cells as immune effectors in encephalitis and demyelination induced by Theiler's virus. J. Immunol. 143: 95.

46. Blakemore, W.F., C.J.R. Welsh, P. Tonks, and A.A. Nash. 1988. Observations on demyelinating lesions induced by Theiler's virus in CBA mice. Acta Neuropath. 76: 581.

47. Fabry, Z., M.M. Waldschmidt, D. Hendrickson, J. Keiner, L. Love-Homan, F. Takei, and M.N. Hart. 1992. Adhesion molecules on murine brain microvascular endothelial cells: expression and regulation of ICAM-1 and Lgp 55. J. Neuroimmunol. 36: 1.

14

IDIOTYPE MANIPULATION IN DISEASE MANAGEMENT

Heinz Kohler, Malaya Bhattacharya-Chatterjee, Sybille Muller, and
Kenneth A. Foon

Markey Cancer Center,
University of Kentucky
Lexington, Kentucky

INTRODUCTION: (NIELS K. JERNE, 1911-1994)

On October 7, 1994, Niels Jerne died. Niels Jerne is the father of the idiotypic network which provides a unifying concept for understanding the immune system. The network hypothesis makes certain predictions which lead themselves to therapeutic exploitation. Indeed, a large body of vaccine experiments in animals and a handful of clinical trials are based on this theory.

Niels Jerne received the Nobel Prize in 1984 for his contribution to a better understanding of the immune system. He will be remembered as a "gentleman" scientist who enjoyed the beauty of logical arguments regarding biological phenomena. A frequent favored phrase of his was, "Doesn't it make sense, does it?" Much of his thinking was based on his intuitive thinking paired with a keenly mathematical approach. His lectures often started with numbers of lymphocytes, antibodies, genes, etc.

As director and founder of the Basel Institute for Immunology, he presided over the golden 10 years of the Institute when three members of the Institute were awarded the Nobel Prize (Niels Jerne, George Kohler, Susumu Tonegawa). Niels Jerne's intellectual contributions have led to a better understanding of the immune system and new approaches to harness the immune system for therapy of diseases.

THE IDIOTYPE VACCINE CONCEPT

The first mention of anti-idiotypic antibodies as potential vaccines was made by Eichmann and Rajensky in an article published in 1975 (1) describing their finding of B and T cell priming using rabbit anti-idiotypic antibodies. The idea laid dormant for five years until 1980 when Nisonoff (2) and Roitt (2) formally proposed to use anti-idiotypic antibodies as vaccines. Their proposal was based on Jerne's "internal image" concept for anti-idiotypic antibodies. Internal images of conventional antigens are represented in the immune network

Immunobiology of Proteins and Peptides VIII
Edited by M. Z. Atassi and G. S. Bixler, Jr., Plenum Press, New York, 1995
117

as idiotype and anti-idiotype. The immense diversity of CDR conformations would permit mimicry of virtually any antigenic structure. The challenge for the experimental immunologist was to find those internal image anti-idiotypes in polyclonal or hybridoma generated antibody repertoires.

It was Niels Jerne who clearly provided a conceptual explanation for the difficulties to generate effective anti-idiotypic vaccines. He divided anti-idiotypes into alpha and beta forms (i.e. Ab2a and Ab2β) (4). Ab2β was the functional internal image antigen and could be formulated into a vaccine. Ab2a was thought to be less interesting for therapeutic or vaccine applications as it did not mimic the antigen based on the observation that binding of Ab1 to the original antigen could not be inhibited by Ab2a. Another Ab2 that produced partial inhibition of Ab1 binding to original antigen was designated as Ab2g by C. Bona and one of us (5).

However, it was subsequently reported that the so called non-image anti-idiotype (Ab2a) could also induce specific immunity (6, 7, 8, 9). The explanation for the immune responses induced by Ab2 was that they acted as an anti-clonotypic antibody for B cell idiotype markers and perhaps also for T cells. We proposed (10) the term "network antigen" to describe anti-idiotypes which had the potential to induce specific immune responses. "Network antigens" would be either Ab2β, a, or g that induce an Ab1 like Ab3 response and stimulate so called silent clones in immune therapy.

We review here the experimental evidence for idiotype-specific mechanisms underlying tumor immunotherapy and the immune response to infection with HIV-1.

CLINICAL TRIALS WITH ANTI-IDIOTYPE ANTIBODIES

In our first clinical trial we enrolled four patients with advanced cutaneous T cell lymphoma. The selection of patients was on the basis of positive reactivity with the monoclonal antibody SN-2 (11) which detects a gp37 T cell lymphoma associated antigen. Only a subset of T cell lymphomas react with SN-2 and no normal cells react with SN2 including T cells. Patients were treated with our monoclonal anti-idiotype antibody 4DC6, which was raised against SN-2 and shown in preclinical studies to be able to induce anti-gp37 anti-tumor antibodies in several species (12). Patients received a minimum of four injections every other week of 1 mg of aluminum hydroxide precipitated 4DC6. Three patients responded with human anti-mouse antibody (HAMA) and anti-anti-idiotype responses. The second patient in the trial had a remarkable clinical response. After the second immunization the skin tumor lesions began to diminish in size. This continued over the immunization course and eventually all nine measureable skin lesions disappeared. Only one lesion demonstrates residual tumor cells in skin biosy. The patient was given a total of eight injections and therapy was stopped. The skin tumors were stable for 11 months. After this time, some of the lesions began to regrow.

Biopsy of the recurrent tumors demonstrated negative staining with SN-2, thus indicating that the tumor relapsed with an immune escape variant. This also indicates that the idiotype vaccine was successful in eliminating all epitope (gp37) positive tumors in this patient.

Our second trial is based on the use of a monoclonal anti-idiotype vaccine which targets carcinoembryonic antigen (CEA) positive tumors. In the initial phase I trial 24 patients were treated with the 3H1 monoclonal antibody. 3H1 was generated against a monoclonal anti-CEA antibody, 8019. 3H1 induces anti-CEA antibodies in several species including non-human primates (13 and manuscript submitted). Patients with advanced disease and having failed conventional therapies were eligible. Patients also were required to respond to at least one antigen among a panel of antigens following intracutaneous injection. Typically these patients had an expected survival of 6-12 months.

Twelve patients have been immunologically evaluated. Ten responded with HAMA and anti-anti-idiotype titers (Ab3). Nine of these patients produced anti-CEA antibodies. Interestingly, in seven patients a proliferative response to the immunizing anti-idiotype and purified CEA was observed. The responding T cells were primarily CD4+. This finding is one of the first in humans demonstrating activation of T cells by an anti-idiotype vaccine. Clinically, tumor regression was not observed in any of the treated patients. Tumor regression in these advanced patients cannot be expected and are likely to be seen only in patients with limited or minimal disease. Therefore, a follow-up trial has been initiated treating patients with the 3H1 anti-idiotype following surgical removal of all measureable disease but limited to those with high risk of recurrence.

Another idiotype vaccine trial in breast cancer patients has been initiated. Here, the therapy target is a milk fat globule protein which is preferentially expressed on 80% of breast carcinomas. An anti-idiotype has been generated, as designated 11D10, and has been shown to induce tumor-specific antibodies in mice, rabbits and monkeys (13 and manuscript submitted).

To summarize our clinical experience with anti-idiotype tumor vaccines the following conclusions can be drawn: 1) No major toxicity was observed, mild side-effects consisted of local inflammation, fever and chills of short duration. 2) The majority of treated patients produced the desired tumor-specific antibody response. This is particularly remarkable for the CEA, since CEA is known to be tolerogenic in man and other tumor vaccines have consistently failed to induce anti-CEA antibodies in cancer patients. 3) In about half of the immunologically responding patients a T cell response could be demonstrated. 4) In one patient a long-lasting near-complete tumor remission was observed. 5) The encouraging immunological response and minimal toxicity in the two trials warrant treating patients in the adjuvant setting.

POTENTIAL CLONOTYPE-SPECIFIC ANTI-IDIOTYPES FOR AIDS

Anti-idiotype antibodies may have therapeutic and diagnostic utilities in infectious diseases induced by bacteria or viruses (14, 15, 16). For some time we and others have studied the humoral response to HIV-1 in seropositive individuals (17, 18, 19, 20). One of the hallmarks of HIV infection is the vigorous immune response at both the T cell and antibody level. Shortly after infection, virus-specific cytolytic T lymphocytes (CTL) are induced followed by neutralizing antibodies against the viral envelope and antibodies against other viral proteins. This initial response coincides with a drastic decline in viremia to low levels which are often difficult to measure. This low grade infection continues over many years without causing AIDS. Evidently the immune system can control the virus for long periods of time, up to 10-12 years in some HIV-1 infected individuals. Therefore, the disease (AIDS) onset is highly variable in infected individuals. The other hallmark of HIV-1 infection is the failure of the immune system to clear the viral infection which in all observed cases was lethal except long-term survivors. This has led to the conclusion that the immune response may be abnormal in HIV-1 infection.

The nature of this B and T cell abnormality, however, is unclear and is subject to speculation about the immunopathology of AIDS (21, 22). Together with Peter Nara from the NCI, the "clonal dominance" hypothesis was proposed (23). In summary, this hypothesis states that an initially established dominant antibody response to the infecting HIV-1 remains stable over years, while the virus mutates and generates immune escape variants. The process leads to a dominant primary immune response and is described by us as 'Deceptive

Imprinting' (24). Key for this process is the inability of virgin B cells to recognize newly arising viral variants.

Without discussing details of this model, we would like to direct attention to another characteristic phenomenon of HIV immunity which may be related to the immunopathology of AIDS.

Because of the persistence of initially recruited B cell anti-HIV clones, we wanted to generate an anti-idiotype which specifically recognizes such primary response B cell clones. Such anti-idiotypes for early or late responding B cells have been made in other experimental model, such as Arsobenzoate (ARS) (25) and hen egg lysozyme (HEL) (26). We used a pool of IgG containing polyclonal anti-HIV antibodies from infected donors with high anti-HIV titers, called HIVIG (27). Mice immunized with this anti-HIV preparation were used to generate hybridomas. Negative selection was performed on normal human IgG (IVIG) and clones were positively selected by an antigen capture ELISA. One hybridoma, designated 1F7, was selected and used in a series of experiments. A large panel of seroposi- tive and negative sera were screened for binding to 1F7: about 70% of seropositive sera reacted with 1F7 while none of the sera from uninfected donors, including autoimmune patients, was 1F7 positive (18, 19). This finding, subsequently and independently confirmed, shows that 1F7 is a HIV infection-associated idiotype. Putative disease-related idiotypes are not uncommon and have been described in SLE patients (28) and on antibodies against a variety of infectious diseases (29, 30). For example, antibodies against Hemophilus influenza share an idiotype marker in the majority of individuals infected with Hemophilus (29). Recently, an encephalitogenic epitope related and recognized by Id-bearing autoreactive T cells has been identified (31).

Next, we asked which anti-HIV antibodies express the shared 1F7 marker. Different recombinant HIV-1 proteins, gp120, p24, RT, were used in capture ELISA and Western Blot assays to detect expression of the 1F7 idiotype. Surprisingly, 1F7 was expressed by antibodies against the viral envelope, the core protein and the reverse transcriptase. Sub- sequent studies showed that 1F7 is expressed on anti-gp120 and anti-gp160 of different HIV laboratory strains. We also tested sera from HIV-1 infected chimpanzees and found 1F7 positive antibodies. Finally, sera from vaccine volunteers, vaccinated with recombinant gp160 and gp120, were also recently found to have 1F7 expressing anti-gp120 antibodies.

Sharing of idiotype markers on antibodies with different specificities has been known for some time as the "Oudin-Cazenave enigma" (32). In the case of anti-HIV antibodies in seropositive individuals the 1F7 sharing is specific for anti-HIV antibodies and not seen on antibodies against non-HIV antigens. Because of the shared expression the molecular basis of 1F7 was of great interest. We tested a number of human monoclonal anti-HIV antibodies (19) for 1F7 expression and found several to be 1F7 positive. These 1F7 positive antibodies belong to different VH and Vk families excluding the possibilities that 1F7 is a V gene family specific marker as previously anticipated. In collaboration with Ed Blalock from the University of Alabama, we scanned the sequence of the variable regions of 1F7 and several 1F7 positive antibodies for expression of inverse hydropathicity (33). A sequence region stretching from the FR3 to CDR3 of VH of 1F7 expressing antibodies was identified as the molecular target for 1F7 binding. Peptides from this region were synthesized and tested for binding to 1F7 and inhibition of 1F7-anti-HIV antibody binding. The peptide was active in both assays. Thus, the heavy chain region from the 3rd framework (FR3) extending to a unique CDR3 sequence is evidently involved in the expression of the 1F7 idiotope marker. Interestingly, we also found in sera from infected individuals antibodies which bind to the above described idiotope peptide. This indicates that in the antibody response to HIV a 1F7-like antibody appears which is an auto-anti-id for anti-HIV antibodies.

The biological significance for this auto-anti-Id remains unclear, but such auto-anti- Id titers are known to occur in chronically stimulated antibody responses (34). However, it might be possible to take advantage of the widely shared 1F7 antibody in the sense of an AIDS idiotype vaccine.

1F7 clearly is not an internal image type anti-idiotype. Rather, 1F7 may be used as an anti-clonotypic antibody to stimulate pre-committed B cells to produce anti-HIV antibodies. The key advantage of this approach would be that 1F7 could induce a wide spectrum of anti-HIV antibodies, including virus neutralizing antibodies against different strains. The disadvantage of current recombinant gp120 subunit vaccines is the narrow strain specificity which does not induce protective immunity against viral field isolates and not only cell line adapted HIV strains (35). 1F7 induced anti-HIV antibodies would not be restricted to neutralize a given strain but could neutralize a wide spectrum of variants, as the majority of infected individuals produce 1F7 anti-HIV antibodies. However, the utility of using a shared HIV infection specific anti-Id as vaccine remains to be demonstrated. One of us described a monoclonal Ab2 (3C9) generated by injection of polyclonal gp120-affinity purified anti-HIVgp120-IgG into mice, that induced an antibody response to the CD4 attachment site of gp120 and broadly neutralizing antibodies in cynomolgus monkeys (36). However, the 3C9-Id was shared only by approximately 5% of HIV-1 infected individuals and 11 out of 18 possible HIV-1 infected individuals who received 3C9 in a controlled phase I clinical study developed an Ab3 response to immunization (37) without significant increase of neutralizing activity. A French group (38) tested in a recent study anti-idiotypic antibodies as molecular mimics of neutralizing epitope of HIV and presented evidence that a private anti-id raised against HIV-specific antibodies may substitute viral antigens for induction of a broader humoral immune response to HIV. The authors, therefore, suggested that the HIV epitope variability may be, in part, overcome with the use of anti-idiotypic reagents.

CONCLUSION

In this article, we covered the beginning of idiotype vaccines as concept and described some practical clinical applications. We discussed the Jerne network concept as an intellectual framework for the functioning of the immune system without attempts to take advantage for immunotherapy. More than ten years had to pass before the idiotype vaccine concept was reduced to clinical practice. Examples of such clinical application are our idiotype vaccine trials against two cancers which showed encouraging clinical and immunological responses. We also discussed our finding with an infectious disease specific idiotype in HIV infected individuals. Such widely shared Ids could be used as targets for clonotypic stimulation to prime for broadly neutralizing antibodies.

DEDICATION

This article is dedicated to the memory of Niels K. Jerne.

REFERENCES

1. Eichmann, K. and Rajewsky. *Eur. J. Immunol.* 5:661, (1975)
2. Nisonoff, A. and Lamoyi, E. *Clin. Immunol. Immunopathol.* 21:397, (1981)
3. Roitt, I.M., Male, D.K., Guamotta, G., De Carvhalo, L.D., Cooke, A., Hay, F.C., Lydard, Y., Thanalava, Y., and Ivanyi, J. *Lancet* 1:10411, (1980)
4. Jerne, N.K., Roland, J. and Cazanave, P.A. *EMBO J.* 1:243, (1982)
5. Bona, C. and Kohler, H. In Monoclonal and Anti-idiotypic Antibodies: Probes for Receptor Structure and Function. (J.C. Venter, Ed.) Vol. 4 pp. 141-149, A.R. Liss, New York, (1984).

6. Schick, M.R., Dreesman, G.R., and Kennedy, R.C. *J. Immunol.* 138:3419, (1985)
7. Huang, J-H., Ward, R.E., and Kohler, H. *J. Immunol.* 770:(1986)
8. Raychaudhuri, S., Saeki, Y., Fuji, H., and Kohler, H. *J. Immunol.* 137:1743, (1986)
9. Zhou, E.M., Lohman, K.L. and Kennedy, R.C. *Virology* 174:9, 1990.
10. Kohler, H., Kaveri, S., Kieber-Emmons, T., Morrow, J.W., Muller, S., and Raychaudhuri, S. *Meth Enzymol* 178:1, (1989)
11. Seon, B.K., Negoro, S., Barcos, M.P., Tebbi, C.K., Chervinsky, D., and Fukkawa, T. *J. Immunol* 132:2089, (1984)
12. Bhattacharya-Chatterjee, M., Pride, M.W., Seon, B.K., and Kohler, H. *J. Immunol.* 139:1354, (1987)
13. Bhattacharya-Chatterjee, M., Chatterjee, S.K., Vasile, S., Seon, B.K., and Kohler, H. *J. Immunol.* 141:1398, (1988)
14. McNamara, M., Ward, R., and Kohler, H. *Science* 226:1325, (1984)
15. Kennedy, R.C., Dreesman, G.R., Butel, J.S., and Landford, R.E. *J. Exp. Med.* 161:1432, (1985)
16. Nepom, J.R., Weiner, H.L., Dichter, M.A., Tandieu, M., Spriggs, D.R., Gramm, C.F., Powers, L.M., Fileds, B.N. and Greene, M.I. *J. Exp. Med.* 155:155, (1982)
17. Morrow, W. J. W., Gaston, I., Anderson, T., Haigwood, N., McGrath, M.S., Rosen, J., and Steimer, K.S. (1990). *Viral Immunol.* 3:99-109.
18. Muller, S., Wang, H-T., Kaveri, S., Chattoppadhyay, S. and Kohler, H. *J. Immunol.* 147:933, (1991)
19. Wang, H.-T., Muller, S., and Kohler, H. *Eur. J. Immunol.* 22:1749, (1992)
20. Muller, S., Wang, H-T., Silverman, G., Bramlet, G., and Kohler, H. *Scand. J. Immunol.* 38:327, (1993)
21. Weiss, R. *Science* 260:1273-1279, (1993)
22. Salk, J., Bretscher, P.A., Salk, P.L., Clerici, M. and Shearer, G.M. *Science* 260:1270-1272, (1993)
23. Kohler, H., Goudsmit, J. and Nara, P. *J. AIDS*, 1158-1168, (1992)
24. Kohler, H., Muller, S., and Nara, P. *Immunology Today* 15:479, (1994)
25. Gurish, M. and Nisonoff, A. in "Idiotypy in Biology and Medicine" 64, (ed. Kohler, Urbain and Cazanave) Academic Press (1984)
26. Metzger, D.W., Furman, A., Miller, A., and Sercarz, E.E. *J. Exp. Med.* 154:701, (1981)
27. Prince, A.M., B. Horowitz, L. Bakers, R.W. Shulman, H. Ralph, J. Volinsky, A. Cundell, B. Brotman, W. Boehle, F. Rey, M. Piet, H. Reesink, N. Lelie, M. Tersmette, F. Miedema, L. Barbosa, G. Nemo, C. L. Nastala, J. S. Allan, D. R. Lee, and J. W. Eichberg. *Proc. Natl. Acad. Sci. USA* 85:6944, (1988)
28. Davidson, A., A. Smith, H. Katz, J.-L. Preude-Homme, A. Solomon, and B. Diamond. *J. Immunol.* 143:179, (1989)
29. Lucas, A.H. *J. Clin. Invest.* 81:480, (1988)
30. Schreiber, J.R., M. Patawaran, M. Tosi, J. Lennon, and G. B. Pier. *J. Immunol.* 144:1023, (1990)
31. Zhou, S-R., Whitaker, J.N., Han, Q., Maier, C.C., and Blalock, J.E. *J. Immunol.* 153:2340, (1994)
32. Oudin, J. and Cazenave, P. *Proc. Natl. Acad. Sci. USA* 68:2616, (1971)
33. Muller, S., Wang, H.-T., Maier, C.C., Blalock, J.E. and Kohler, H. Abstract: Vaccines New Technologies & Applications. CMI Organizers. Alexandria, VA, March 21-23, 1994
34. Kluskens, L. and Kohler, H. *Proc. Natl. Acad. Sci.* 71:5083, (1974)
35. Kohler, H., Muller, S., and Nara, P. *The Immunologist*, in press, (1994)
36. Kang, C.-Y., Nara, P., Chamat, S., Caralli, V., Chen, A., Nguye, M.-L., Yoshiyama, H., Morrow, W.J.W., Ho, D.D., and Kohler, H. *Proc. Natl. Acad. Sci. USA* 89:2546-2550, (1992)
37. Kang, C.-Y., Hambrick, R.H., Jacobson, C., Cardelli, V., McCutchen, J.A., Meritt, J. Abstract WS-B28L3: IXth International Conference on AIDS abstract book, volume I, p. 71, Berlin, Germany, June 6-11, 1993.
38. Boudet, F., Theze, J., and Zouali, M. *Virology* 200:176, (1994)

15

INTRAVENOUS IMMUNOGLOBINS (IVIGS) TO PREVENT AND TREAT INFECTIOUS DISEASES

A. S. Cross

University of Maryland Cancer Center
22 South Greene Street, S9D05
Baltimore, Maryland 21201

INTRODUCTION

Long before the current concerns about pathogens that are resistant to multiple antibiotics, there was considerable interest in the implementation of therapies that would supplement conventional antibiotics and supportive care. The impetus for such treatments was driven in part by the suboptimal clinical results obtained with conventional care in many clinical conditions, such as sepsis where a 30-50% mortality persists despite the introduction of a progression of potent new antimicrobials. With the development of immunoglobulins that were sufficient safe to be given intravenously (IVIG), this form of adjunctive therapy has been implemented not only for infectious but also for non-infectious conditions, often in the absence of convincing data to establish its efficacy.

BACKGROUND

The passive administration of antisera for the treatment of disease was initiated over a century ago, before the concept of antibodies was proposed by Ehrlich. Animal antisera was found to be beneficial when administered to people with diphtheria, rabies and measles. During the 1920's and early 1930's antisera was being routinely given to patients with pneumococcal disease. Heffron's classic text on pneumococcal disease devotes nearly 100 pages to a review of serum therapy. With this widespread use of antisera it became clear that anaphylactic and pyrogenic reactions were considerable risks of such treatment, as was the risk of acquiring hepatitis. Consequently, attempts were made to purify an active antibody portion of the serum away from more toxic fractions. Early efforts towards antibody purification included ammonium sulfate precipitation of human placental extracts, as well as alcohol and/or acetone treatment of animal serum. Some of these products were administered on a large scale to patient populations and while there was a decrease in anaphylactic

Immunobiology of Proteins and Peptides VIII
Edited by M. Z. Atassi and G. S. Bixler, Jr., Plenum Press, New York, 1995

123

reactions, such therapy was associated with severe pyrogenic reactions, consistent with what would be expected if these preparations were contaminated by endotoxic pyrogens.

These early forms of passive immunotherapy were soon supplanted by the discovery in the 1930's and 1940's of antibiotics with activity against common infectious diseases. The continued reactions to passive immunotherapy as well as the promise of these new antibacterial agents soon led to the eclipse of antisera therapy. During the next decade, however, Cohn and colleagues devised a simpler method of fractionating sera on a large scale based on alcohol treatment of plasma under carefully controlled conditions of pH, temperature and ionic strength. This process actually led to a gamma globulin fraction enriched in antibodies, free from widespread analyphylactic activity typical of whole serum and free of potentially infectious virus (particularly hepatitis). Indeed, Cohn showed that in his hands the gamma globulin obtained from his process was safe even for intravenous use (he gave it to himself without any incident). In subsequent studies gamma globulin was shown to be effective in preventing or attenuating the development of infectious hepatitis, polio and measles among subjects at risk following exposure. With the identification of patients who had abnormally low levels of gamma globulin, it was soon shown in a series of studies by the British Medical Research Council that the monthly administration of the immune serum globulin (ISG) obtained by the Cohn technique could reduce the development of infection. Thus, by the 1950's it was possible to obtain large amounts of gamma globulin that could be given safely, and in a number of clinical situations it was shown that the prophylactic administration of ISG was clearly efficacious. ISG was not suitable for intravenous administration because as a 16% solution it was too viscous, and move importantly IgG aggregates were associated with the activation of both complement and bradykin cascades.

With the development of techniques for the treatment of ISG that would prevent the severe reactions that followed intravenous administration of gamma globulins, there was a renewed interest in the use of immunoglobulins, not only for prophylaxis against disease but also for therapy of active disease. Initially it was demonstrated that the use of immunoglobulins for intravenous use (IVIG) in patients with hypogammaglobulinemia was as efficacious as was the use of ISG intramuscularly. In addition, the IVIG was better tolerated such that it has become the preferred form of antibody replacement. The use of IVIG to prevent infections has also been expanded to other patient sub-populations that are characterized by a functional (or physiologic) hypogammaglobulinemia. Perhaps the most intensively studied area in this regard is the administration of IVIG to premature neonates who are delivered before receiving both the full complement of the mother's antibody and before their own antibody-producing capability matures. Controlled studies of IVIG for the prevention of nosocomial infection in neonates has yielded conflicting results, with some studies demonstrating a reduction in infection not reproducible in other studies. Most recently Baker and colleagues reported a placebo-controlled study of 588 premature infants (< 1500 gms) at high risk of nosocomial infection given IVIG at a dose to maintain mean IgG levels above 400 to 500 mg/dl (1). They observed a significant reduction in documented nosocomial infections, prolongation in time to first infection and a decrease in multiple episodes of infection; however, such prophylaxis did not reduce overall morbidity or mortality. In contrast, no beneficial effect from IVIG administration results when given to full-term neonates. More recently, the prophylactic administration of IVIG was shown to benefit an additional population with functional hypogammaglobulinemia, children with AIDS (2). In a placebo controlled, double-blind study of 372 HIV-infected children, the prophylactic use of IVIG at 400 mg/kg every 28 days significantly decreased the time free from serious bacterial infections for those patients whose CD4 lymphocyte counts at entry exceeded 0.2 x 10^9 per liter. There is no data, however, demonstrating that IVIG has a similar beneficial effect in the adult population with AIDS. Finally, patients with chronic lymphocytic leukemia (CLL) are known to have a functional hypogammaglobulinemia that puts them at risk of recurrent infections. In one study in which IVIG was given to patients with CLL, a modest, but significant decrease in respiratory infections was observed (3), however, some have questioned whether this strategy of infection prevention is cost effective.

Therapeutic Uses of IVIG

There is little data to suggest that IVIG is efficacious for the treatment of established infection. With recognition during the last 20 years that the introduction of potent, broad-spectrum antimicrobial agents has had relatively little impact on the overall mortality from sepsis, the greatest efforts towards studying the potential efficacy of passive immunotherapy has been for that clinical condition. A number of studies had been performed to assess the efficacy of various non-IVIG strategies for the treatment of sepsis (Table 1). Ziegler and colleagues culminated a decade of preclinical studies with antisera to Gram-negative bacteremia and shock. This antisera, directed against an epitope in the core region of lipopolysaccharide (LPS) that was common to a wide range of Gram-negative bacteria, was generated in normal volunteers after immunization with a boiled whole bacterial vaccine. Although a beneficial effect was shown from such treatment, these investigators were unable to show that it was the antibody fraction of the antisera that conferred the protection (5). Additional studies were performed with non-IVIG antisera or plasma. In one, the investigators screened plasma for increased levels of natural antibody to a panel of Gram-negative bacterial antigens and used this plasma to treat obstetrical patients who appeared septic. Here, too, an improvement in mortality was observed; however the numbers were too small for statistical validity (6).

More recently, plasma enriched in antibody to the J5 core LPS epitope showed no ability to decrease mortality in children with meningococcal sepsis (8). All of these studies should be viewed as clinical experiments and not phase III trials of potential products. None of the preparations studied were prepared in a manner that would satisfy current standards: i.e. defined products prepared from large pools of donors in a manner that would guarantee lot-to-lot reproducibility of the therapeutic agent. Consequently, with the availability of IVIG there was a renewed interest in the use of adjunctive antibody therapy for sepsis (Table 2). Cometta and colleagues compared the ability of standard IVIG, IVIG prepared from plasma that was screened for increased levels of antibody to core LPS epitopes and placebo (albumen) for the treatment/prophylaxis of Gram-negative sepsis (12). Surprisingly, these investigators found that standard, but not core LPS antibody-enriched IVIG was effective in decreasing the incidence of lower respiratory tract infection, but had little effect on sepsis. The investigators did not explain why the core LPS antibody-enriched IVIG, which should have been identical to the standard IVIG. In contrast, one study demonstrated that use of an IVIG preparation enriched in anti-LPS core has highly beneficial therapeutic effect. This IVIG product was manufactured in a manner that resulted in an enrichment of IgM isotype antibody. Several investigators have proposed that this fraction may have more activity against Gram-negative bacteria.

Of particular interest is the study by Calandra and colleagues who proposed that if antibody to a core region in the LPS of Gram-negative bacteria (the "J5 epitope") was the active element in the earlier study by Ziegler and colleagues, then preparation of an IVIG from the plasma of volunteers immunized with a similar boiled, whole cell (J5) vaccine should be as effective as the whole sera (9). These investigators immunized donors, prepared IVIG from the plasma of these volunteers, and administered it in a placebo-controlled trial to patients who were septic. They failed to find any benefit from the J5 IVIG. There are several possible explanations for this failure of J5 IVIG: for example, they achieved only a 2-3 fold increase in anti-J5 antibody following immunization or they may have modified the antibody during the preparation of IVIG.

Table 1. Passive immunotherapy: plasma and blood products - early trials

Study	Design (No. pts)	Preparation/dose		Exptal	Control
				Mortality	
				(No. dead/Total)	
Baumgartner et al. [4]	Randomized, D-B (262)	immune plasma	shock	6/126	15/136
Lancet, 1985	Placebo-controlled (P)	1 unit q 5 d	mortality	2/126	9/136
			infection	16/126	23/136
Ziegler et al. [5]	Randomized, D-B (212)	immune serum	mortality if:		
NEJM, 1982	Placebo-controlled (T)	1 unit	bacteremia	23/103	42/109
			shock	18/41	30/39
Lachman et al. [6]	Randomized, no blind (33)	immune plasma	mortality	1/14	9/19
Lancet, 1984	no placebo (T)	3 units			
Jones et al. [7]	Randomized, (60 children)	immune plasma	mortality:		
Lancet, 1980	Controlled (P) (91 adults)		children	0/18	9/42
			adults	3/30	22/61

Note: P = prophylaxis; T = treatment; Exptal = experimental; GN = gram negative; D-B = double-blinded.

Thus, to date, there is a mixed record for IVIG in the treatment and prophylaxis in infectious diseases. One important variable may be in both the levels, and perhaps quality (of affinity) of antibodies in the IVIG preparations. Studies that failed to show significant benefit following use of standard IVIG for the treatment of infectious diseases may have had insufficient levels of antibody against the specific causative agent. Givner has documented, for example, that there is considerable lot-to-lot variation in antibody levels to group B streptococcus in IVIG made by the same manufacturer (13). This undoubtedly reflects variations in the antibody levels in the plasma donor pools used to make the IVIG. Another important variable is whether the study is designed to assess efficacy for prophylaxis or treatment of an infectious disease. In the former instance, the IVIG is active against a much smaller inoculum than is the case during active, replicating infection.

To address the possibly low, and variable levels of specific antibody in the IVIG, there has been interest in IVIG products that are enriched in antibody levels to pathogens of particular interest (hyperimmune immunoglobulins). Santosham et al prepared in ISG from volunteers who were first immunized with vaccines licensed for pneumococcus, meningo-coccus and H. influenzae, type b (14). This ISG hyperimmune to these bacterial polysaccharide antigens (BPIG) was administered on a prophylactic basis to Apache Indian children who are at increased risk from serious infection with these pathogens. These investigators found a statistically decreased mortality from infection with pneumococcus and hemophilus influenzae, type B in those who received the BPIG. This effect was observed only during the first 90 days after receipt of BPIG, when the levels of type-specific antibody were presumably sufficiently high.

In another study of interest, Shigeoka and colleagues administered whole blood to neonates septic with group B streptococcus (GBS) (15). They found that 9 of 9 neonates survived if they received units of blood that had antibody to GBS, but 3 of 6 neonates survived if the units of blood were devoid of antibody to GBS. They estimated that such transfusions replaced 40% of the blood volume of the neonates.

The studies of Santosham et al and Shigeoka et al, while both being performed with non-IVIG products for passive administration, are nonetheless instructive. They demonstrate that if there are adequate levels of antibody against the infection of interest, then a protective effect may be anticipated. The inability to date in achieving such high levels of antibody in adults may explain why it has been more difficult to demonstrate efficacy. Shigeoka's data also suggest that even if it were possible to give lots of IVIG, if levels of specific antibody

Table 2. IVIG for the treatment and prophylaxis of sepsis: Recent studies

Study	Design	No. pts	Mortality[a]		Comments
			Experimental	Control	
Calandra et al., 1988[9]	Placebo, D-B Treatment	71	15/30	20/41	No reduction in systemic complications or delay in death
De Simone et al., 1988[10]	Open, randomized Treatment	24	7/12	9/12	Faster defervescence and more frequent culture sterilization with IVIG.
Schedel et al., 1991[11]	No placebo or blind Treatment	55	1/27	9/28 (p<0.01)	IVIG enriched in IgM & IgA Anti-core LPS specificity Reduction in circulating LPS

Study	Design	No. pts		Placebo	H-IVIG	IVIG	Comments
Cometta et al., 1992[12]	3-arm D-B, placebo-controlled Prophylaxis	319	Mortality[a]: Infections: Pneumonia:	22/112 53/112 30/112	20/108 50/108 25/108	15/109 (NS) 36/109 15/109	H-IVIG from screened plasma No difference in risk or systemic infection, shock or mortality

Note: D-B = double-blind; H-IVIG = hyperimmune IVIG.
[a]Mortality = No. dead/Total.

were low, it may not be sufficient to treat an infectious disease. The study of Santosham et al might also suggest that beyond the first weeks of life, that prophylaxis of infection may be more amenable to passive immunotherapy than is treatment of active infection.

With these studies in mind our own laboratory has been involved in the development of hyperimmune immunoglobulins. In collaboration with investigators at the Swiss Serum and Vaccine Institute (SSVI), we have developed polyvalent vaccines against those serotypes of Gram negative enteric bacilli that most commonly cause bacteremia. While the presence of over 150 different O serogroups of E. coli, 80 capsular polysaccharide types of Klebsiella and 17 serogroups of Pseudomonas have made the prospect of type-specific coverage of enteric bacilli not feasible, bacteriologic surveys of clinical isolates have shown that a far more limited range of bacteremic strains causes disease. Consequently, we have prepared an IVIG that is enriched in IgG antibody to 24 capsular types of Klebsiella and to 8 serotypes of Pseudomonas. This IVIG has increased levels of type-specific antibody by ELISA, and can mediate opsonophagocytosis in vitro and protection in animal models of infection. This hyperimmune IVIG product was evaluated in a multicenter study involving hospitals in the Veterans' Administration and in the Defense Department. Patients entering intensive care units (ICU) were given one dose of either hyperimmune IVIG (100 mg/kg) or placebo (albumen) prophylactically. They were then followed for 6 weeks for the development of type-specific infection or the amelioration of infection. The study was terminated after enrolling 1500 patients because there was no decrease in overall (ie, Pseudomonas and Klebsiella) infection; however, a trend towards decrease in Klebsiella infection was apparent (16). Since one possible explanation for the lack of effect was inadequate antibody administration, a follow-up study was initiated in which the IVIG was tripled (to 300 mg/kg). This study is currently in progress with over 1500 patients enrolled. These two studies represent the largest studies done with IVIG for the evaluation of its efficacy in preventing infection.

Our laboratory has also been intrigued with the possibility that the J5 common epitope in the core of Gram negative bacilli may be an appropriate target for IVIG therapy. We hypothesized that the earlier inability of J5 IVIG to protect patients from septic complications may have resulted from the meager antibody response in the donor population to the J5 vaccine. Consequently, we prepared a similar whole cell, boiled J5 vaccine, hyperimmunized rabbits and made a purified antibody from the sera. This J5-specific antibody was used as therapy at the first sign of fever in a neutropenic rat model of Pseudomonas sepsis. We found that J5 LPS-affinity purified IgG antibody was able to afford significant protection against septic death compared to IgG prepared from pre-immune serum fraction that did not stick to the J5 LPS affinity column (ie, low in anti-J5, but high in anti-lipid A antibody) (17). Based on these studies showing that IgG antibody to J5 LPS provides heterologous protection against septic death in this animal model, we prepared a vaccine composed of detoxified J5 LPS complexed to group B meningococcal outer membrane protein (OMP). This vaccine, like the whole cell vaccine, induces anti-J5 LPS IgG that is similarly protective when given as therapy in the neutropenic model of infection. It is not clear why this anti-J5 IgG is protective in the animal model of infection whereas the J5 IVIG was not effective in the human clinical trial. In addition to the explanations offered above, the affinity of the J5 antibody in the animal studies may have been greater since the sera from which the antibody was made was from animals that were hyperimmunized with several immunization boosts after the primary series.

Hyperimmunize Immunoglobulins

While many ISGs enriched in antibody to specific pathogens or toxins (eg. tetanus and rabies immune globulin) have been available for many years, only recently have there been studies with IVIG enriched in antibody to specific pathogens. Hyperimmune IVIGs would have an advantage of having less lot-to-lot variation in antibody levels to pathogens of particular interest. Moreover, studies of IVIGs enriched in antibody to selected pathogens based on an increased antibody levels determined by screening plasma might not be as effective as IVIGs prepared from the plasma of volunteers immunized with vaccines designed to induce antibodies of interest. Antibodies induced following immunization may have higher affinities than those found in screening protocols; however, if hyperimmune IVIGs are shown to have efficacy in the treatment of specific diseases, then it would be of great importance to have methods to rapidly diagnose infections so that therapy with specific hyperimmune globulins could be rapidly initiated.

Synergy of Immunoglobulins with Antibiotics

The synergistic bactericidal effect of the combination of immunoglobulins with antibiotics was described by Alexander nearly 50 years ago. In those studies she demonstrated that the addition of sulfa antibiotics to antibody directed against H. influenzae increased the lethal dose (ie, was less virulent) nearly 100-fold. She followed those experimental studies with a clinical study of 75 babies in which there was a large reduction in septic mortality from a disease that historically had > 90% mortality (18). Similar experimental studies were conducted by Fisher who demonstrated the addition of experimental studies were conducted by Fisher who demonstrated the addition of chloramphenicol to ISG markedly enhanced bactericidal activity (19). Clinical studies by Waisbren during the 1950's suggested the possibility of an improved clinical effect from the combination of antibiotics and ISG when used as treatment for staphylococcal infections. More recently, many laboratories have reported that a mechanism for a possible synergy between antibiotics and anti-endotoxin core antibodies may be the ability of antibiotics to expose previously obscured structures in the bacterial outer membrane for binding by antibody.

SUMMARY

The availability of IVIGs for the prevention and treatment of bacterial diseases has presented may challenges. While a role for prophylaxis against infection has been suggested for some conditions characterized by hypogammaglobulinemia (or physiologic hypogammaglobulinemia), it has been difficult to demonstrate a convincing effect when IVIG is used as treatment for infectious diseases. The development of IVIGs enriched in antibody against specific pathogens may yet show therapeutic efficacy, but cost effective strategies for generating such reagents must be defined.

REFERENCES

1. Baker CJ, Melish ME, Hall RT, et al. 1992: Intravenous immune globulin for the prevention of nosocomial infection in low-birth-weight neonates. N. Engl. J. Med. 327:231-9.
2. The National Institute of Child Health and Human Development Intravenous Immunoglobulin Study Group. 1991: Intravenous immune globulin for the prevention of bacterial infections in children with symptomatic human immunodeficiency virus infection. N. Engl. J. of Med. 325:73-80.
3. Cooperative Group for the Study of Immunoglobulin in Chronic Lymphocytic Leukemia. 1988: Intravenous immunoglobulin for the prevention of infection in chronic lymphocytic leukemia, a randomized, controlled clinical trial. N. Engl. J. Med. 319:902-907.
4. Baumgartner J-D, McCutchan JA, Van Melle G, et al. 1985: Prevention of gram-negative shock and death in surgical patients by antibody to endotoxin core glycolipid. Lancet 2:59-63.

5. Ziegler EJ, McCutchan A, Fierer J, Glasuer MP, Sadoff JC, Douglas H, Braude AI. 1982: Treatment of gram-negative bacteremia and shock with human antiserum to a mutant *Escherichia coli*. N. Engl. J. Med. 307:1225-30.

6. Lachman E, Pitsoe SB, Gaffin SL. 1984: Antilipopolysaccharide immunotherapy in management of septic shock of obstetric and gynecologic origin. Lancet 1:981-3.

7. Jones RJ, Roe EA, Gupta JL. 1980: Controlled trial of pseudomonad immunoglobulin and vaccine in burn patients. Lancet 2:1263-1265.

8. J5 Study Group. 1992: Treatment of severe infectious purpura in children with human plasma from donors immunized with *Escherichia coli* J5: a prospective double-blind study. J. Infect. Dis. 165:695-701.

9. Calandra T, Glauser P, Schellekens J, Verhoef J, and the Swiss-Dutch J5 Immunoglobulin Study Group. 1988: Treatment of gram-negative septic shock with human IgG antibody to *Escherichia coli* J5: A prospective, double-blind, randomized trial, J. Inf. Dis. 158:312-318.

10. DeSimone C, Delogu G, and Corbetta G. 1988: Intravenous immunoglobulins in association with antibiotics: A therapeutic trial in septic intensive care unit patients. Crit. Care Med. 16:23-26.

11. Schedel I, Dreikhausen U, Nentwig B, Hockenschnieder M, Rauthmann D, Balikcioglu S, Coldewey R, Deicher H. 1991: Treatment of gram-negative septic shock with an immunoglobulin preparation: a prospective, randomized clinical trial. Crit. Care Med. 19:1104-1118.

12. The Intravenous Immunoglobulin Collaborative Study Group. 1992: Prophylactic intravenous administration of standard immune globulin as compared with core-lipopolysaccharide immune globulin in patients at high risk of postsurgical infection. N. Engl. J. Med. 327:234-40).

13. Givner LB. 1990: Human immunoglobulins for intravenous use: comparison of available preparations for group B streptococcal antibody levels, opsonic activity, and efficacy in animal models. Pediatrics 86:955-62.

14. Santosham M, Reid R, Ambrosino DM, et al. 1987: Prevention of *Haemophilus influenzae* type B infections in high-risk infants treated with bacterial polysaccharide immune globulin. N. Engl. J. Med. 317:923-9.

15. Shigeoka AO, Hall RT, Hill HR. 1978: Blood-transfusion in group-B streptococcal sepsis, The Lancet, p. 636-638.

16. Donta ST, Cross AS, Sadoff J, Peduzzi P. 1993: Federal Hyperimmune Immunoglobulin Study Group, Immunoprophylaxis of Klebsiella and Pseudomonas Infections. Infectious Diseases Society of American Annual Meeting. Abstract #202. Clin. Infect. Dis. 117:564.

17. Bhattacharjee AK, Opal SM, Palardy JE, Drabick JJ, Collins H, Taylor R, Cotton A, Cross AS. 1994: Affinity-purified *Escherichia coli* J5 lipopolysaccharide-specific IgG protects neutropenic rats against gram-negative bacterial sepsis, J. Inf. Dis. 170:622-629.

18. Alexander HE. 1946: Experimental basis for treatment of *Haemophilus influenzae* infections, Am. J. Dis. Child 66:172-87.

19. Fisher MW. 1957: Syngergism between human gamma globulin and chloramphenicol in the treatment of experimental bacterial infections. Antibiot. Chemo. 7:315.

16

QUALITATIVE AND QUANTITATIVE IMMUNE RESPONSE TO BACTERIAL CAPSULAR POLYSACCHARIDES AND THEIR CONJUGATES IN MOUSE AND MAN

Ali Fattom

W. W. Karakawa Microbial Pathogenesis Laboratory
Univax Biologics Inc.
Rockville, Maryland 20852

INTRODUCTION

Capsular polysaccharides (CP), antigens present on the surfaces of Gram-negative and Gram-positive bacteria that cause invasive diseases, serve as bacterial virulence factors and protective antigens. Capsules have been shown to interfere with antibody binding and complement-mediated phagocytosis. In contrast, non-encapsulated bacteria activate complement and are rapidly cleared from the blood without the need for specific antibodies [28]. CP vaccines against several different bacterial pathogens have been successfully developed and are being used in sub-groups of the population who are at risk for infection [27]. Currently, several different CP conjugate vaccines are in development and clinical evaluation for active immunization and for the generation of antibodies for passive immunization. Development of these vaccines has involved the use of different conjugation methodologies and different carrier proteins. Further development of these vaccines will eventually result in the expansion of vaccination programs to include larger populations and the use of multivalent combined vaccines requiring a *multiple* injection vaccination regimen. In this paper, an overview of CP vaccines will be presented, as will a discussion of the issues that may arise as a result of the development of new vaccines. Possible strategies to resolve such issues will also be discussed.

BACKGROUND

CP are of polymeric nature and their structures are usually simple, being comprised of repeating units containing one to a few monosaccharides. Many of these CP are acidic molecules that are negatively charged at physiological pH and which may induce a repulsive force between the bacteria and the negatively charged surface of mammalian cells. However,

Immunobiology of Proteins and Peptides VIII
Edited by M. Z. Atassi and G. S. Bixler, Jr., Plenum Press, New York, 1995

a few CP, such as pneumococcal types 14 and 7F, are neutral at physiological pH. In addition, CP may have other groups, such as acetyl groups (O- or N-linked) or pyruvates, attached to a saccharide backbone. Such side groups may also play a role in the virulence promoting nature of the CP and their immunogenicity. Moreover, some CP are homopolymers comprised of sialic acid, a molecule also present on gangliosides. Representatives of these capsules are presented in table 1.

Several mechanisms may be suggested to explain how capsules interfere with host defense mechanisms. As noted above, capsules are frequently negatively charged. The opposing negative charges of the bacterial capsule and surface of mammalian cells may block the interactions normally involved in phagocytosis [4]. Alternatively or in addition, bacterial capsules may also prevent the deposition of complement proteins (e.g., C3b) or interfere with the recognition of opsonins by Fc or C3b receptors on phagocytic cells [24, 33]. Consequently, opsonophagocytosis, a critical host defense mechanism, may be circumvented. Other capsules, such as those containing sialic acid as an immunodeterminant, may camouflage the bacteria leading the immune system of the host to consider the bacteria as self antigens [table 1].

ANTIGENICITY AND IMMUNOGENICITY OF CAPSULAR POLYSACCHARIDES

In contrast to globular proteins, CP are compositionally simple structures with relatively few immunodeterminants. Their immunodeterminants may be related to or dominated by the presence of a side chain such as O-acetyl groups. Alternatively, immunodominant determinants may be focused on and around the glycosidic bonds that link the monosaccharides [13]. The CP of S. aureus types 5 and 8 are comprised of the nearly identical monosaccharides in their repeat unit. They differ, however, in the glycosidic bonds and the site of O-acetyl groups (see table 1). In spite of the similarities in the structure of these two CP, no immunologic cross-reactivity has been observed between the type 5 and type 8 CP [11,12, 20]. In addition, S. aureus type 5 CP shares a structure in common with pneumococcal type 12 CP, while type 8 CP shows linkage similarities to pneumococcal type 4 CP. In both these cases, immunological cross-reactivity has been observed. While there is structural similarity between type III Group B streptococcus and pneumococcal type 14 [21], the absence of immunological cross-reactivity in this case is attributable to the immunological dominance of a single epitope on the type III Group B streptococcus CP. These finding indicate that the few immunodeterminants present on CP are dominated by their chemical structure and the specific side groups that are attached to the backbone.

Problematically, CP and polysaccharides in general, are poor immunogens in humans. Their immunogenicity in man is age related; children under age two years and the elderly generally respond poorly to immunization with polysaccharide vaccines [25]. There is also a relationship between immunogenicity and molecular size. Polysaccharides of lower molecular size are generally less immunogenic than those with high molecular size. Furthermore, the immune response to polysaccharides is T-independent. As a result, only a humoral response is elicited with the polysaccharide and there is typically no booster effect on antibody levels upon repeated immunizations.

The search for possible ways to potentiate the immune response to CP was initiated in the late 1920's by Geobel and Avery [14] and subsequently by Robbins and colleagues in the 1970's. By covalently binding CP to carrier proteins, the immunogenicity of the CP was shown to be enhanced [30]. Conjugation of CP or other polysaccharides to carrier proteins is now well established and has significantly advanced the development of polysaccharide-

Table 1. Representative bacterial CP and their structures

Bacterium type	Capsular polysaccharide structure	References
Staphylococcus aureus type 5	→4)-β-D-Man$p$ANAc3Ac-(1→4)-α-L-Fuc$p$NAc-(1→3)-β-D-Fuc$p$NAc-(1→	Moreau, M. *et al.* 1990
Staphylococcus aureus type 8	→3)-β-D-Man$p$ANAc4Ac-(1→3)-α-L-Fuc$p$NAc-(1→3)-β-D-Fuc$p$NAc-(1→	Fournier, J.-M. *et al.* 1984
Meningococcus type B	→8)-α-D-Neup5Ac-(2→	Bhattacharjee, A.K. *et al.* 1975
Meningococcus type C	→9)-α-D-Neup5Ac-(2→	Bhattacharjee, A.K. *et al.* 1975
Group B *Streptococcus* type III	→4)-β-D-Glc$p$-(1→6)-β-D-Glc$p$NAc-(1→3)-β-D-Gal$p$-(1→ 4 ↑ 1 α-D-Neup5Ac-(2→3)-β-D-Gal$p$	Wessels, M.R. *et al.* 1987
Pneumococcus type 14	→4)-β-D-Glc$p$-(1→6)-β-D-Glc$p$NAc-(1→3)-β-D-Gal$p$-(1→ 4 ↑ 1 β-D-Gal$p$	Lindberg, B. *et al.* 1977
Haemophilus influenzae type b	→3)-β-D-Rib$f$-(1→1)-Rib-ol-(5→O)-P-(O→	Crisel, R.M. *et al.* 1975

based vaccines [19,26,30]. These efforts culminated in the introduction and licensing of *Haemophilus influenzae b* (Hib) conjugates vaccines and continuing research in the field has resulted in the introduction of several different coupling methodologies, the use of various spacers and reaction with different functional groups on the CP [3,7,30,31].

S. AUREUS VACCINES

Production of S. aureus Vaccines

Of the 11 recently discovered *S. aureus* CP, types 5 and 8 comprise >90% of clinical isolates [17]. As is the case with other invasive bacteria, the capsule of *S. aureus* plays a role in protecting the microorganisms from phagocytosis by polymorphonuclear cells (PMN's). Addition of homologous immune sera to an *in vitro* opsonophagocytosis assay containing human PMNs and bacteria results in the enhancement of opsonophagocytosis and the killing of the bacteria. The enhancement of opsonophagocytosis was found to be capsular type specific and was mediated by homologous polyclonal, as well as monoclonal capsular specific antibodies [16].

S. aureus type 5 and type 8 capsular polysaccharides have been isolated, purified and their chemical and physical characterizations have been elucidated [11,20]. These CP were found to be of small molecular size compared to the CP of other polysaccharide encapsulated bacteria and smaller than currently licensed CP vaccines. Consequently, S. *aureus* CP were expected to be poor immunogens in humans.

Initially, *S. aureus* CP type 8 was chosen as a model for the evaluation of two different conjugation methodologies utilizing the carboxyl group on the uronic acid component of the CP. The first method used adipic acid dihydrazide (ADH) as a spacer. CP was derivatized with ADH through activation with 1-ethyl-3-(3-dimethlaminopropyl)carbodiimide (EDAC), then conjugated to the carrier protein by further addition of EDAC. For the second method, cystamine was used as the spacer. CP was activated with EDAC and subsequently derivatized with cystamine. The purified CP-cystamine derivative was reduced and conjugated to N-succinimidyl-3-(2-pyridyldithio) propionate (SPDP) derivatized protein. Figure 1 illustrates the two conjugation methodologies and the chemical structure of the spacers. Two

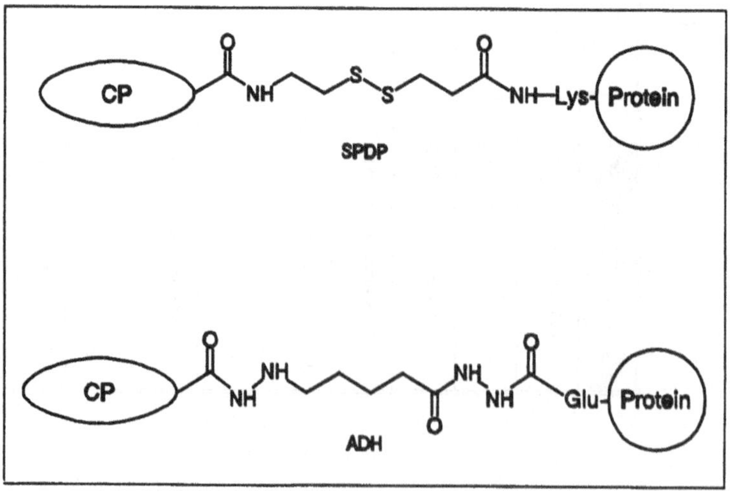

Figure 1. A scheme of the two conjugate types produced by ADH or SPDP conjugation methods.

carrier proteins were evaluated with each method, diphtheria toxoid (DT) and recombinant exoprotein A (rEPA). rEPA is a non-toxic, genetic mutant of *Pseudomonas aeruginosa* exotoxin A which is antigenically identical to the exotoxin [23]. Figure 1 also illustrates the two conjugate types produced by these methods.

Immune Response to *S. aureus* Vaccines in Mice

The immunogenicity of these conjugates was evaluated in 6-8 wk old, outbred ICR mice. Groups of ten mice were immunized subcutaneously once, twice, or three times at two week intervals with a conjugated CP dose containing 2.5ug of polysaccharide. Groups of mice were exsanguinated one week after each injection. Sera from individual mice were evaluated for CP specific antibodies in an avidin-biotin ELISA and geometric mean titers (GMT) were calculated [9]. As expected, unconjugated CP did not elicit antibodies following either one, two or three injections. A single injection of any of the conjugate also failed to elicit significant titers ($P>0.05$). However, the levels of antibodies increased dramatically following the second injection of each of the conjugates ($P<0.001$). Following a third injection , all of the conjugates elicited a detectable, but not significant, rise in antibody levels compared to levels achieved after only two injections. Moreover, a single injection of conjugate into mice previously injected with carrier protein alone elicited an immune response comparable to two injections of the conjugate [10]. These data indicate that conjugation of CP to carrier protein, regardless of the conjugation chemistry or carrier protein used, conferred on the CP properties which are associated with T-dependent antigens; i.e., a booster response and a carrier priming effect. rEPA conjugates made with the ADH as a spacer were more immunogenic than those made with the cystamine-SPDP method (244ug/ml vs 74ug/ml IgG, $P<0.001$) . This was attributed to a higher degree of cross-linking obtained using this conjugation method [9].

It has been established that the nature of an antigen is one factor which determines the isotype pattern of the immunoglobulins elicited in response to the antigen. T-cell dependent antigens usually induce an immune response with IgG_1 subclass as the major component. In contrast, T-cell independent antigens such as unconjugated CP antigens, routinely induce antibodies of the IgG_3 subclass. To determine whether the conjugation methodology affects the IgG subclass distribution, *S. aureus* type 8 CP (CP8) specific IgG antibodies, elicited by the different conjugates following the third injection, were evaluated in a sandwich ELISA using subclass specific secondary reagents [34]. Greater than 90% of the type 8 CP IgG elicited by the four different conjugates was of the IgG_1 subclass. The levels of the other IgG subclasses were so low as to be barely detectable. These data indicate that upon conjugation, CP antigens behave as T-dependent antigens, and the enhancement of their immunogenicity is apparently through the stimulation of T helper cells of the TH2 subset [15].

Within the context of the conjugation methods examined, these data show that once a conjugate is formed, the beneficial effects of coupling a CP to a carrier protein are independent of the method used. Regardless of the method, higher immune responses and T-dependent properties were achieved with conjugated CP. Variations in antibody levels observed with the different conjugates were not significant, but may have been due in part by as yet undetermined parameters relating to the variabilities inherent in the conjugation reactions involved. The data also showed a similarity among the conjugation methods in the IgG subclass distribution achieved as a measure of the quality of the immune response observed.

Evaluation of *S. aureus* Types 5 and 8 Conjugate Vaccines in Humans

The cystamine-SPDP method was used to produce several lots of vaccines to be evaluated in human volunteers. This method was chosen because both types 5 and 8 CP do not react in any of the known colorimetric assays for the quantitation of polysaccharides. This conjugation method potentially allowed direct quantitation of conjugated CP through reopening the disulfide bond in the conjugates, thus generating free CP which could be quantified by absorption after high performance liquid chromatography. rEPA was chosen as carrier instead of the more conventional diphtheria toxoid or tetanus toxoid, in an attempt to develop an additional carrier for combinational vaccines which may itself be both clinically relevant and acceptable for human studies. The conjugates were evaluated in normal, healthy, adult volunteers for safety and immunogenicity.

Both *S. aureus* type 5 and 8 r-EPA conjugates vaccines were found to be safe. No significant systemic reactions were observed and the few local reactions noted generally lasted less than 48 hours. Since native exotoxin A targets the liver, liver functions were evaluated following injection of the rEPA-containing vaccines. Liver enzymes did not rise above normal levels nor did they exceed normal variability limits at 1, 3, 5, 7, or 14 days following injection with the vaccine. It was noted that all volunteers had pre-existing *S. aureus* CP 5 and CP8 specific antibody levels ranging from 7-10ug/ml. Following a single injection of the vaccine, all volunteers injected responded with 10-20 fold increased levels of specific IgM and IgG, reaching maximum values at 6 weeks post immunization. A second injection of vaccine, however, did not elicit a further boost in antibody levels. The lack of a booster response in human adults has also been observed with other conjugate vaccines (29). IgG_1, IgG_2, or both IgG subclasses were the major specific immunoglobulin components in all volunteers. Few volunteers responded with low levels of IgG_3 or IgG_4. Six months after the second boost, only a 20% decrease in the geometric mean of specific IgG was observed. Thus, the vaccinees maintained high levels of specific serum IgG for extended periods of time. Furthermore, antibodies elicited by both vaccines were found to mediate type-specific opsonophagocytosis by human polymorphonuclear cells in an *in vitro* opsonophagocytosis assay.

S. aureus Types 5 and 8 Conjugate Vaccines in Target Populations

S. aureus is termed an opportunistic pathogen since it rarely causes disease in otherwise healthy people. Patients with even mildly immunocompromising conditions, however, are more susceptible to *S. aureus* infections and are at a overall higher risk for infection. One such group are patients with end stage renal disease (ESRD) on dialysis. In collaboration with Dr. P. Welch from Walter Reed Army Medical Center in Washington DC, we evaluated a *S. aureus* type 5-rEPA conjugate vaccine in volunteers on hemodialysis. Although the vaccine elicited a significant increase in IgG levels (16 fold) over pre-existing titers, the overall antibody response achieved in these patient volunteers was lower than that elicited by the same conjugate vaccine in healthy adults (GMT of 180ug/ml in patients compared to 318ug/ml in healthy volunteers, P<0.05) [Manuscript in preparation]. Contrary to the responses observed in healthy adult volunteers, we also observed a number of non-responders (defined as a response of < 4 fold increase in specific IgG levels over pre-vaccination levels) among the dialysis patients, and a more dramatic and rapid decline in IgG levels at six month following the second immunization (P<0.05).

Another group at risk for infections with *S. aureus* are hospitalized, low birth-weight neonates. Active immunization is not practical for this group of patients since they typically do not respond well to vaccines due to their immature immune systems. Our approach for this and other immunolgically compromised patient populations is based on the use of passive immunization. In this approach, normal, healthy, adult plasma donors are vaccinated and their plasma collected. The pooled plasma is used to produce polyclonal immunoglobulin containing high titers of specific antibodies. This immunoglobulin can then be infused into these high risk patients to provide immediate protection against infection. A bivalent CP conjugate vaccine is currently in a pilot "donor stimulation" clinical study.

Achieving high levels of both type 5 and type 8 specific IgG antibodies in plasma donors is a primary goal of the donor stimulation program. A multivalent vaccine formulation would be one potential approach to achieving antibody responses to both type 5 and type 8 in plasma donors. In addition, other vaccines could be used in the same donor population to generate multivalent human intravenous immunoglobulin (H-IVIG) products which could be used to address multiple pathogens. This combined vaccine approach would simplify immunization schedules and reduce the manufacturing costs. However, a multivalent vaccine formulation may be at disadvantage because of possible interference among the monovalent components [6].

Evaluation of Conjugate Vaccines Formulated with Adjuvants

To further enhance the immunological response to these conjugates in immunocompromised patients, to overcome possible immunological interference phenomena, and to achieve high levels of specific antibodies in plasma donors, formulation of the vaccines with adjuvants or in alternative delivery systems may prove to be advantageous. Therefore, we evaluated *S. aureus* CP conjugate vaccines and their corresponding non-conjugated CP in combination with monophosphryl lipid A (MPL), Stimulon™ (QS21), and a liposome-like delivery system (Novasomes™) in mice [Fattom, A., et. al. 1995, *Vaccine* 13: in press].

MPL, QS-21, and Novasomes™ significantly enhanced the immune response to the CP conjugates (>5 fold increase) compared to a non-adjuvanted formulation. The immunogenicity of the unconjugated CP, however, was not enhanced by any of the adjuvants or by formulation in Novasomes™. Clearly conjugation of CP to carrier protein is required to elicit an anti-CP immune response. It is also clear, however, that when used with conjugate vaccines, adjuvants can further enhance the immune response and possibly overcome some of the aforementioned issues which may obstruct the full expression of the potency of these vaccines. When the IgG subclasses in the sera from mice receiving the non-adjuvanted and adjuvanted formulations were compared, we observed that IgG_1 continued to account for the the majority of the immune response to the CP conjugates regardless of the formulation. However, significant increases in the levels of the other IgG subclasses, namely IgG_{2a} and IgG_{2b} were observed. The extent of these IgG sub-class shifts and the magnitude of the responses were comparable but not necessarily equal in all cases (MPL> QS21> Novasomes™). IgG sub-classes IgG_{2a} and IgG_{2b} are thought to have a higher capacity to fix complement than does IgG_1. As a result of increased complement fixation, these IgG_2 subclasses are thought to facilitate opsonophagocytosis of microorganisms, thus conferring higher degree of protection against infections [15]. Ultimately, the usefulness of these and other adjuvants in humans and the significance of their effect on the subclass composition of human immune response needs to be investigated. Results expected from planned clinical trials with these adjuvants will be great value in determining the possible use of the adjuvants especially in immunocompromised patients.

In conclusion, the development of CP conjugate vaccines shows great promise for the prevention of bacterial infections. Through the work of a large number of investigators over the years, a great deal has been learned about capsular polysaccharides and their role in bacterial pathogenesis. Continued studies of such vaccines and the immunoglobulins induced by them are addressing pathogens such as *S. aureus*, *S. epidermidis*, and the enterococci which are of growing concern due to their prevalence, especially among nosocomial infections, and their growing resistance to other antimicrobial agents.

ACKNOWLEDGMENT

I am grateful to Garvin Bixler, Robert Naso, and Sara Shepherd, for assisting in the preparation and reviewing this manuscript.

REFERENCES

1. Bhattacharjee, A.K., Jennings, H.J., Kenny, C.P., Martin, A., Smith, I.C.P., 1975, Structural determination of the sialic acid polysaccharide antigens of *Neisseria meningitidis* serogroup B and serogroup C with carbon 13 nuclear magnetic resonance. J. Biol. Chem. 250:1926-1932.
2. Bishop, C. T., and Jennings, H.J., 1982, Immunology of polysaccharides. *Polysaccharides* 1: 291-230.
3. Carlsson, J., Drevin, H. and Axen, R., 1978, Protein thiolation and reversible protein-protein conjugation. N-succinimidyl 3- (2-pyridyldithio) propionate, a new heterofunctional reagent. *Biochem. J.* 173:723-737.
4. Chu, C.Y., Schneerson, R., Robbins, J. B. and Rastogi, S. C., 1983, Further studies on the immunogenicity of *Haemophilus influenza* type b and pneumococcal type 6A polysaccharide-protein conjugates. Infect. Immun. 40:245-256.
5. Crisel, R.M., Baker, R.S., and Dorman, D.E., 1975, Capsular polymer of *Heamophilus influenzae* type b. Part 1 Structural characterisation of the capsualr polymer of the strain Egan. *J. Biol. Chem.* 250:4926-4930.
6. Cross, A., Artenstein, A., Que, J., Fredeking, T., Furer, E., Sadoff, J.C., and Cryz, S.J., Jr., 1994, Safety and immunogenicity of a polyvalent *Escherichia coli* vaccine in human volunteers. *J. Infect. Dis.* 170:834-840.
7. Cryz, S. J., Jr., Sandoff, J.C. and Furer, E., 1988, Immunization with a *Pseudomonas aeruginosa* immunotype 5 O-polysaccharide-toxin A conjugate vaccine: Effect of a booster dose on antibody levels in humans. *Infect. Immun.* 56:1829-1830.
8. Fattom, A., Schneerson, R., Karakawa, W.W., Fitzgerald, D., Pastan, I., Li, X., Shiloach, J., Bryla, D.A. and Robbins, J.B., 1993, Laboratory and Clinical evaluation of conjugate vaccines composed of *Staph aureus* types 5 and 8 capsular polysaccharides bound to *Pseudomonas aeruginosa* recombinant exoprotein A. *Infect. Immun.* 61:1023-1032.
9. Fattom, A., Shiloach, J., Bryla, D., Fitzgerald, D., Pastan, I., Karakawa, W., Robbins, J.B. and Schneerson, R., 1992, Comparative immunogenicity of conjugates composed of *S. aureus* type 8 capsular polysaccharide bound to carrier proteins by adipic acid dihydrazide or N-succinimidyl-3-(2-pyridyldithio) propionate. *Infect. Immun.* 60:584-589 (15)
10. Fattom, A., Schneerson, R., Szu, S.C., Vann, W.F., Shiloach, J. and Robbins, J.B., 1990, Synthesis and immunologic properties in mice of vaccines composed of *S. aureus* type 5 and type 8 capsular polysaccarides conjugated to *Pseudomonas aeruginosa* exotoxin A. *Infect. Immun.* 58:2367-2374.
11. Fournier, J-M., Hannon, K., Moreau, M., Karakawa, W.W., and Vann, W.F., 1987, Isolation of type 5 capsular polysaccharide from *Staphylococcus aureus*. *Ann. Inst. Pasteur/Microbiol.* 138:561-567.
12. Fournier, J-M., Vann, W.F., and Karakawa, W.W., 1984, Purification and characterization of *Staphylocccus aureus* type 8 capsular polysaccharide. *Infect. Immun.* 45:87-93.
13. Glode, M.P., Lewin E.B., Sutton, A., Le, C.T., Gotschlich, E.C., and Robbins, J.B., 1979, Comparative immunogenicity of vaccines prepared from capsular polysaccharides of group C *Neisseria meningitidis* o-acetyl-positive and o-acetyl-negative variants and *Escherichia coli* K92 in adult volunteers. *J. Infect. Dis.* 139:52-59.
14. Goebel, W. F., 1929, Chemo-immunological studies on conjugated carbohydrate- proteins. XII. The immunological properties of an artificial antigen containing cellobiuronic acid. *J. Exp. Med.* 50:469-520.
15. Golding B., 1991, Cytokine regulation of humoral immune response. In: *Topics in Vaccine Adjuvant Research* (Eds Spriggs, D. R. and Koff, W. C.) CRC Press, Boca Raton, Ann Arbor, Boston, pp:25-37.
16. Karakawa, W.W., Sutton, A., Schneerson, R., Karpas, A., and Vann, W.F., 1988, Capsular antibodies induce type-specific phagocytosis of capsulated *Staphylococcus aureus* by human polymorphonuclear leukocytes. *Infect. Immun.* 56,1090-1095.

17. Karakawa, W.W., Fournier, J-M., Vann, W.F., Arbeit, R., Schneerson, R.S., and Robbins, J.B., 1985, Method for the serological typing of the capsular polysaccharides of *Staphylococcus aureus. J. Clin Micro.* 22:445-447.

18. Kenne, L., and Lindbergh, B., 1984, Bacterial polysaccharides. *Polysaccharides* 2:262-290.

19. Manclark,C. R., Meade, B. D. and Burstyn, D. G., 1986, Serological response to *Bordetella pertussis*. In: *Manual of Clinical Laboratory Immunology, 3rd ed.* (Eds Rose, N. R., Friedman, H., and Fahey, J. L.) American Society for Microbiology, Washington, D.C. pp:388-394.

20. Moreau, M., Richards, J.C., Fournier, J-M., Byrd, R.A., Karakawa, W.W., and Vann, W.F., 1991, Structure of the type 5 capsular polysaccharide of *Staphylococcus aureus. Carb. Res.* 201:285-297.

21. Lee, C-J., 1987, Bacterial capsular polysaccharides-biochemistry, immunity and vaccine. *Molecular Immunol.* 24:1005-1019.

22. Linberg, B., Lonngren, J., and Powell, D.A., 1977, Structural studies on the specific type 14 pneumococcal polysaccharide. *Carbohydr. Res.* 58: 177-186.

23. Lucak, M., Pier, G. B. and Collier, R. J., 1988, Toxoid of *P. aeruginosa* exotoxin A generated by deletion of an active-site residue. *Infect. Immun.* 56:3095-3098.

24. Peterson, P. K., Kim, Y., Wilkinson, B. J., Schmeling, D., Michael, A.F., and Quie, P. G., 1978, Dichotomy between opsonization and serum complement activation by encapsulated staphylococci. *Infect. Immun.* 20:770-775.

25. Reingold , A. L., Broome, C. V., Hightower, A. W., Ajello, G. W., Bolan, G. A., Adamsbaum, C., Jones, E. E., Phillips, C., Tiendrebeogo, H. and Yada, A., 1985, Age specific differences in duration of clinical protection after vaccination with meningococcal polysaccharide A vaccine. *Lancet* ii:114-118.

26. Robbins, J.B. and Schneerson, R., 1990, Polysaccharide-protein conjugates: A new generation of vaccines. *J. Infect. Dis.* 161:821-832.

27. Robbins, J.B., Schneerson, R., Egan, W. B., Vann, W. F. and Liu, D. T., 1980, Virulence properties of bacterial capsular polysaccharides-unanswered questions. In: *The Molecular Basis of Microbial Pathogenecity* (Eds Smith, H., Skehel, J. J. and Turner, M. J.) Dahlem Konferenzen, Verlag Chemie GmbH, Weinheim, Ferderal Republic of Germany, pp: 115-132.

28. Robbins, J. B., 1978, Vaccines for the prevention of encapsulated bacterial diseases: Current status, problems and prospects for the future. *Immunochemistry* 15:839-854.

29. Schneerson, R., Robbins, J. B., Parker, J. C. Jr., Sutton, A., Wang, Z., Schlesselman, J.J., Schiffman, G., Bell, C., Karpas, A. and Hardegree, M. C., 1986, Quantitative and qualitative analyses of serum *Haemophilus influenzae* type b, pneumococcus type 6A and tetanus toxin antibodies elicited by polysaccharide-protein conjugates in adult volunteers. *Infect. Immun.* 52:519-528.

30. Schneerson, R., Barrera, O., Sutton, A. and Robbins, J. B., 1980, Preparation, characterization and immunogenicity of *Haemophilus influenzae* type b polysaccharide-protein conjugates. *J. Exp. Med.* 152:361-376.

31. Szu, S.C., Stone, A. I., Robbins, J. D., Schneerson, R. and Robbins, J. B., 1987, Vi capsular polysaccharide-protein conjugates for prevention of typhoid fever. *J. Exp. Med.* 166:1510-1524.

32. Wessels, M.R., Pozsgay, V., Kasper, D.L., and Jennings, H.J., 1987, Structure and immunochemistry of an oligosaccharide repeating unit of the capsular polysaccharide of the type III group B Streptococcus. *J. Biol. Chem.* 262:8262-8267.

33. Wilkinson, B. J., Sisson, S. P., Kim, Y., and Peterson, P. K., 1979, Localization of the third component of complement on the cell wall of encapsulated *S. aureus* M: Implications for the mechanism of resistance to phagocytosis. *Infect. Immun.* 26:1159-1163.

34. Zollinger, D. W. and Boslego, J. W., 1981, A general approach to standardization of the solid-phase radioimmunoassay for quantitation of class-specific antibodies. *J.Immunol. Methods.* 46:129-140.

17

AUTOIMMUNE RECOGNITION OF ACETYLCHOLINE RECEPTOR AND MANIPULATION OF THE AUTOIMMUNE RESPONSES BY SYNTHETIC PEPTIDES

M. Zouhair Atassi and Minako Oshima

Department of Biochemistry
Baylor College of Medicine
Houston, Texas 77030

INTRODUCTION

A major activity of this laboratory has been focused on the determination and synthesis of the functional sites and regions of immune and autoimmune recognition of protein receptors and the use of the synthetic peptides for manipulation of receptor function. These studies have included insulin receptor (1,2), thyroid-stimulating hormone receptor (3-5) and acetylcholine receptor (AChR). This paper describes some of our studies on the autoimmune recognition of AChR and the use of synthetic peptides corresponding to antibody and/or T-cell autoimmune recognition regions on AChR.

The nicotinic AChR is a membrane protein which has a principal role in post-synaptic neuromuscular transmission because it mediates ion flux across the cell membrane in response to binding of acetylcholine (ACh) (6,7). The receptor is a pentamer comprising four subunits ($\alpha_2\beta\gamma\delta$). The primary structures of the four AChR subunits of *Torpedo californica* (*t*) (8-11) and mouse (*m*) (12-15) and the α-subunits of human (*h*) and bovine (16) have been deduced from the respective cDNA or mRNA sequences. Most functional studies have focused on the α-subunit because it is responsible for binding ACh (17,18) and α-neurotoxin (19,20). Studies from this laboratory have enabled the localization of the full profile of the binding regions for long [α-bungarotoxin (BTX) and cobratoxin] and for short (erabutoxin and cobratoxin) neurotoxins on the α-subunit of *t*AChR and *h*AChR (21-25). Conversely, the binding regions for AChR on BTX were mapped by synthetic peptides representing each of the toxin loops (26,27). Identification of the binding regions on BTX for AChR and on the receptor for the toxin allowed the construction of a 3-D model of the binding-site cavity for the toxin on AChR (28) and provided a molecular explanation for the observed differences between long and short α-neurotoxins in their actions on AChR (25).

From the primary structure of each AChR subunit, it was possible to identify transmembrane hydrophobic regions and the extracellular parts of the chain (10,29,30).

Immunobiology of Proteins and Peptides VIII
Edited by M. Z. Atassi and G. S. Bixler, Jr., Plenum Press, New York, 1995

Immunological and toxin-binding studies with intertransmembrane synthetic peptides confirmed (31) the model postulating five transmembrane regions (29,30).

Thus, we now obtained a plan for the transmembrane organization of the AChR subunits as well as a working 3-D model for the α-neurotoxin and the ACh binding cavity of AChR. The secondary structure of the AChR subunits was estimated from the sequence by standard algorithms (10). But this proposed secondary structure lacks experimental verification and there is no information about the folding of the extracellular part and whether insertion of the receptor in the membrane and on a live cell affects this folding. To study the organization of the polypeptide chain of the main extracellular domain of the *t*AChR α-subunit, we examined the ability of free *t*AChR and membrane-bound (*mb*) AChR fractions to bind a panel of antibodies against overlapping synthetic peptides which encompassed the entire domain (32,33). These binding studies, which enabled a comparison of the accessible regions in *mb*AChR and free AChR, revealed that the receptor undergoes changes in conformation upon removal from the cell membrane (33). Very recently, our experiments were designed (34) to give a functional view of exposed regions on the extracellular part of the α-chain of AChR in a *live* muscle cell line. These studies revealed that receptor conformation is somewhat different in the membrane of a *live* cell relative to that of AChR in the membrane fraction. Mapping of the accessible regions on the extracellular part of the AChR subunits should provide a structure-function correlation for autodeterminant recognition. Furthermore, knowledge of the surface areas should permit the investigation of the role played by antibodies against various accessible regions in inducing myasthenia gravis (MG).

Autoimmune Recognition Of Acetylcholine Receptor IN Myasthenia Gravis

Myasthenia gravis is a disabling autoimmune disease in which autoantibodies are produced against AChR and inhibit its regulatory activity (35-37). Sera of mg patients contain autoantibodies which compete with cholinergic agents for binding to the α-subunit (38,39). By employing eighteen synthetic overlapping peptides encompassing the entire extracellular part (residues α1-210) of the α-chain of *h*AChR and a 19th peptide (residues α262-276) corresponding to an extracellular connection between two transmembrane regions (Figure 1), we have found (40) that antibodies in sera of mg patients recognize only a limited number of the synthetic peptides. The regions recognized resided predominately within the areas of *h*α10-30, *h*α111-145 and *h*α175-198 and, less frequently, *h*α45-77. Differences in the recognition profile of peptides from patient to patient indicated that the autoantibody responses were under genetic control. However, by using a mixture of the appropriate peptides, it was possible to determine autoantibodies in all 15 myasthenia sera and to distinguish between these, normal human sera and other neurological or autoimmune diseases. The mapping of the continuous regions recognized by autoantibodies on the α-chain of *h*AChR has permitted a comparison of the regions recognized by autoantibodies (40) and autoimmune T-cells (41) from the same donor. It also provided a peptide-based direct antibody binding method for diagnosis of MG (40) which enabled the determination of the autoantibodies, even when mg sera gave a false negative by the standard method using ^{125}I-labeled BTX-receptor complex (38,39,42). A major toxin binding region was found (22-24) to reside within the peptide *h*α122-138 and two minor regions occur within peptides *h*α34-49 and *h*α194-210. The finding that peptide *h*α122-138 also contains an autoantigenic region in most MG sera would explain (40) the false negatives obtained with the method that relies on the precipitation by autoantisera of the ^{125}I-BTX-AChR complex (42) because

```
Peptide
Position   Species        Structure

α1-16      Human      S E H E T R L V A K L F K D Y S
           Torpedo    - - - - - - - - N - L S N - M

α12-27     Human      F K D Y S S V V R P V E D E R Q
           Torpedo    L E N - N K - I - - - - N - T N

α23-38     Human      E D E R Q V V E V T V G L Q L I
           Torpedo    - N - T N F - D I - - - - - - -

α34-49     Human      G L Q L I Q L I N V D E V N Q I
           Torpedo    - - - - - - - - S - - - - - - -

α45-60     Human      E V N Q I V T T N V R L K Q Q V
           Torpedo    - - - - - - E - - - - - R - - -

α56-71     Human      L K Q Q W V D Y N L K W N P D D
           Torpedo    - R - - - I - V R - R - - - A -

α67-82     Human      W N P D D Y G G V K K I N I P S
           Torpedo    - - - A - - - - I - - - R L - -

α78-93     Human      I N I P S E K I W R P D L V L Y
           Torpedo    - R L - - D D V - L - - - - - -

α89-104    Human      D L V L Y N N A D G D F A I V K
           Torpedo    - - - - - - - - - - - - - - - N

α100-115   Human      F A I V K F T K V L L Q Y T G N
           Torpedo    - - - - N M - - L - - D - - - K

α111-126   Human      Q Y T G N I T W T P P A I F K S
           Torpedo    D - - - K - M - - - - - - - - -

α122-138   Human      A I F K S Y G E I I V T N F P F D
           Torpedo    - - - - - - - - - - - - - - - - -

α134-150   Human      N F P F D E Q N G S M K L G T W T
           Torpedo    - - - - - Q - - - - T - - - - I - -

α146-162   Human      L G T W T Y D G S V V A I N P S
           Torpedo    - - I - - - - - T K - S - S - - -

α158-174   Human      I N P S S D Q P D L S N F N E S G
           Torpedo    - S - - - - R - - - - T - - - - -

α170-186   Human      F N E S G E W V I K E S R G W K N
           Torpedo    - - - - - - - - M - D Y - - - - -

α182-198   Human      R G W K N S V T Y S G G P D T P Y
           Torpedo    - - - - - W - Y - T - - - - - - -

α194-210   Human      P D T P Y L D I T Y H F V M Q R L
           Torpedo    - - - - - - - - - - - - I - - - I

α262-276   Human      E L I P S T S S A V P L I G K
           Torpedo    - - - - - - - - I - - - - - -
```

Figure 1. Covalent structures of the synthetic overlapping peptides representing the main extracellular part (residues 1-210) of the α-chain of hAChR and a 19th peptide (residues α267-276) corresponding to an extracellular connection between two transmembrane regions. Under the hAChR peptides, are shown the sequence differences in the corresponding regions of tAChR α chain. Segments in boldface represent the 5-residue overlaps between consecutive peptides.

the binding of BTX to the region would block the binding of the antibody to the same region. The region hα122-138 carries contact residues of the ACh (18). The finding that autoantibodies also bind to this region provides a molecular explanation for dysfunction of AChR in mg (40). In order to define the epitopes of tAChR that are recognized by AChR-specific T-cells, we generated T-cell populations and T-cell hybridoma clones from tAChR-immunized C57BL/6(B6) mice and tested their reactivity to the synthetic uniform-sized overlapping peptides representing the entire extracellular portion of the α-chain of tAChR (43). We found that the reactivity of the T-cell clones and the parent lines was directed against peptides tα111-126, tα146-162 and tα182-198, with the immunodominant response being against tα146-162 (43). This data is consistent with a highly limited recognition of AChR determinants in murine experimental autoimmune MG (EAMG) tAChR-specific T-cell (43).

The localization of a specific region of AChR that is recognized immunodominantly by antigen-specific T-cells represented an advance toward the modification of the autoimmune process in a antigen-specific manner. This was in fact achieved and those studies are described below.

MHC Control of EAMG

Myasthenia gravis has been associated with certain HLA antigens (44,45). As already mentioned, immunization of mice with *t*AChR produces EAMG. EAMG susceptibility has been mapped to the I-A subregion of the mouse major histocompatibility complex (mhc) (46-48). The development of EAMG is primarily influenced by the class-II genes (46-48). Class II-restricted AChR-reactive Th cells are activated in MG patients and appear to contribute to the postsynaptic pathology (41,43,49,50).

Profile of the α-Chain Regions Recognized by Antibodies and by T-cells in EAMG Susceptible and Non-susceptible Mouse Strains after Different Periods of Immunization. The T-cell recognition profiles of *t*AChR α-chains were initially determined at 1 week after immunization in B6, SJL and other mouse strains (43,50). On the other hand, B6 mice develop EAMG after a second immunization (at week 5) or more with *t*AChR. To determine whether EAMG is related to recognition of particular region(s) on the main extracellular domain of the α-chain (residues α1-210) in prolonged immunization, we have examined the differences in the antibody and T-cell recognition profiles of B6 and SJL (a strain that does not develop EAMG) mice after different periods and a number of immunizations with *t*AChR (51). In a given strain, antibodies and T-cells recognized immunodominant regions, which may coincide or maybe uniquely B-cell or T-cell determinants. Both B6 and SJL exhibited similar antibody recognition profiles after the 2nd and through the 4th immunization with *t*AChR. Major differences between the two strains were found (51) in their T-cell recognition of regions in the second part (residues *t*α100-210) of the main extracellular domain of the α-chain (Figures 2 and 3). T-cells of SJL recognized consistently only one region (*t*α111-126) within this part of the α-chain, whereas in B6, T-cell recognition of three peptides (*t*α111-126, *t*α146-162 and *t*α182-198) and next neighbor regions to them persisted throughout the period (51). Of these three peptides, *t*α146-162, was immunodominant peptide unique to B6, as the other two peptides (*t*α111-126 and *t*α182-198) were also recognized by either T-cells or antibodies in SJL.

To study the role of T-cells recognizing region *t*α146-162 in EAMG, a T-cell line was generated against this region and the cells transferred into B6 mice followed by one *t*AChR injection (51). The antibody responses to peptides *t*α12-27, *t*α111-126 appeared in these mice and the response to peptide *t*α134-150 increased substantially as an influence of T-cell transfer compared to antibody profiles at 1 week. In addition, one out of three mice examined showed signs of EAMG. These results suggest the importance of T-cells recognizing residues *t*α146-162 in EAMG. It is concluded (51) that the presence of persistent T-cell responses to the second half (residues *t*α100-210) of the main extracellular domain of the α-chain is associated with the development of EAMG in B6 mice, while absence of these responses in SJL mice may enable them to escape the disease (51). The preservation of the immunodominance of peptide *t*α146-162 in the T-cell recognition of B6 is probably most important for the pathogenesis of EAMG in this strain (51).

Comparison of the Responses of B6 and bm12 Mouse Strains. A gene conversion event between I-Eβ^b and I-Aβ^b in the B6.C-H-2^{bm12} (bml2) strain (52,53), which altered three amino acids in the C-terminal half of the first domain of I-Aβ^b (Ile-67 → Phe; Arg-70 → Gln; Thr-71 → Lys) (54), resulted in resistance (of bml2) to EAMG development (46) and lower cellular and humoral immune responses to *t*AChR. Very recently, we studied the effect of the gene conversion at I-Aβ^b positions 67, 70, 71 on the T-cell responses to epitopes of

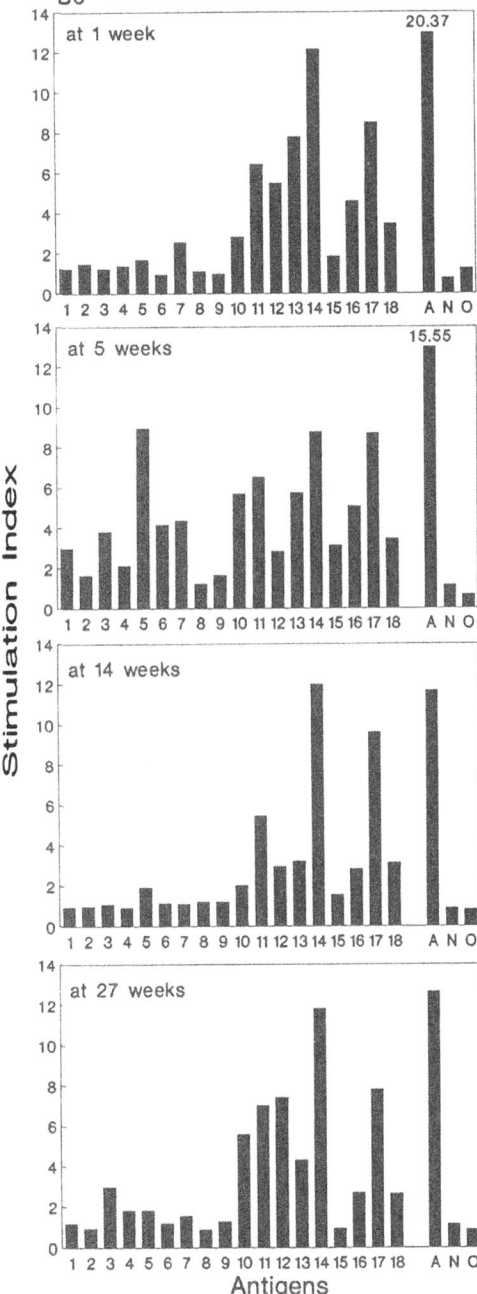

Figure 2. A bar diagram summarizing the *in vitro* proliferative responses to the synthetic *t*AChR peptides of LNC from B6 mice after different periods of immunization and numbers of injections with *t*AChR. In each experiment, LNC (5×10^5 cells/well) were challenged with various doses of the *t*AChR peptides (10-40 μg/ml) or *t*AChR (0.4-12 μg/ml). Ovalbumin, lysozyme (100 μg/ml) and unrelated peptide (10-40 μg/ml) were used as negative controls. Con A (1 μg/ml) and LPS (500 μg/ml) were used as positive controls. The following values are cpm for unstimulated cells and negative controls (in parenthesis): 1 week, 6637 (6570); 5 weeks, 7419 (4451); 14 weeks, 4901 (4067); 27 weeks, 3070 (2670). The diagram shows the stimulation index at the optimum challenge dose of each peptide: (1) peptide 1-16, (2) peptide 12-27, (3) peptide 23-38, (4) peptide 34-49, (5) peptide 45-60, (6) peptide 56-71, (7) peptide 67-82, (8) peptide 78-93, (9) peptide 89-104, (10) peptide 100-115, (11) peptide 111-126, (12) peptide 122-138, (13) peptide 134-150, (14) peptide 146-162, (15) peptide 158-174, (16) peptide 170-186, (17) peptide 182-198, (18) peptide 194-210. Additional antigen letter symbols are: A, *t*AChR; N, unrelated (to AChR) nonsense peptide; O, ovalbumin. (Figure is from ref. 51).

*t*AChR α-subunit in B6 and bml2 mice (55). The mice were primed with *t*AChR, and the profiles of T-lymphocyte proliferation were determined (55) with synthetic overlapping peptides encompassing the entire extracellular portion of the *t*AChR α-subunit (Figure 1). The proliferative responses of *t*AChR-primed bml2 lymphocytes (Figure 4) were markedly reduced in two (*t*α146-162 and *t*α182-198) of the three *t*AChR peptides (*t*α111-126, *t*α146-162, and *t*α182-198) that are immunodominant in B6 mice (43,50). Thus, the I-Aβ[b]

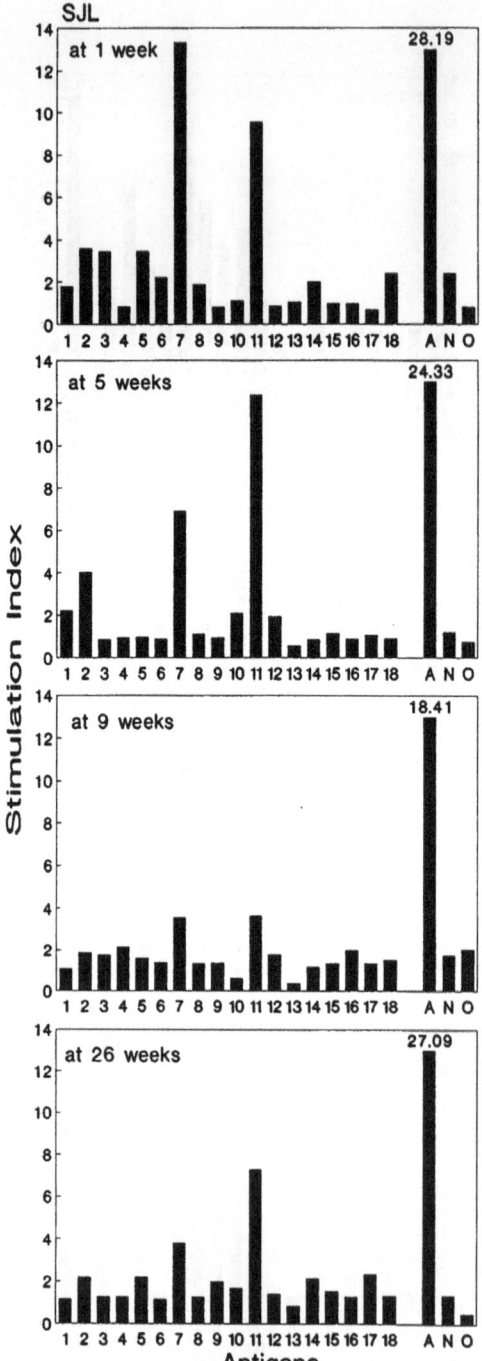

Figure 3. Presentation showing the changes with time after immunization of the *in vitro* proliferative responses to the synthetic *t*AChR peptides of LNC from SJL mice. The LNC were obtained from mice after different periods of immunization and numbers of injections with *t*AChR. Details are as in Figure 2, except that the cell numbers per well used in the assays were as follows: 1 week, 6.5×10^5; 5 weeks, 6.5×10^5; 9 weeks, 8×10^5; 26 weeks, 5×10^5. Results are expressed in stimulation index. Unstimulated cells and negative controls (in parenthesis) gave following cpm values: 1 week, 1081 (951); 5 weeks, 7187 (6576); 9 weeks, 6075 (5179); 26 weeks, 6096 (3091). Numbers and symbols of antigens are as in Figure 2. (Figure is from ref. 51).

residues encompassing the region 67-71 determine the immunogenicity of two of the AChR α-subunit T-cell epitopes in *t*α146-162 and *t*α182-198. In determinant selection, the Ia molecule is believed to specifically bind some, but not all peptide antigens during presentation, thereby selecting the determinants that are to be presented by the antigen presenting cells (APC) (56,57). Therefore, the reduced lymphocyte proliferative response seen in bml2 to peptide *t*α146-162 and *t*α182-198 may be due to inefficient I-A^{bml2} binding and/or

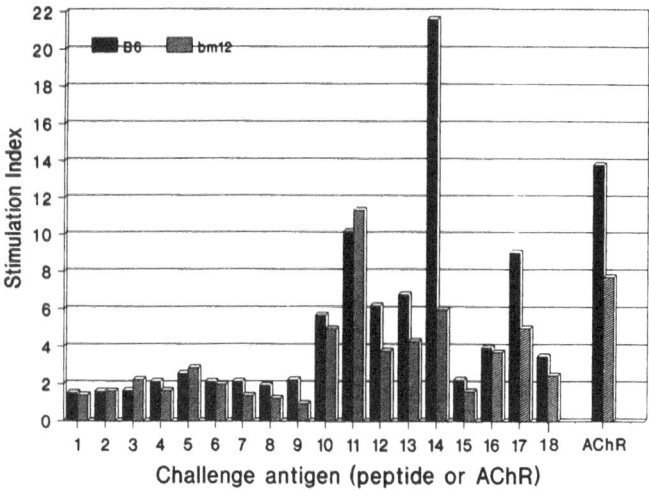

Figure 4. Peptide recognition profiles of B6 and bm12 AChR-specific T cells. Proliferative responses to the synthetic tAChR α-chain peptides of LNC from B6 and bm12 mice. Unstimulated cells gave the following cpm: B6, 6208; bm12, 6872. The diagram shows the stimulation index at the optimum challenge dose of each peptide. Peptide numbers are as in Figure 2. Results were expressed as stimulation index (stimulation index = mean cpm incorporated by stimulated cells / mean cpm incorporated by unstimulated cells). (Plotted from data in Shenoy *et al.*, ref.55).

expression of $A\beta^{bm12}$:$A\alpha^b$, by directly decreasing the efficiency of α:β heterodimer formation and/or surface membrane expression (58), and this might contribute to the suppressed response to the dominant epitopes.

Binding of AChR α-chain peptide to I-Aβ peptides

As mentioned in the preceding section, a mutant mouse of B6, $B6.C-H2^{bm12}$ (bm12) (52,53) which has 3 amino acid changes (at residues 67, 70 and 71) in the I-Aβ^b subunit of MHC class II molecules (54) is resistant to EAMG development (46). This mutant is, therefore, useful for clarifying the mechanism of EAMG pathogenesis. These data suggested that non-susceptibility of bm12 mice to EAMG might be due to the reduced T cell response against peptide $t\alpha$146-162. Helper T cells recognize the antigen (peptide) in the context of MHC class II molecules. The three amino acid substitutions between the I-Aβ^b and I-A^{bm12} are located within one of the two top α-helices (antigen-binding groove) in MHC class II molecule (59). In this location, even a single amino acid substitution may alter the shape of peptide-binding groove to change its specificity. The reduced response in bm12 to peptide $t\alpha$146-162, relative to the parental B6 strain, could be affected by a lower affinity of this peptide to I-A^{bm12} molecule.

Recent studies (60) were carried out to investigate whether the reduced T cell response in bm12 to peptide $t\alpha$146-162 is a result of low binding of this peptide to I-A^{bm12} and, at the same time, to assess which amino acid(s) replacement plays an importnat role in this reduction. We synthesized region 62-76 of I-Aβ^b (peptide b6) and I-Aβ^{m12}(peptide bm), and three additional peptides of the same region in which the amino acids at positions 67, 70 or 71 were altered one at a time to that of peptide bm (Figure 5), and investigated the binding affinities of these five peptides (60) to tAChR peptide $t\alpha$146-162. In order to verify that the binding of MHC peptides to antigenic peptides reflects the binding of antigenic peptides to intact MHC II molecules, we also conducted (60) a similar binding

Peptide			Sequence	
Abbreviation	Full name	62		76
b6	I-Aβ^{b}62-76	Asn-Ser-Gln-Pro-Glu-Ile-Leu-Glu-Arg-Thr-Arg-Ala-Glu-Leu-Asp		
b6(67F)	I-Aβ^{b}62-76(67F)	Asn-Ser-Gln-Pro-Glu-**Phe**-Leu-Glu-Arg-Thr-Arg-Ala-Glu-Leu-Asp		
b6(70Q)	I-Aβ^{b}62-76(70Q)	Asn-Ser-Gln-Pro-Glu-Ile-Leu-Glu-**Gln**-Thr-Arg-Ala-Glu-Leu-Asp		
b6(71K)	I-Aβ^{b}62-76(71K)	Asn-Ser-Gln-Pro-Glu-Ile-Leu-Glu-Arg-**Lys**-Arg-Ala-Glu-Leu-Asp		
bm	I-Aβ^{bm12}62-76	Asn-Ser-Gln-Pro-Glu-**Phe**-Leu-Glu-**Gln**-**Lys**-Arg-Ala-Glu-Leu-Asp		

Figure 5. Amino acid sequences of the synthetic I-Aβ peptides and analogs. Letters in bold represent the residues that are different from the I-Aβ^b sequence in region 62-76. (Figure is from ref. 60).

investigated the binding affinities of these five peptides (60) to tAChR peptide $t\alpha$146-162. In order to verify that the binding of MHC peptides to antigenic peptides reflects the binding of antigenic peptides to intact MHC II molecules, we also conducted (60) a similar binding study of peptide $t\alpha$146-162 to MHC molecules, I-A^b and I-A^{bm12}. Secondly, we compared the T cell response to B6 and bm12 mice when peptide $t\alpha$146-162 alone was used as an immunogen (60) to evaluate if low recognition of peptide still occurs in bm12 mice without competition from other peptide regions on tAChR. Finally, we cross-examined the capability of B6 and bm12 spleen cells to present peptide $t\alpha$146-162, tAChR to peptide-, or tAChR-specific T-cell lines derived from B6 and bm12 mice (60) to determine whether differences in the aforementioned affinities correlated with the presentation capability of the respective APC.

Peptide $t\alpha$146-162 bound with a significantly higher affinity to peptide b6 than to peptides bm or b6(71K) (Table 1), suggesting that the lower affinity of peptide $t\alpha$146-162 to I-A^{bm12} is a factor in the reduced response to this peptide by bm12 T cells. This was confirmed by measurement of the K_d values of the binding of peptide $t\alpha$146-162 to the I-A molecules of B6 and bm12. The average K_d values from independent assays were: I-A^b, $8.4 \pm 2.1 \times 10^{-7}$ [M]; I-A^{bm12} $1.75 \pm 0.61 \times 10^{-6}$ [M] (Table 1). Thus, peptide $t\alpha$146-162 bound to I-A^{bm12} with an affinity that was 2 times lower than its affinity to I-A^b (60). Furthermore, APC of bm12 presented the peptide, or tAChR, poorly to peptide-specific (Figure 6) or to tAChR-specific B6 T cells. The major effect is caused by

Table 1. Dissociation constants (K_d) of the binding of ^{125}I-labeled peptide $t\alpha$146-162 to the mouse I-Aβ peptides, analogs, and I-A molecules

Peptide abbreviation	Full name	K_d	$\pm$ SD	[M]	K_a	[M^{-1}]
Peptides						
b6	I-Aβ^{b}62-76	3.6	± 0.3	x 10^{-7}	2.8	x 10^6
b6(67F)	I-Aβ^{b}62-76(67F)	5.9	± 0.4		1.7	
b6(70Q)	I-Aβ^{b}62-76(70Q)	8.5	± 0.7		1.2	
b6(71K)	I-Aβ^{b}62-76(71K)	13.8	± 2.9		0.72	
bm	I-Aβ^{bm12}62-76	14.8	± 3.4		0.68	
I-A molecules						
I-A^b	—	8.4	± 2.1	x 10^{-7}	1.19	x 10^6
I-A^{bm12}	—	1.75	± 0.61	x 10^{-6}	0.57	x 10^6

Dissociation constants (K_d) of the binding of ^{125}I-labeled peptide α146-162 to adsorbents of each of the five peptides were determined as described in ref. 17. K_a, is the association constant $=1/K_d$ [M^{-1}]. Table is from Oshima and Atassi (71).

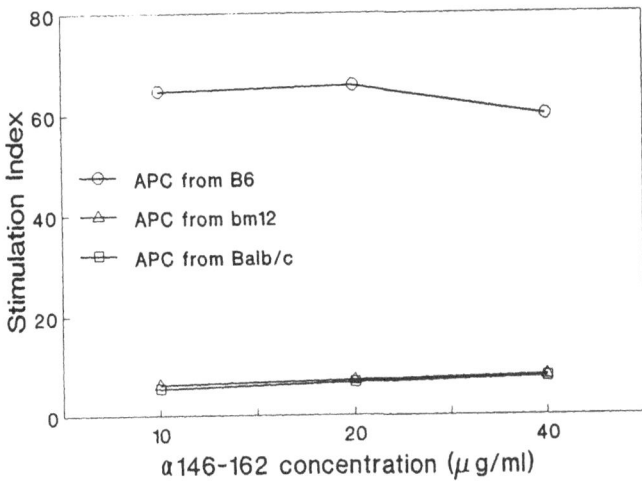

Figure 6. Dose response curves of the antigen presentation capability of APC from B6 and bm12 of peptide $t\alpha$146-162 to the peptide-specific T cell line from B6 at 3 challenge doses of the peptide. APC from Balb/c were used as a control. Results are expressed in stimulation index. Unstimulated cells and negative controls of unrelated challenge antigens (in parenthesis) gave the following cpm values: B6, 2651 (1157); bm12, 995 (416); Balb/c, 1498 (867). (Figure is from ref. 60).

the change of Thr-71 in I-Aβ^b to lysine in I-Aβ^{bm12}. However, APC of B6 also presented peptide $t\alpha$146-162 much less efficiently to peptide-specific T-cells of bm12 (Figure 7). These findings indicate (60) that the lower recognition of peptide $t\alpha$146-162 by bm12 T cells is affected, at least in part, by a less efficient binding of this peptide to I-A^{bm12}. This does not rule out contributions by other factors such as: (1) Change or failure in interaction of MHC molecules and/or peptide with T cell receptor; (2) Absence of T cells responding to MHC-bound peptide due to clonal elimination or anergy; (3) The lower

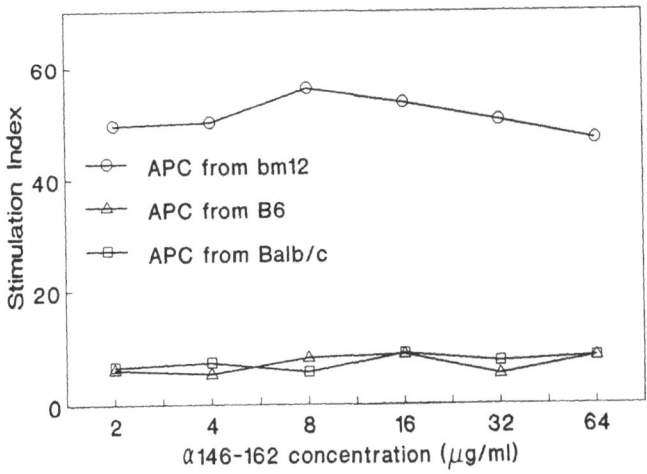

Figure 7. Dose response curves of antigen presentation by APC from bm12 and B6 of peptide $t\alpha$146-162 to the peptide-specific T cell line from bm12 at 6 challenge doses of the peptide. APC from Balb/c were used as a control. Results are expressed in stimulation index. Unstimulated cells and negative controls of unrelated challenge antigens (in parenthesis) gave the following cpm values: bm12, 1644 (1616); B6, 3918 (6440); Balb/c, 3984 (4545). (Figure is from ref. 60).

Table 2. Effect of neonatal tolerance to *t*AChR and *t*AChR α-chain peptides in the development of clinical EAMG

Neonatal injection	Muscle weakness (grade)				Total (%)	p^a
	0	1	2	3		
—	4	3	4	0	6/10 (60)	
*t*AChR	8	0	1	0	1/9 (11)	0.04[b]
*t*α146-162	10	2	0	0	2/12 (16)	0.04[b]
*t*α182-198	4	2	2	1	5/9 (55)	

[a]Fisher exact probability test was used to determine the statistical significance of the data obtained.
[b]The difference between the incidence of muscle weakness in non-tolerized and *t*AChR- or *t*α146—162-tolerized mice is statistically significant at $p = 0.04$. The difference between the incidence of muscle weakness in non-tolerized and *t*α182—198-tolerized mice did not attain statistical significance. (From Shenoy *et al.*, ref 55).

affinity of peptide *t*α146-162 to I-A^{bm12} may be associated with a different conformation than the one it acquires in the binding to I-A^b.

MANIPULATION OF THE AUTOIMMUNE RESPONSE

Suppression of EAMG by Epitope-Specific Neonatal Tolerance to an AChR Peptide

Because bml2 resistance to EAMG correlates with reduced proliferative response to the otherwise immunodominant epitopes within peptides *t*α146-162 and *t*α182-198 on the *t*AChR-α-subunit, the data implicated the importance of these epitopes in EAMG pathogenesis (55). We had previously achieved epitope-specific neonatal tolerance with synthetic peptides (61). This capability has been exploited in the animal model of experimental autoimmune encephalomyelitis (62). To test the involvement of *t*AChR α-chain epitopes within peptide *t*α146-162 in EAMG pathogenesis, B6 mice were neonatally tolerized with soluble peptide *t*α146-162 and subsequently immunized with *t*AChR in complete Freund's

Table 3. Effect of neonatal tolerance to *t*AChR and *t*AChR α-chain peptides on serum autoantibody to *m*AChR and lymphocyte proliferative response

Neonatal injection	Antibody to $mAChR^a(x10^{-10}m)$	[^{3}H] Thymidine uptake^b (ΔCPM ± SEM)	
		*t*AChR	*t*α146–162
—	6.89 ± 1.85		
*t*AChR	1.68 ± 1.85		
*t*α146-162 (Torpedo)	3.14 ± 1.31		
*t*α182-198 (human)	4.87 ± 1.73		
—		18.4 ± 4.7	2.8 ± 0.5
*t*α146-162 (Torpedo)		15.4 ± 10.3	1.3 ± 0.5

[a]Expressed as bungaroxin binding sites precipitated per liter of serum.
[b]Values represent mean [^{3}H] thymidine uptake of triplicate cultures (x10^{-3} ± SEM). [^{3}H] Thymidine incorporation is expressed as cpm, with background (cells without antigen) values subtracted (Δcpm). (From Shenoy *et al.*, ref 55).

adjuvant (55). Neonatal tolerance to tAChR or to peptide $t\alpha$146-162 reduced the incidence of clinical MG (Table 2) and suppressed serum anti-AChR antibodies and the T-cell response to peptide $t\alpha$146-162 (Table 3) (55). This indicates the involvement of T-cell epitope(s) within region $t\alpha$146-162 in EAMG pathogenesis. Neonatal tolerance to peptide $t\alpha$146-162 could have caused specific clonal deletion, and/or clonal anergy, and/or recruited suppressor cells to prevent clinical EAMG. Presumably, epitope(s) within $t\alpha$146-162, in the context of the I-Aβ^b region 67-71, stimulate specific T-helper cells which interact with specific B cells to produce pathogenic antibodies which cause the end plate lesion in patients with MG.

Epitope-Specific Suppression of Antibody Response in EAMG by a Monomethoxypolyethylene Glycol (mPEG) Conjugate of a Synthetic Peptide

The majority of the autoantibodies in EAMG is directed against the main extracellular part of the α-subunit of AChR. As described in the preceding sections, the mapping of the complete antibody recognition profile of the entire extracellular part of the tAChR α-chain demonstrated that the peptide $t\alpha$125-148 contains a major antigenic site (63), is a potent region for induction of EAMG (64) and is involved in the binding of ACh to AChR (18). We have shown in recent studies (65) that a peptide $t\alpha$125-148-mPEG conjugate suppresses the development of the tAChR-induced EAMG in B6 mice by electrophysiological criteria.

Previous studies have shown that conjugation of mPEG, or PVA to protein antigens causes a loss of most of the antigenicity of the antigens (66-70) and prior injection of animals with antigen-mPEG conjugates leads to the development of tolerance to subsequent immunization with the native antigen (67,69) In more recent studies, we described the synthesis of mPEG-peptide conjugates (71) and have shown that treatment with an mPEG-peptide conjugate prior to immunization with the intact protein resulted in suppression of the antibody response specific for the respective region without effect on the antibody responses to other antigenic sites of the protein (65). Injection of mice with mPEG-$t\alpha$125-148 and subsequent immunization with whole tAChR suppressed the development of EAMG by electrophysiological criteria (Figure 8). In anti-tAChR sera from these animals, the antibody response against unconjugated $t\alpha$125-148 was decreased, while the antibody responses against whole tAChR and other epitopes were not altered (Figure 9) (65). There were no

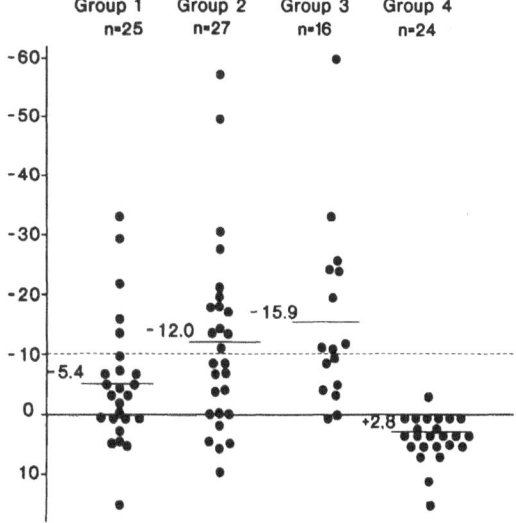

Figure 8. Effects of prior administration of mPEG-$t\alpha$125-148 on the development of electrophysiological EAMG. Note that in group 1 mice, both the mean amplitude change and the proportion of mice showing >10% decrement were smaller than in group 2 or 3 (p <0.05) but greater than in group 4 (p<0.05), suggesting that the mPEG-$t\alpha$125-148 conjugate suppresses development of electrophysiological EAMG although the suppression is incomplete. Prior to immunization with tAChR, the mice received: Group 1, mPEG-$t\alpha$125-148; Group 2, mPEG-nonsense peptide conjugate; Group 3, unaltered free peptide $t\alpha$125-148; Group 4, none. (From Atassi et al., ref. 65).

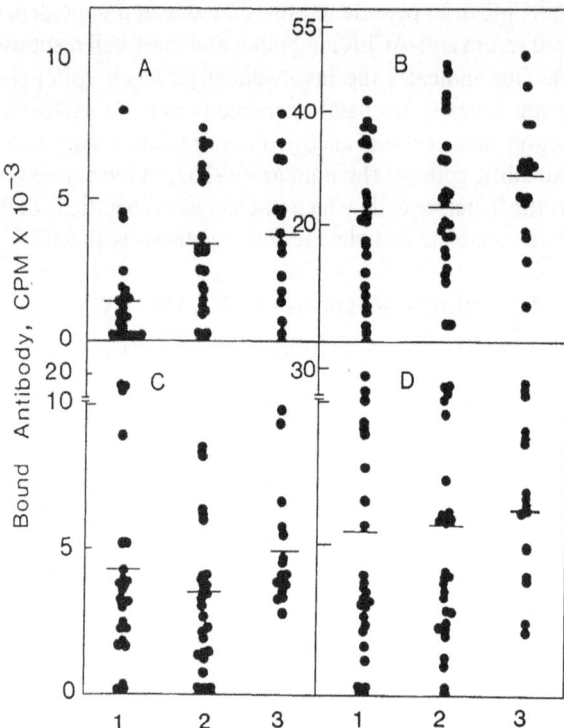

Figure 9. Effects of preadministration of mPEG-$t\alpha$125-148 on the antibody response to immunization with tAChR. The anti-tAChR antisera from the three groups of mice (groups 1-3) were studied for antibody binding to $t\alpha$125-148 (A), whole tAChR (B), $t\alpha$45-60 (C), and $t\alpha$182-198 (D). In A, group 1 mice showed significant suppression of the antibody population that binds to $t\alpha$125-148 (mean net cpm ±standard deviation = 1414 ± 1801) compared with the mice in group 2 (3334 ± 2318, $p < 0.05$) and group 3 (3626 ± 2214, $p < 0.05$: by the t test). Antibodies against whole receptor (B) ($p > 0.5$), $t\alpha$45-60 (C) ($p > 0.1$-0.5) and $t\alpha$182-198 (D)($p > 0.1$-0.5) showed no significant suppression in group 1 compared with the control groups. The groups of mice are described in Figure 8. (From Atassi *et al.*, ref. 65).

detectable changes in the T-cell proliferative responses to $t\alpha$125-148 or to whole tAChR in these animals. Prior injections with a "nonsense" peptide-mPEG conjugate had no effect on responses to the subsequent immunization with whole tAChR. The results indicated that the mPEG-$t\alpha$125-148 conjugate has epitope-specific tolerogenicity for antibody responses in EAMG and that the tAChR α-subunit region $t\alpha$125-148 plays an important pathophysiological role in EAMG (65). The epitope-directed tolerogenic conjugates may be useful for future immunotherapies of human MG. The strategy of specific suppression of the antibody response to a predetermined epitope by using a synthetic mPEG-peptide conjugate may prove useful in manipulation and suppression of unwanted immune responses such as autoimmunity and allergy (65).

CONCLUSIONS

Studies from this laboratory have mapped and synthesized the functional (i.e. ACh and α-neurotoxin binding) sites as well as the sites of immune and autoimmune T- and B-cell recognition of tAChR and hAChR. Antibodies against tAChR and mAChR α-chain peptides have been used to map the surface accessibility of the α-chain extracellular domain in soluble (i.e. free) and in membrane-bound tAChR as well as in mAChR on a live muscle cell line in culture. We have also performed studies, which have been quite successful, on the manipulation of the autoimmune antibody and T cell responses by synthetic peptides. These studies should open up important avenues for the use of synthetic peptides in the manipulation and control of unwanted immune responses.

REFERENCES

1. Sakata, S., Kobayashi, M., Miura, K. and Atassi., M. Z.(1988) Molecular recognition of human insulin receptor by autoantibodies in a human serum. *Immunological Investigations* **17**, 237-242.

2. Nakamura, S., Sakata S. and Atassi, M. Z. (1990) Localization and synthesis of an insulin-binding region on human insulin receptor. *J. Prot. Chem.* **9**, 229-233.

3. Atassi, M. Z., Manshouri, T. and Sakata, S. (1991) Localization and synthesis of the hormone-binding regions of the human thyrotropin receptor. *Proc. Natl. Acad. Sci. USA* **88**, 3613-3617.

4. Sakata, S., Ogawa, T., Matsui, I., Manshouri, T. and Atassi., M. Z. (1992) Biological activities of rabbit antibodies against synthetic human thyrotropin receptor peptides representing thyrotropin binding regions. *Biochem. Biophys. Res. Commun.* **182**, 1369-1375.

5. Atassi, M. Z., Manshouri, T. and Sakata, S. (1992) Synthesis, biological activity and autoimmune recognition of the hormone-binding regions of the human thyrotropin receptor. *J. Prot. Chem.* **11**, 417-419.

6. Karlin, A. (1980) Molecular properties of nicotinic acetylcholine receptors. In *Cell Surface and Neuronal Function* (Edited by Colman, C.W., Poste, G. and Nicolson, G.L.), pp. 191-260. Elsevier/North-Holland Biomedical Press, New York.

7. Changeux, J.P., Devillers-Thiery, A. and Chemouilli, P. (1984) Acetylcholine receptor: an allosteric protein. *Science* **225**, 1335-1345.

8. Noda, M., Takahashi, H., Tanabe, T., Toyosato, M., Furutani, Y., Hirose, T., Asai, M., Inayama, S., Miyata, T. and Numa, S. (1982) Primary structure of α-subunit precursor of *Torpedo californica* acetylcholine receptor deduced from cDNA sequence. *Nature (London)* **299**, 793-797.

9. Noda, M., Takahashi, H., Tanabe, T., Toyosato, M., Kikyotani, S., Hirose, T., Asai, M., Takashima, H., Inayama, S., Miyata, T. and Numa, S. (1983) Primary structures of ß- and α-subunit precursors of *Torpedo californica* acetylcholine receptor deduced from cDNA sequences. *Nature (London)* **301**, 251-255.

10. Noda, M., Takahashi, H., Tanabe, T., Toyosato, M., Kikyotani, S., Miyata, T. and Numa, S. (1983) Structural homology of *Torpedo californica* acetylcholine receptor subunits. *Nature (London)* **302**, 528-532.

11. Claudio, T., Ballivet, M., Patrick, J. and Heinemann, S. (1983) Nucleotide and deduced amino acid sequences of *Torpedo californica* acetylcholine receptor subunit. *Proc. Natl. Acad. Sci. USA* **80**, 1111-1115.

12. Isenberg, K.E., Mudd, J., Shah, V. and Merlie, J.P. (1986) Nucleotide sequence of the mouse muscle nicotinic acetylcholine receptor α subunit. *Nucleic Acids Res.* **14**, 5111-5111; Boulter, J., Evans, K., Goldman, D., Martin, G., Treco, D., Heinemann, D. and Patrick, J. (1986) Isolation of a cDNA clone coding for a possible neural nicotinic acetylcholine receptor α-subunit. *Nature (London)* **319**, 368-374.

13. Buonanno, A., Mudd, J., Shah, V. and Merlie, J.P. (1986) A universal oligonucleotide probe for acetylcholine receptor genes: Selection and sequencing of cDNA clones of the mouse muscle β subunit. *J. Biol. Chem.* **261**, 16451-16458.

14. Yu, L., LaPolla, J. and Davidson, N. (1986) Mouse nicotinic acetylcholine receptor γ subunit: cDNA sequence and gene expression. *Nucleic Acids Res.* **14**, 3539-3555.

15. LaPolla, R.J., Mayne, K.M. and Davidson, N. (1984) Isolation and characterization of a cDNA clone for the complete protein coding region of the δ subunit of the mouse acetylcholine receptor. *Proc. Natl. Acad. Sci. USA* **81**, 7970-7974.

16. Noda, M., Furutani, Y., Takahashi, H., Toyosato, M., Tanabe, T., Shimizu, S., Kikyotani, S., Kayano, T., Hirose, T., Inayama, S. and Numa, S. (1983) Cloning and sequence analysis of calf cDNA and human genomic DNA encoding alpha-subunit precursor of muscle acetylcholine receptor. *Nature (London)* **305**, 818-823.

17. Sobel, A., Weber, M. and Changeux, J.P. (1977) Large-scale purification of the acetylcholine-receptor protein in its membrane-bound and detergent-extracted forms from *Torpedo marmorata* electric organ. *Eur. J. Biochem.* **80**, 215-224.

18. McCormick, D.J. and Atassi, M.Z. (1984) Localization and synthesis of the acetylcholine-binding site in the α-chain of the *Torpedo californica* acetylcholine receptor. *Biochem. J.* **224**, 995-1000.

19. Tzartos, S.J. and Changeux, J.P. (1983) High affinity binding of α bungarotoxin to the purified α subunit and its 27,000-dalton proteolytic peptide from *Torpedo marmorata* acetylcholine receptor. Requirements for sodium dodecil sulfate. *EMBO. J.* **2**, 381-387.

20. Lee, C.Y. (1979) Recent advances in chemistry and pharmacology of snake toxins. *Adv. Cytopharmacol.* **3**, 1-16.

21. Mulac-Jericevic, B. and Atassi, M.Z. (1986) Segment α182-198 of *Torpedo californica* acetylcholine receptor contains a second toxin-binding region and binds anti-receptor antibodies. *FEBS Lett.* **199**, 68-74.

22. Mulac-Jericevic, B. and Atassi, M.Z. (1987) α-Neurotoxin binding to acetylcholine receptor: localization of the full profile of the cobratoxin-binding regions in the α-chain of *Torpedo californica* acetylcholine receptor by a comprehensive synthetic strategy. *J. Prot. Chem.* **6**, 365-373.

23. Mulac-Jericevic, B. and Atassi, M.Z. (1987) Profile of the α-bungarotoxin binding regions on the extracellular part of the α-chain of *Torpedo californica* acetylcholine receptor. *Biochem. J.* **248**, 847-852.

24. Mulac-Jericevic, B., Manshouri, T., Yokoi, T. and Atassi, M.Z. (1988) The regions of α-neurotoxin binding on the extracellular part of the α-subunit of human acetylcholine receptor. *J. Prot. Chem.* **7**, 173-177.

25. Ruan, K.-H., Stiles, B.G. and Atassi, M.Z. (1991) The short-neurotoxin binding regions on the α-chain of human and *Torpedo californica* acetylcholine receptors. *Biochem. J.* **274**, 849-854.

26. McDaniel, C.S., Manshouri, T. and Atassi, M.Z. (1987) A novel peptide mimicking the interaction of α-neurotoxins with acetylcholine receptor. *J. Prot. Chem.* **6**, 455-461.

27. Atassi, M.Z., McDaniel, C.S. and Manshouri, T. (1988) Mapping by synthetic peptides of the binding sites for acetylcholine receptor on α-bungarotoxin. *J. Prot. Chem.* **7**, 655-666.

28. Ruan, K.-H., Spurlino, J., Quiocho, F.A. and Atassi, M.Z. (1990) Acetylcholine receptor α bungarotoxin interactions: determination of the region-to-region contacts by peptide-peptide interactions and molecular modeling of the receptor cavity. *Proc. Natl. Acad. Sci. USA* **87**, 6156-6160.

29. Guy, H.R. (1983) A structural model of the acetylcholine receptor channel based on partition energy and helix packing calculations. *Biophys. J.* **45**, 249-261.

30. Finer-Moore, J. and Stroud, R.M. (1984) Amphipathic analysis and possible formation of the ion channel in an acetylcholine receptor. *Proc. Natl. Acad. Sci. USA* **81**, 155-159.

31. Atassi, M.Z., McDaniel, C.S. and Manshouri, T. (1988) Mapping by synthetic peptides of the binding sites for acetylcholine receptor on α-bungarotoxin. *J. Prot. Chem.* **7**, 655-666.

32. Atassi, M.Z., Mulac-Jericevic, B. and Ashizawa, T. (1994) Mapping of the polypeptide chain organization of the main extracellular domain of the α-subunit in membrane-bound acetylcholine receptor by anti-peptide antibodies spanning the entire domain. *Adv. Expt. Med. Biol.* **347**, 221-228.

33. Atassi, M.Z. and Mulac-Jericevic, B. (1994). Mapping of the extracellular topography of the α-chain in free and in membrane-bound acetylcholine receptor by antibodies against overlapping peptides spanning the entire extracellular parts of the chain. *J. Prot. Chem.* **13**, 37-47.

34. Jinnai, K., Ashizawa, T. and Atassi, M.Z. (1994) Analysis of exposed regions on the main extracellular domain of mouse acetylcholine receptor a subunit in *live* muscle cells by binding profiles of antipeptide antibodies: *J. Prot. Chem.,* **13**, 715-722.

35. Appel, S.H., Almon, R.R. and Levy, N. (1975) Acetylcholine receptor antibodies in myasthenia gravis. *N. Engl. J. Med.* **293**, 760-761.

36. Berman, P.W. and Patrick J. (1980) Linkage between the frequency of muscle weakness and loci that regulate immune responsiveness in murine experimental gravis. *J. Exp. Med.* **152**, 507-520.

37. Engel, A.G. (1984) Myasthenia gravis and myasthenic syndromes. *Ann. Neurol.* **16**, 519-533.

38. Lindstrom, J.M., Seybold, M.E., Lennon, V.,A., Whittingham, S. and Duane, D.D. (1976) Antibody to acetylcholine receptor in myasthenia gravis. Prevalence, clinical correlates, and diagnostic value. *Neurology* **26**, 1054-1059.

39. Falpius, B.W., Miskin, R. and Riche, E. (1980) Antibodies from myasthenic patients that compete with cholinergic agents for binding to nicotinic receptors. *Proc. Natl. Acad. Sci. USA* **77**, 4326-4330.

40. Ashizawa, T., Ruan, K.H., Jinnai, K. and Atassi, M.Z. (1992) Profile of the regions on the α-chain of human acetylcholine receptor recognized by autoantibodies in myasthenia gravis. *Mol. Immunol.* **29**, 1507-1514.

41. Oshima, M., Ashizawa, T., Pollack, M.S. and Atassi, M.Z. (1990) Autoimmune T cell recognition of human acetylcholine receptor: the sites of T cell recognition in myasthenia gravis on the extracellular part of the α subunit. *Eur. J. Immunol.* **20**, 2563-2569.

42. Vincent, A. and Newson-Davis, J. (1985) Acetylcholine receptor antibody as a diagnostic test for myasthenia gravis: results in 153 validated cases and 2967 diagnostic assays. *J. Neurol. Neurosurg. Psych.* **48**, 1246-1252.

43. Pachner, A.R., Kantor, F.S., Mulac-Jericevic, B. and Atassi, M.Z. (1989) An immunodominant site of acetylcholine receptor in experimental myasthenia gravis mapped with T lymphocyte clones and synthetic peptides. *Immunol. Lett.* **20**, 199-204.

44. Safwenberg, J., Hammerstrom, L., Lindbluom, J.B., Matell, G., Moller, E., Osterman, P.O. and Smith, S.I.E. (1978) HLA-A, -B, and -D antigens in male patients with myasthenia gravis. *Tissue Antigens* **12**, 136-142.

45. Bell, J., Rassenti, L., Smoot, S., Smith, K., Newby, C., Hohlfeld, R., Toyka, K., McDevitt, H. and Steinman, L. (1986) HLA-DQ beta-chain polymorphism linked to myastenia gravis. *Lancet* **i**, 1058-1060.

46. Christadoss, P. (1989) Immunogenetics of experimental autoimmune myasthenia gravis. *Crit. Rev. Immunol.* **9**, 247-278.

47. Christadoss, P., David, C.S., Shenoy, M. and Keve, S. (1990) Ea^k transgene in B10 mice suppresses the development of myasthenia gravis. *Immunogenetics* **31**, 241-244.

48. Christadoss, P., David, C.S. and Keve, S. (1992) I-Aα^k transgene pairs with I-Aβ^b gene and protects C57BL10 mice from developing autoimmune myasthenia gravis. *Clin. Immunol. Immunpathol.* **62**, 235-239.

49. Hohlfeld, R., Toyka, K.V., Heininger, K., Grosse-Wilde, H. and Kalies, I. (1984) Autoimmune human T lymphocytes specific for acetylcholine receptor. *Nature (London)* **310**, 244-246.

50. Yokoi, T., Mulac-Jericevic, B., Kurasaki, J. and Atassi, M.Z. (1987) T lymphocyte recognition of acetylcholine receptor: localization of the full T cell recognition profile on the extracellular part of the α chain of *Torpedo californica* acetylcholine receptor. *Eur. J. Immunol.* **17**, 1697-1702.

51. Oshima, M., Pachner, A.R. and Atassi, M.Z. (1994) Profile of the regions of acetylcholine receptor α chain recognized by T lymphocytes and by antibodies in EAMG-susceptible and non-susceptible mouse strains after different periods of immunization with the receptor. *Mol. Immunol.* **31**, 833-843.

52. Widera, G., and Flavell, R.A. (1984) The nucleotide sequence of the murine I-Eβ^b immune response gene: evidence for gene conversion events in class II genes of the major histocompatibility complex. *EMBO J.* **3**, 1221-1225.

53. Denaro, M., Hammerling, U., Rask, L. and Peterson, P.A. (1984) The Eβ^b gene may have acted as the donor gene in a gene conversion event generating the Aβ^{bm12} mutant. *EMBO J.* **3**, 2029-2032.

54. McIntyre, K.R. and Seidman, J.G. (1984) Nucleotide sequence of mutant I-Aβ^{bm12} gene is evidence for genetic exchange between mouse immune response genes. *Nature* (London) **308**, 551-553.

55. Shenoy, M. Oshima, M., Atassi, M.Z. and Christadoss, P. (1993) Suppression of experimental autoimmune myasthenia gravis by epitope-specific neonatal tolerance to synthetic region α146-162 of acetylcholine receptor. *Clin. Immunol. Immunopathol.* **66**, 230-238.

56. Rosenthal, A.S. (1978) Determinant selection and macrophage function in genetic control of the immune response. *Immunol. Rev.* **40**, 136-152.

57. Benacerraf, B. (1978) A hypothesis to relate the specificity of T lymphocytes and the activity of I region-specific Ir genes in macrophages and B lymphocytes. *J. Immunol.* **120**, 1809-1812.

58. Ronchese, F., Brown, M.A. and Germain, R.N. (1987) Structure-function analysis of the Aβ^{bm12}. *J. Immunol.* **139**, 629-638.

59. Brown, J. H., Jardetsky, T. S., Gorga, J. C., Stern, L. J., Urban, R. G., Strominger, J. L. and Wiley, D. C. (1993) Three-dimensional structure of the human class II histocompatability antigen HLA-DR1. *Nature (London)* **364**, 33-39.

60. Oshima, M. and Atassi, M. Z. (1995) Effect of amino acid substitutions within the region 62-76 of I-Aβ^b on binding with and antigen presentation of *Torpedo* acetylcholine receptor α chain peptide 146-162. *J. Immunol., in press.*

61. Young, C.R. and Atassi, M.Z. (1983) T lymphocyte recognition of sperm whale myoglobin specificity of T cell recognition following neonatal tolerance with either myoglobin or synthetic peptides of an antigenic site. *J. Immunogenet.* **10**, 161-169.

62. Clayton, J. Gammon, G.M., Ando, D.G., Kono, D.H., Hood, L. and Sercarz, E.E. (1989) Peptide-specific prevention of experimental allergic encephalomyelitis: Neonatal tolerance induced to the dominant T-cell determinant of myelin basic protein. *J. Exp. Med.* **169**, 1681-1691.

63. Mulac-Jericevic, B., Kurisaki, J. and Atassi, M.Z. (1987) Profile of the continuous antigenic regions on the extracellular part of the α chain of an acetylcholine receptor. *Proc. Natl. Acad. Sci. USA* **84**, 3633-3637.

64. Lennon, V.A., McCormick, D.J., Lambert, E.H., Griesmann, G.E. and Atassi, M.Z. (1985) Region of peptide 125-147 of acetylcholine receptor α-subunit is exposed at neuromuscular junction and induces experimental autoimmune myasthenia gravis, T-cell immunity and modulating autoantibodies. *Proc. Natl. Acad. Sci. USA* **82**, 8805-8809.

65. Atassi, M.Z., Ruan, K.H., Jinnai, K., Oshima, M. and Ashizawa, T. (1992) Epitope-specific suppression of antibody response in experimental autoimmune myasthenia gravis by an monomethoxypolyethylene glycol conjugate of a myasthenogenic synthetic peptide. *Proc. Natl. Acad. Sci. USA* **89**, 5852-5856.

66. Abuchowski, A., van Es., T., Palczuk, N.C. & Davis, F.F. (1977) Alteration of immunological properties of bovine serum albumin by covalent attachment of polyethylene glycol. *J. Biol. Chem.* **252**, 3578-3581.

67. Lee, W.Y. and Sehon, A.H. (1977) Abrogation of reaginic antibodies with modified allergens. *Nature (London)* **267**, 618-619.

68. Davis, F.F., Abuchowski, A., van Es., T., Palczuk, N.C., Savoca, K., Chen. R.H.-L. and Pyatuk, P. (1980) in: Biochemical Polymers: Polymeric Materials and Pharmaceuticals for Biomedical Use, E.P. Goldberg, and A. Nakajima, eds., pp. 441-452, Academic, New York.

69. Savoca, K.V., Davis, F.F. and Palczuk, N.C. (1984) Induction of tolerance in mice by uricase and monomethoxypolyethylene glycol-modified uricase. *Int. Arch. Allergy Appl. Immunol.* **75**, 58-66.
70. Sehon, A.H. and Lang, G.M. (1986) The use of nonionic, water soluble polymers for the synthesis of tolerogenic conjugates of antigens, in: Mediators of Immune Regulation and Immunotherapy, S.K. Singal and T.L. Delovitch, eds., pp. 190-203, Elsevier, New York.
71. Atassi, M.Z. and Manshouri, T. (1991) Synthesis of tolerogenic monomethoxypolyethylene glycol and polyvinyl alcohol conjugates to peptides. *J. Prot. Chem.* **10**, 623-687.

AMELIORATION OF AUTOIMMUNE REACTIONS BY ANTIGEN-INDUCED APOPTOSIS OF T CELLS

Hugh I. McFarland,[1] Jeffrey M. Critchfield,[2] Michael K. Racke,[3]
John P. Mueller,[4] Steven H. Nye,[4] Stefen A. Boehme,[1] and
Michael J. Lenardo[1]

[1] Laboratory of Immunology, National Institute of Allergy and Infectious
   Diseases
National Institutes of Health, Bethesda, MD 20892
[2] Department of Medicine, University of California at San Francisco,
San Francisco, California 94143
[3] Department of Neurology, Washington University School of Medicine,
St. Louis, Missouri 63110
[4] Alexion Pharmaceuticals, Inc.
25 Science Park, suite 360, New Haven, Connecticut 06511

INTRODUCTION

Exposure of mature T cells to their specific antigen normally results in proliferation, but in the presence of high and repeated doses of antigen we have found that these same T cells undergo programmed cell death. The conditions required for antigen to induce the death of mature T cells include: an initial mitogenic stimulus that induces T cell responsiveness to growth lymphokines such as interleukin 2 (IL-2) or interleukin 4 (IL-4), T cell progression through the cell cycle, and finally strong T cell receptor (TCR) restimulation while still under the influence of growth lymphokine. Our data suggest that antigen-induced T cell death is part of a feedback regulatory loop which senses and controls the magnitude of the T cell response to antigen, that we have termed, "Propriocidal regulation." Propriocidal regulation provides an explanation for the phenomenon of high dose suppression, and a mechanism for the peripheral deletion of T cells resulting in antigen-specific immunological tolerance. In this review, we will discuss data demonstrating the effeciveness of an immunotherapy based on the propriocidal mechanism to delete autoantigen-reactive T cells in experimental allergic encephalomyelitis (EAE), a mouse model for multiple sclerosis (MS).

PROPRIOCIDAL REGULATION

The Role of Growth Lymphokines

Interleukin 2. The initial finding implicating IL-2 and antigen in the death of mature T cells came unexpectedly in studies involving A.E7 cells, a CD4$^+$ T$_H$1 clone (1). A.E7 cells recognize a pigeon cytochrome c peptide (PCC) amino acids 81-104 and constitutively express high affinity IL-2 receptors. Stimulation of these cells first in 100 U/ml IL-2 for 2 days followed by 1 µM PCC 81-104 with antigen presenting cells resulted in the death, as determined by trypan blue exclusion, of 82% of these cells relative to controls that received no antigen. This observation suggests that the order of antigen and IL-2 stimulation of T cells may be important in determining whether the outcome of the stimulation is proliferation or death. Additional studies were perfomed using plate-bound anti-CD3ε in place of antigen, in order to avoid endogenous IL-2 production. The A.E7 cells were cultured in doses of IL-2 ranging from 0-50 U/ml for 2 days followed by 48 hours on anti-CD3ε. Again, cells exposed first to IL-2 and then to TCR stimulation died, but here the death was shown to be dependent on the inital dose of IL-2: 74% of cells that received no IL-2 survived (relative to control cells that were plated without anti-CD3ε), while 2 U/ml IL-2 resulted in 51% survival and treatment with 10 U/ml IL-2 dropped the survival to 14%. Agarose gel electrophoresis of DNA from IL-2 and anti-CD3ε-treated cells revealed the oligonucleosome-sized fragments characteristic of DNA from apoptotic cells. IL-2 levels, therefore, appear to regulate the susceptibility of T cells to death following TCR stimulation, and this death is apoptotic.

Interleukin 4. Similar studies were done to determine whether IL-4, another T cell growth lymphokine, can predispose T cells to die by apoptosis (2). As A.E7 cells will upregulate IL-4 receptors following antigenic stimulation, and can then proliferate in response to IL-4 as well as to IL-2, the cells were pre-incubated in various dilutions of IL-2 or IL-4 for 48 hours prior to stimulation on plate-bound anti-CD3ε for an additional 48 hours.

As with IL-2, pretreatment with IL-4 caused dramatic T cell loss (84% cell loss with 1000 U/ml IL-4). Neutralizing antibody to IL-2, present during the IL-4 pretreatment, did not prevent the death, indicating that endogenous IL-2 production was not involved, and IL-4 alone is sufficient to predispose T cells to death. Again, apoptotic death was shown to be the cause of the cell loss.

The Cell Cycle. As growth lymphokines are involved in T cell entry into the cell cycle, it was envisioned that the predisposition of T cells to apoptosis involves continuous cycling or entry into an apoptosis-vulnerable stage of the cell cycle. The approach taken was to evaluate the effects of a battery of cell cycle blockers on the susceptibility of A.E7 T cell clones to apoptosis (2). Mimosine, deferoxamine, and dibutyryl cAMP were used as G1 blockers, and aphidicolin and excess thymidine were used to block cells in S phase. Antigen-stimulated A.E7 cells were incubated with IL-2 or IL-4 for 32 hours, one of the cell cycle blockers was then added, and the cells were incubated a further 16 hours (excess thymidine, when used, was present throughout the 40- hour incubation period). The A.E7 cells blocked in S remained fully susceptible to apoptosis, but a significant degree of protection was achieved by the G1 blockers. This suggests that T cells in G1 are resistant to apoptosis. Thus, the ability of IL-2 and IL-4 to predispose T cells to die appears to be due to their ability to drive cells into stage(s) of the cell cycle where they are susceptible to apoptosis following TCR stimulation.

The Role of Antigen

High Dose Suppression. Propriocidal regulation provides a mechanism for the phenomenon of "high dose suppression" or "high zone tolerance" which has been long discussed in the immunological literature (3-5). In these studies, as well as in studies of viral infection (6,7), high antigen load paradoxically suppresses the immune response. Our data suggest that the fate of a stimulated population of T cells is dependent on an interplay of growth cytokine levels and hence the proportion of the population in S phase at any given time, and the intensity of TCR stimulation as a result of antigen load and TCR affinity. Strong antigenic stimulation of activated, cycling T cells leads to their programmed death.

Antigen Dose. To examine the effect of antigen dose on the induction of apoptosis in mature T cells, A.E7 cells were incubated in the presence of a range of antigen concentrations from 0-10 μM (8). Peak proliferation occurred at 0.01 μM which coincided with peak viable cell recovery (Fig. 1A). Greater concentrations of antigen (from 1-10 μM) showed reductions in viability (up to 90%) compared to the maximal proliferation dose. IL-2 production by the A.E7 cells as well as the fraction of cells in S phase, peaked and remained on a plateau at antigen concentrations greater than 0.01 μM. This experiment illustrates the concept of propriocidal regulation, in which a critical interplay between IL-2 and antigen

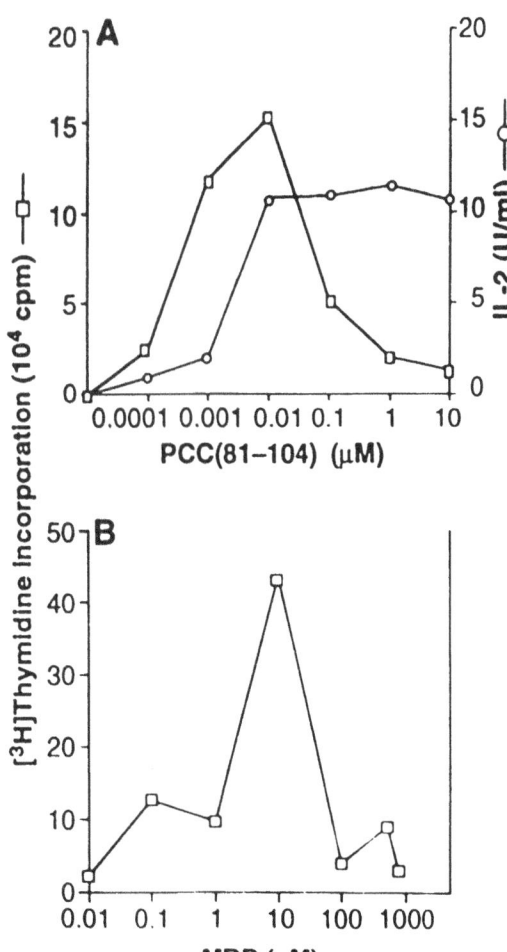

Figure 1. High Dose Suppression in : (A) A.E7 T cells and (B) Vα2+ Vβ8+ TCR Transgenic LNTC. (³H) Thymidine incorporation assay (squares) and IL-2 bioassay (circles). (from: Science 263:1140 (1994) with permission).

load controls T cell proliferation. As antigen increases, IL-2 production increases until a maximum level of proliferation is achieved and then exceeded. As antigen receptor reengagement of cycling cells occurs, T cells begin to die, and IL-2 production and consequently the proportion of cells in cycle becomes limited. Further increases in antigen only result in greater cell death without a subsequent increase in IL-2 production or proliferation. Thus, the magnitude of the T cell response as a function of growth lymphokine and antigen levels are "sensed" by T cells as the intensity of cycling and TCR stimulation.

Repetitive Dosing. Insights as to the usefulness of propriocidal regulation as a potential immunotherapy were gained in studies with the superantigen staphylococcal enterotoxin B (SEB). Short-term in vitro exposure of T cells to SEB results in the activation and proliferation of TCR Vβ8 T cells (9). In order to analyze the involvement of the propriocidal mechanism in SEB-induced T cell deletion, a protocol of repetitive SEB treatment was devised to maximize TCR restimulation by SEB at times of maximal cell cycling (1). Mice were injected with 250 μg SEB i.v. on day 0 and 125 μg on days 2 and 4. Lymph node cells (LNC) were harvested on day 8, and the proportion of Vβ8 and Vβ6 T cells determined by flow cytometry. It was hypothesized that cells activated and proliferating in response to the initial SEB dose would then be killed by TCR reengagement on day 2. Vβ8 cells not killed on day 2, either because they were not activated by the first SEB treatment or did not receive sufficient TCR restimulation to undergo apoptosis, would then be activated and proliferate in response to the second treatment. These cells should then be deleted by the third SEB treatment on day 4. This treatment regimen reduced the Vβ8$^+$ LNC from 23.3% to 7.5%, while Vβ6$^+$ LNC were not deleted. The inclusion of IL-2 receptor blocking antibody but not anti-IL-4 prevented most of the Vβ8 deletion caused by SEB, demonstrating the critical role of growth lymphokine and cell cycling in SEB-induced T cell deletion. Additional experiments showed progressively increasing loss of Vβ8 T cells with each dose of SEB. These data suggest that repetitive high-dose antigen treatment might be used in vivo to eliminate T cell subsets reactive to a specific antigen.

Bystander Death? Additional information about the specificity of the T cell deletion by antigen-induced apoptosis was provided by an in vitro experiment using plate-bound TCR Vβ-specific monoclonal antibodies (1). Concanavalin A-stimulated LNTC were incubated with 3 to 100 U/ml IL-2 for 2 days prior to stimulation with anti-Vβ8 or anti-Vβ6. In low dose IL-2 the cells proliferated, but with high dose IL-2 a selective elimination of Vβ8$^+$ T cells was observed in the LNTC plated on anti-Vβ8, and Vβ6+ T cells were specifically deleted from the LNTC treated with anti-Vβ6 antibody. No death of T cells bearing the inappropriate Vβ was observed (i.e. no Vβ6 elimination from anti-Vβ8 treated LNTC), and both CD4$^+$ and CD8$^+$ cells were depleted. This indicates that only T cells that receive a TCR challenge and not bystander cells undergo programmed death.

HIGH DOSE ANTIGEN THERAPY FOR EXPERIMENTAL ALLERGIC ENCEPHALOMYELITIS

Overview

Multiple sclerosis, myasthenia gravis, autoimmune uveitis, and graft rejection are examples of disorders in which T cells mediate a destructive immune response. At present, therapies for these disorders involve general immunosuppression with drugs such as cyclosporin and steroids. Here we will review studies aimed at the development of an immunotherapy for EAE based on the antigen-induced death of autoreactive T cells. The specific elimination of only autoreactive T cells by this approach, while sparing T cells needed for normal immune function, could avoid the potentially dangerous immunosuppression caused by conventional therapies.

The EAE Model

EAE in mice is an experimental demyelinating disease induced by immunization with myelin basic protein (MBP) or by adoptive transfer of MBP-activated T cells. The disease is mediated by CD4[+] T cells, and is characterized by a perivascular inflammatory cell infiltrate and demyelination within the central nervous system. The clinical and pathological features of EAE are very similar in most regards to the human demyelinating disease, multiple sclerosis (MS) (10). We have used two EAE models: one is a model utilizing MBP epitope-specific TCR transgenic (Tg) mice (11), the other is the relapsing mouse model of EAE (R-EAE) as originally described by Pettinelli and McFarlin (12). In the R-EAE model, mice of susceptible strains are immunized with GP-MBP and 10 days later LNC are obtained and stimulated in vitro for 4 days with MBP. Disease is induced by the adoptive transfer of 3×10^7 MBP-activated LNC into naive recipients of the same strain. Onset occurs within 7-10 days and the disease follows a relapsing-remitting course similar to human MS. The transgenic model is very similar, with the exception that splenocytes from the Vα2 Vβ8 transgenic mice are activated with their specific peptide MBP Ac1-11 for 4 days in vitro prior to adoptive transfer of these cells into B10.PL mice.

MBP-Induced Apoptosis in vitro

The first step in the development of an a therapy for EAE based on specific T cell deletion, was to demonstrate that MBP-reactive T cells could be killed in vitro by high dose antigen (8). Vα2, Vβ8 transgenic LNTC were cultured for 120 hours with doses of MBP from 0.01 to 1000 μM with syngeneic splenocytes to provide antigen presenting cells (Fig. 1B). Peak proliferation occured at 10 μM MBP. Proliferation sharply declined at higher MBP doses. This reduction in proliferation was accompanied by a loss of cell viability of greater than 90%, and microscopic observation showed that apoptosis was responsible for the cell loss. In recent experiments we have found that, MBP-specific human CD4[+] T cell lines exhibit a similar susceptibility to death at high doses of MBP (Clara Pelfrey, unpublished observations).

Prevention of EAE with High Dose MBP

Treatments in the Transgenic Model. The next step was to develop a protocol for in vivo high dose MBP treatment of EAE. Based on the successful specific deletion of Vβ8 T cells in mice by repetitive administration of SEB as discussed earlier, a similar repetitive high dose MBP strategy was developed. On days 0,2,4, and 6 after the adoptive transfer of encephalitogenic T cells, B10.PL mice received 400 μg of MBP or 1400 μg of Ac1-11 twice daily at 6-8 hour intervals. As the cycling of the T cells activated by the early doses of MBP were likely to be asynchronous, the twice daily doses were intended to provide a broad window of high dose TCR restimulation to capture as many of the autoreactive T cells as possible at vulnerable stages of the cell cycle. By day 10, when the animals were sacrificed, untreated EAE mice exhibited severe hind limb paralysis, but animals receiving the described MBP or peptide treatment showed little evidence of clinical disease. The fraction of CD4[+] LNTC expressing the transgenic Vα2.3·Vβ8.2 TCR decreased from 23.7% in untreated EAE animals to 3.8% and 4.5% in animals treated with MBP or Ac1-11 peptide respectively. Sham treatment of mice with ovalbumin had little effect on clinical disease or the number of $V\alpha2.3^+$ $V\beta8.2^+$ LNTC as compared to untreated controls. Of $CD4^+$ LNTC from normal PL/J mice which received no transgenic T cells, 2.9% were $V\alpha2.3^+$ and $V\beta8.2^+$. The adoptive transfer of transgenic cells increased this percentage to 11.9 by day 10 post-transfer. Eight injections of MBP on days 0,2,4, and 6 after adoptive transfer resulted in a 90% deletion of the transfered $V\alpha2.3^+$ $V\beta8.2^+$ cells. A single 400μg dose of MBP on day 0 post-transfer resulted in a 49% deletion of transgenic T cells. Mice adoptively transferred with resting transgenic T cells followed by similar MBP treatments showed no deletion of $V\alpha2.3^+$ $V\beta8.2^+$ cells. The observation that resting cells are not killed is consistent with the in vitro observation that cycling is essential for the susceptibility of T cells to antigen-induced death.

Treatments in the R-EAE Model. These studies with the transgenic EAE model demonstrate that high dose antigen treatment can delete antigen-reactive T cells in vivo, and prevent the onset of clinical disease 10 days after adoptive transfer. Further studies were done using the nontransgenic R-EAE model, in part, because donor and recipient in the transgenic model were not completely syngeneic. (PL x SJL)F_1 mice received MBP-activated syngeneic LNTC on day 0. The mice were given 2-400 μg doses of MBP on days 0,2 and 4. This treatment reduced the incidence and severity of clinical disease, observed over a period of 65 days, from 100% with severe disease in the untreated group to 20% in the MBP-treated group with minimal disease (Fig. 2). Histological staining of spinal cord sections from untreated animals revealed demyelination and fibrillary astrogliosis with a general loss of spinal cord architecture, while spinal cords from MBP-treated animals were essentially normal with only mild inflammation. Determination of the degree of deletion of transferred T cells proved more difficult than in the transgenic EAE model. In short-term experiments, MBP-activated LNC were stained with the fluorescent vital membrane dye DiI prior to intravenous administration. Untreated mice exhibited severe disease. MBP treatment twice daily on days 0.2 and 4 resulted in complete abrogation of clinical signs and an 80% reduction

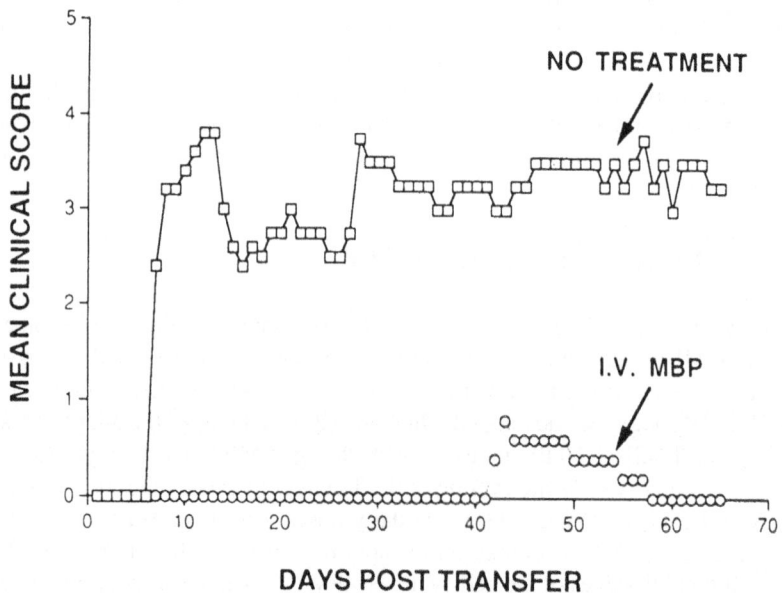

Figure 2. High Dose Treatment Abrogates EAE. MBP administered i.v. on days 0,2 and 4 after adoptive transfer of encephalitogenic LNTC. (from: Science 263:1139 (1994) with permission).

in the fraction of CD4$^+$ LNTC that were DiI$^+$. Attempts to measure the deletion of transferred cells at later times were unsuccessful, probably due, in part, to dilution of the dye through cell division. In long-term experiments, the extent of deletion was determined by limiting dilution analysis.

Protocol Variations in Timing and Dose. The effectiveness of variations in the MBP treatment protocol were examined in a series of experiments (13). A single 400µg dose of MBP on day 0 post-transfer was completely ineffective in treating EAE. A double dose (800 µg) of MBP given on day 4 delayed the onset of disease by 11-21 days, but did not affect the severity of the clinical symptoms, and low dose MBP (100 µg) given twice per day on days 0,2 and 4 resulted in only a slight delay in onset. These data demonstrate that repetitive high dose treatments are necessary to achieve a substantial reduction of clinical symptoms.

Incorporation of IL-2 in the Treatment Protocol. As previously described, in vitro studies have demonstrated the essential role of growth lymphokines such as IL-2 in predisposing T cells to undergo programmed cell death (PCD) following TCR restimulation. Thus, it was thought that the addition of IL-2 to the standard MBP treatment protocol might enhance the cycling of encephalitogenic T cells, rendering them more vulnerable to antigen-induced death (13). The addition of 30,000 U of IL-2 administered twice daily on days 0-5, to the MBP treatment on days 0,2 and 4, slightly reduced the effectiveness of the treatment. In one experiment the incidence increased from 20% with MBP alone to 40% with MBP and IL-2. At the same time, the precursor frequency for MBP-reactive T cells showed reductions for both MBP alone and MBP + IL-2 treatments. the frequencies were: 1/356,799 and 1/322,394 for MBP alone, 1/435,210 and 1/228,415 for MBP + IL-2, as compared to 1/69,156 and 1/138,307 for ovalbumin-treated controls. The reasons for the discrepancy between clinical effects and MBP-reactive T cell deletion are not entirely clear. IL-2 is known to have stimulatory effects on other inflammatory cell types, and this may cause clinical symptoms. The identical IL-2 treatment in the absence of MBP resulted in 100% mortality by day 7, probably due to heightened activation of the encephalitogenic T cells or other cells in the absence of antigen-induced apoptosis.

High Dose MBP Therapy in Established Disease

The previous experiments demonstrated that repetitive high dose MBP treatment is effective in preventing the onset of the clinical and pathological signs of EAE. To determine the efficacy of this therapy in established disease, the standard treatment regimen of twice daily doses of 400 µg of MBP every other day for a total of 6 doses, was begun on day 9, 17 or 31 post-transfer of MBP-activated LNC (13). MBP treatment initiated on day 9, at the onset of clinical signs, was as effective as treatment on day 0,2 and 4 in ameliorating the disease, as well as reducing the precursor frequency of MBP-reactive T cells. Identical treatment beginning on day 17 reduced the clinical severity of EAE as observed over the following 34 days. Mean clinical scores were also reduced in mice treated with MBP at days 31, 33 and 35. This clinical improvement was accompanied by similar reductions in the precursor frequency for MBP-reactive T cells as described for treatments earlier in the course of the disease.

Relapses

High dose, repetitive MBP treatment, though effective in reducing the severity of EAE, has been unable to prevent relapses in some animals particularly late in the observed disease course (>40 days post-transfer). Even a few encephalitogenic T cells not killed during the treatment, could proliferate and eventually result in a relapse. The argument against this is obviously that the precursor frequency for MBP-reactive T cells, at least in the periphery, continues to be reduced 56 days after treatment. Additional cycles of MBP treatment might prevent the outgrowth of undeleted encephalitogenic T cells. A second possibility is that epitope spreading (14,15) may result in the activation and proliferation of encephalitogenic T cells reactive to myelin proteins other than MBP, such as proteolipid protein (PLP), myelin oligodendroglia glycoprotein (MOG), or myelin associated glycoprotein (MAG) (10). In preliminary studies, PLP-reactive T cells could not be identified among T cells from MBP-treated mice late in the course of the disease (M. Racke, unpublished observations). High dose therapy incorporating one or more of these other myelin proteins could be effective in the prevention of late relapses. These studies are ongoing.

Toxicity of MBP Preparations

During our studies, we have observed a problem with toxicity of certain purified preparations of MBP(17). Two 400 µg doses of MBP given at the peak of the initial onset of disease symptoms in EAE (day 7-8) often results in 20-100% mortality among treated animals within 48 hours. Necropsy and histopathology suggested terminal circulatory failure consistent with tumor necrosis factor alpha (TNFα)-induced shock. No enhancement of CNS damage was apparent in these animals. Serum cytokine levels measured 2-6 hours following the second MBP dose showed greatly elevated levels of TNFα, IL-2 and interferon-γ (IFNγ). IFNγ has been shown to be essential for the systemic, toxic effects of TNFα (16). Inclusion of anti-IFNγ to the MBP treatment in the previous experiment significantly reduced the mortality, suggesting that cytokine release-induced shock is the probable cause of death. Endotoxin contamination is unlikely to be involved as an inducer of TNFα, as the amount of contaminating endotoxin administered daily to mice in our MBP preparation was less than 5 pg. In vitro studies have implicated macrophages as the primary TNFα-producer. Our preparation of MBP from guinea pig spinal cord, as well as MBP isolated from human brain can directly induce substantial TNFα by various types of macrophages. A recombinant human MBP induced relatively little TNFα from these cells, suggesting that either a CNS contaminant is responsible for the TNFα production, or there is some difference, such as post-translational modifications, between native and recombinant MBP. Whatever the cause, recombinant proteins may be safer to use in high dose therapy. Experiments are under way to determine the identity of the TNFα-inducing agent. It is likely that the toxicity observed in MBP treatment at the onset of EAE indicates that the specific manner of antigen preparation is likely to be important for high dose antigen therapy.

SUMMARY

We have shown that T cells vigorously cycling in response to growth lympokines are driven into apoptosis by potent TCR restimulation. This process, termed propriocidal regulation, appears to be a normal feedback inhibitory mechanism to prevent excessive T cell proliferation and lymphokine production. Exposure of T cells to repeated high dose antigen treatments creates the conditions just described by activating T cells, and stimulating the production of growth lymphokines and their receptors. High growth lymphokine levels induced by the large amount of antigen present, stimulate vigorous cycling. The continued presence of high antigen levels subjects the cycling T cells to strong TCR restimulation as they enter the vulnerable S phase, inducing apoptosis in T cells responsive to the adminis-tered antigen. Thus, simple, repetitive, intravenous administration of high dose antigen may be used to delete potentially destructive clones of T cells, resulting in a state of peripheral tolerance. This has obvious therapeutic potential in disorders where the elimination of pathogenic T cell clones could be beneficial. We have described in EAE, an animal model for MS, that high dose MBP therapy is effective in preventing CNS pathology and the onset of disease as well as reducing the severity of the clinical symptoms of established EAE. We are currently involved in expanding this approach to other animal models of autoimmunity and graft rejection, as well as refining the immunotherapy in the EAE model with the objective of developing a clinical therapy for human demyelinating disease.

ACKNOWLEDGMENTS

We would like to thank Carol Trageser and Laura Quigley for their expert technical assistance, and Kelly Joe and Dr. Louis Matis for discussions and assistance in the preparation of this manuscript. This research was supported by Alexion Pharmaceuticals Inc., New Haven, CT. J.M.C. was a Howard Hughes-NIH Research Scholar. M.J.L. was supported in part by a Cancer Research Institute Investigator Award.

REFERENCES

1. Lenardo, M.J., 1991, Interleukin-2 programs mouse $\alpha\beta$ T lymphocytes for apoptosis, *Nature* 353:858-861.
2. Boehme, S.A., and Lenardo, M.J., 1993, Propriocidal apoptosis of mature T lymphocytes occurs at S phase of the cell cycle, *Eur. J. Immunol.* 23:1552-1560.
3. Mitchison, N.A., 1964, Induction of immunological paralysis in two zones of dosage, *Proc. Roy. Soc. (Lond.)* 161:275.
4. Matis, L.A., Glimcher, L.H., Paul, W.E., and Schwartz, R.H., 1983, Magnitude of response of histocompatibility-restricted T cell clones is a function of the product of the concentrations of antigen and Ia molecules, *Proc. Natl. Acad. Sci. USA* 80:6019-6023.
5. Suzuki, G., Kawase, Y., Koyasu, S., Yahara, I., Kobayashi, Y., and Schwartz, R.H., 1988, Antigen-induced suppression of the proliferative response of T cell clones, *J. Immunol.* 140:1359.
6. Razvi, E., and Welsh, R.M., 1993, Programmed cell death of T lymphocytes during acute viral infection: a mechanism for virus-induced immune deficiency. *J. Virol.* 67:5754-5765.
7. Moskophidis, D., Lechner, F., Pircher, H., and Zinkernagel, R.M., 1993, Virus persistence in acutely infected immunocompetent mice by exhaustion of antiviral cytotoxic effector T cells, *Nature* 362:758-761.
8. Critchfield, J.M., Racke, M.K., Zuñiga-Pflücker, J.C., Cannella, B., Goverman, J., and Lenardo, M.J., 1994, T cell deletion in high antigen dose therapy of autoimmune encephalomyelitis, *Science* 263:1139-1143.
9. Marrack, P., Blackman, M., Kushnir, E., and Kappler, J., 1990, The toxicity of staphylococcal enterotoxin B in mice is mediated by T cells, *J. Exp. Med.* 171:455-464.
10. Martin, R., McFarland, H.F., and McFarlin, D.E., 1992, Immunological aspects of demyelinating diseases, *Ann. Rev. Immunol.* 10:153-187.
11. Goverman, J., Woods, A., Larson, L., Weiner, L.P., Hood, L., and Zaller, D.M., 1993, Transgenic mice that express a myelin basic protein-specific T cell receptor develop spontaneous autoimmunity, *Cell,* 72:551-560.
12. Pettinelli, C.B., and McFarlin, D.E., 1981, Adoptive transfer of experimental allergic encephalomyelitis in SJL/J mice after in vitro activation of lymph node cells by myelin basic protein: requirement for Lyt 1+ 2- T lymphocytes, *J. Immunol.* 127:1420-1423.
13. Racke, M., Critchfield, J.M., and Lenardo, M.J., Manuscript in preparation.
14. McCarron, R.M., Fallis, R.J., and McFarlin, D.E., 1990, Alterations in T cell antigen specificity and class II restriction during the course of chronic relapsing experimental allergic encephalomyelitis, *J. Neuroimmunol.* 29:73-79.
15. Perry, L.L., Barzaga-Gilbert, E., and Trotter, J.L., 1991, T cell sensitization to proteolipid protein in myelin basic protein-induced relapsing experimental allergic encephalomyelitis, *J. Neuroimmunol.* 33:7-15.

16. Doherty, G.M., Lange, J.R., Langstein, H.N., Alexander, H.R., Buresh, C.M., and Norton, J.A., 1992, Evidence for IFN-γ as a mediator of the lethality of endotoxin and tumor necrosis factor-α, *J. Immunol.* 149:1666-1670.
17. McFarland, H.I., and Lenardo, M.J., Manuscript in preparation.

19

BAND 3 AND ITS PEPTIDES DURING AGING, RADIATION EXPOSURE, AND ALZHEIMER'S DISEASE: ALTERATIONS AND SELF-RECOGNITION

Marguerite M. B. Kay,[1,2] Douglas Lake,[1] and Cathleen Cover[1]

[1] Departments of Microbiology and Immunology,
University of Arizona College of Medicine
1501 North Campbell Ave., Tucson, Arizona 85724
[2] Department of Medicine
Veterans Administration Medical Center
Tucson, Arizona

ABSTRACT

An aging antigen, senescent cell antigen, resides on the 911 amino acid membrane protein band 3. It marks cells for removal by initiating specific IgG autoantibody binding. Band 3 is a ubiquitous membrane transport protein found in the plasma membrane of diverse cell types and tissues, and in nuclear, mitochondrial, and golgi membranes. Band 3 in tissues such as brain performs the same functions as it does in red blood cells forming senescent cell antigen. Oxidation is a mechanism for generating senescent cell antigen. The aging antigenic sites reside on human band 3 map residues 538-554, and 812-830. Carbohydrate moieties are not required for the antigenicity or recognition of senescent cell antigen. Anion transport site were mapped to residues 588-594, 822-839, and 869-883. The aging vulnerable site which triggers the antigenic site and the transport sites of band 3 were mapped using overlapping synthetic peptides along the molecule.

Naturally occurring autoantibodies to regions of band 3 comprising both senescent cell antigen and B cells producing these antibodies were demonstrated in the sera of normal, healthy individuals. The presence of these antibodies tend to increase with age. Individuals with autoimmune diseases (rheumatoid arthritis and systemic lupus erythematosus) have increased antibodies to senescent cell antigen peptides. Radiation exposure results in an increase in antibodies to peptides 588-602 which lies in a transport region containing the aging vulnerable site.

Band 3 ages as cells and tissues age. Our studies, to date, indicate, that the anion transport ability of band 3 decreases in brains and lymphocytes from old mice. This decreased transport ability precedes obvious structural changes such as band 3 degradation and

Immunobiology of Proteins and Peptides VIII
Edited by M. Z. Atassi and G. S. Bixler, Jr., Plenum Press, New York, 1995

generation of SCA, and is the earliest change thus far detected in band 3 function. Other changes include a decreased efficiency of anion transport (decreased V_{max}) in spite of an increase in number of anion binding sites (increased K_m), decreased glucose transport, increased phosphorylation, increased degradation to smaller fragments as detected by quantitative binding of antibodies to band 3 breakdown products and residue 812-830, and binding of physiologic IgG autoantibodies *in situ*. The latter 3 findings indicate that post-translational changes occur.

In Alzheimer's Disease (AD), our results indicate that post-translational changes occur in band 3. These include decreased band 3 phosphorylation of a 25-28kD segment, increased degradation of band 3, alterations in band 3 recognized by antibodies, and decreased anion and glucose transport by blood cells. Serum autoantibodies were increased in AD patients compared to controls to band 3 peptide 822-839. This band 3 residue lies in an anion transport\binding region.

INTRODUCTION

Band 3, the anion exchange protein, is ubiquitous [1-12]. It is present in nuclear [1], golgi [9], and mitochondrial membranes [8] as well as in cell membranes. It is present in brain and all other tissues examined [1-12]. The presence of band 3 related molecules in non-erythroid tissues was first demonstrated in 1983 [1]. A protein immunologically related to band 3 was demonstrated in diverse cell types such as isolated neurons, hepatocytes, squamous epithelial cells, alveolar (lung) cells, lymphocytes, neurons in vivo, and fibroblasts using an antibody to band 3 that reacts with the transmembrane, anion transport domain of band 3 [1]. Since then, band 3 has been described in numerous cell types and tissues including fibroblasts, hepatoma cells, and lymphoid cells [1-12]. There appears to be one band 3 gene. Differences in band 3 in different tissues and cells are located primarily in the cytoplasmic segment consisting of 393 amino acids, and appear due to gene splicing and/or alternate promoters acting on the gene [13-15]. Band 3 proteins in nucleated cells participate in band 3 antibody induced cell surface patching and capping [1]. Band 3 maintains acid-base balance by mediating the exchange of anions (e.g. chloride, bicarbonate) [16], is the binding site for glycolytic enzymes [16,17], and is the major structural protein linking the plasma membrane to the cytoskeleton. Because of its central role in CO_2 respiration, band 3 is the most heavily used ion transport system in vertebrate animals and acid base balance. Based upon the translated sequence of cDNA, the human erythrocyte band 3 protein, which is the "proto-type" for all band 3s, consists of 911 amino acids [13]. Band 3 is highly conserved as indicated by derived protein sequences for many species including mouse, chicken, rat, trout, and lamprey [13,18,19]. The N-terminal 393 amino acids are located within the cytoplasm of the cell where they associate with the cytoskeleton by binding to ankyrin [13,20,21]. This so-called "cytoplasmic domain" is not involved in transport or aging [22-25]. Most of the species and tissue variation occurs in this domain as do the few band 3 polymorphisms discovered [19]. Two crucial properties of the band 3 molecule, namely anion transport and formation of the senescent cell antigen map to the membrane-associated "domain" [22-27].

Senescent cell antigen (SCA), an aging antigen, is a degradation product of band 3 that appears on old cells and acts as a specific signal for the termination of that cell by initiating the binding of IgG autoantibody and subsequent removal by phagocytes [28-44]. This appears to be a general physiologic process for removing senescent and damaged cells in mammals and other vertebrates [31]. SCA has been found on all cells examined [31] including neurons [1,2,24,31]. Besides its central role in the removal of senescent and damaged cells, SCA is present in neurofibrillary tangles and senile plaques in Alzheimer's disease (AD) [2,24], is involved in the removal of neural cells [2,24,31,46] and platelets

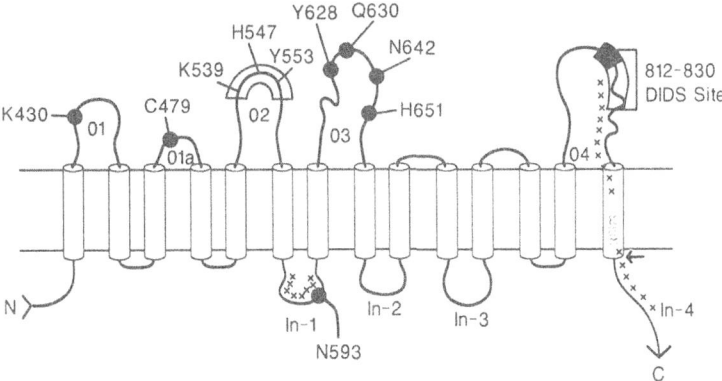

Figure 1. Model of membrane associated and external regions of anion transport protein band 3, approximate residues (R) 400-900 showing aging antigenic sites (speckled area), minimal residues that function as aging epitopes ([ ]), anion transport sites (xxx), and DIDS binding site. The arrow indicates the approximate position of the proline to leucine mutation in high transport band 3 which exhibits anion transport 2-3 times normal. K, lysine; C, cysteine; H, histidine; Y, tyrosine; N, asparagine; the number following a letter indicates residue number; In, inside the cell; O, outside of the cell; arrow C, carboxyl terminus; arrow N, amino terminus. The model was based upon application of the program PEPPLOT of the GCG package to identify membrane spanning nonpolar helices and intervening hydrophilic loops. The location of the hydrophilic loops as extracellular or intracellular was predicted on the basis of established chemical or biological markers, e.g., the demonstration that residues 814-829 contain a DIDS binding site accessible from the outside or external radioiodination of the tyrosine (Y) at position 553. Key residues are identified to facilitate their identification within the sequence. This is a two dimensional representation that does not reflect three dimensional associations of residues that are separated by long stretches of sequence. Results show, however, that close steric association must be maintained by external loops 02 and 04 and In-1 and In-4. The hydrophobicity plots indicate multiple turns/bends in the band 3 molecule in the region of the DIDS binding site. Modified from [26].

[31], and the removal of erythrocytes (rbc) in clinical hemolytic anemias [34,44], sickle cell anemia [35,47,48], and malaria [49]. Oxidation generates SCA *in situ* [33]. Application of hydrophilicity/hydrophobicity algorithms to the derived protein sequence of band 3 indicates that the membrane associated region consists of 11 hydrophobic membrane spanning helical regions and four major external and three major internal hydrophilic loops [19,26]. We have mapped the aging and transport sites along band 3 (Figure 1) using peptide chemistry coupled with cell biology.

Since band 3 differences between tissues and cells are located primarily in the cytoplasmic segment and the membrane domain is highly conserved across tissues and species [13,14,18,19], alterations\mutations in the transport segment, but not the cytoplasmic segment of the band 3 molecule might be expected to be of major importance. We discovered and sequenced the only mutation of band 3 located in the transport domain as part of our ongoing studies [18,50,51]. This mutation of band 3, designated high transport band 3, exhibits anion transport that is 2-3 times normal. A cytosine replaces a thymidine at position 2603 resulting in a proline to leucine alteration in the protein sequence at position 868 in the transport region [18,50,51].

Band 3 ages as cells and tissues age [18,19,24,45,52-55]. Our studies, to date, indicate, that the anion transport ability of band 3 decreases in brains and lymphocytes from old mice [18,24,45,52-55]. This decreased transport ability precedes obvious structural changes such as band 3 degradation and generation of SCA, [2,10,11,19,34,53,56,57] and is the earliest change thus far detected in band 3. The following changes occur in brain band 3 with aging. These changes are: a decreased efficiency of anion transport (decreased V_{max})

in spite of an increase in number of anion binding sites (increased K_m), decreased glucose transport, increased phosphorylation, increased degradation to smaller fragments as detected by quantitative of binding of antibodies to band 3 breakdown products and residue 812-830, and binding of physiologic IgG autoantibodies in situ [23,28-31]. The latter 3 findings indicate that post-translational changes occur in band 3 during aging.

MATERIALS AND METHODS

Tissue Preparation

CBA and BALB/C mice were decapitated. The entire brain was rapidly removed and immediately placed on ice in a 5mM HEPES (pH 8.0) buffered solution containing the protease inhibitors Aprotinin, Leupeptin, and Pepstatin A [11]. The brain was then homogenized for 5 minutes with a Brinkman tissue tearer and subjected to differential centrifugation for 8 minutes at 5,000 rpm. The supernatant was decanted and placed on ice. The remaining pellet was homogenized and the previous steps repeated for a total of three times, saving and combining the supernatant after each centrifugation. The remaining pellet was discarded after the final centrifugation. The supernatant was centrifuged at 20,000 rpm for 30 minutes and the supernatant discarded. The remaining pellet was resuspended in 1 ml 5mM HEPES (pH 8.0), 10mM ATP, and 20% ultrapure glycerol. Protein concentration was determined by the Lowry method and tissues were frozen in isopentane. Preparation of tissue for biochemical assays is as described by Saitoh [58, 59].

Brain Anion Transport

Influx of sulfate was performed essentially according to the method of Elgavish *et al.* [60]. Membranes prepared from brain as described above were washed extensively with 10 mM Tris-HEPES, pH 7.5, with 300mM glucose. Transport was initiated by addition of 10 mM Tris-HEPES, pH 7.5/300mM glucose, 1mM Na2SO4/35SO4 (10 µCi) to membrane vesicles (1.0-5.0 mg/ml protein). Aliquots were taken at 30 seconds and added to 25 mM 4-morpholine propanesulfonic acid (MES)/4mM Tris/pH 5.5 which terminates transport. Membranes were washed extensively with this buffer. Samples were deproteinated with 0.1 N NaOH for 30 min and the amount of radioactivity in a 0.7ml aliquot was counted in a Beckman liquid scintillation counter. Non-band 3 anion transport was determined by measuring the degree to which sulfate exchange is inhibited by inhibiting transport with DIDS.

IgG Binding and Inhibition Assay. Competitive inhibition studies were performed using synthetic peptides to absorb the IgG isolated from senescent erythrocytes [61]. This is the same IgG that initiates phagocytosis *in situ*. The Fc portion IgG is required for binding and phagocytosis of cells by macrophages [28-34]. In order to simulate the physiological situation, intact senescent cell IgG that binds to senescent cells in situ and initiates their removal was used [28-34]. Only IgG isolated from aged erythrocytes binds specifically to senescent cells. For example, IgG eluted from young control erythrocytes did not bind to senescent cells [28-34]. Moreover, the specific binding capacity of the autoantibody was eliminated by absorption with purified senescent cell antigen [31,32]. SCIgG (0.71 µg) was absorbed with synthetic peptides at the concentrations indicated or purified SCA or buffer only for 60 min at room temperature, and incubated with stored red cells for 60 min at room temperature [34]. Storage mimics normal aging in situ immunologically and biochemically [53, 62-73]. After incubation with absorbed IgG, cells are washed four times with 40-50 volumes of phosphate buffered saline containing 0.2% bovine serum albumin (BSA, fraction V, Sigma, St. Louis, MO) and 0.5% glucose. Washed cells are transferred to BSA-coated tubes ($5X10^7$ cells/50 μl) and incubated for 30 min. at 37°C with ^{125}I-Protein A (Amersham, Arlington Heights, IL 30-38 mCi/mg, 10-15 ng/tube). Cells are then washed four times and transferred to new tubes before counting in a gamma counter (Beckman, Gamma 5500). The number of red cell-bound IgG molecules was quantitated before and after absorption using equilibrium binding kinetics [61]. Scatchard analysis was performed. Percent inhibition is calculated from the following formula: 100 [1-(x-b/T-b)] where x = molecules of IgG autoantibody bound per cell; T = total number of IgG antibody molecules bound in the absence of inhibitor; b = background Protein A binding.

Peptides and Antibodies. Peptides were prepared by solid phase synthesis using an Applied Biosystems 430A automatic peptide synthesizer, analyzed for purity, and treated as previously described [19, 26,27,46]. Peptides were synthesized based on the sequence data from Tanner *et al.* [13]. The peptides synthesized were as follows: pNterm (DAGVIQSPR-HEVTEMG), pCDR1 (CKPISGHNSLFWYRQT), pFr3 (KIQPSEPRDSAVXFCA) pCβ (QPLKEQPALNDSRYCL). pTERM served as a negative control based on prior lack of reactivity. Band 3 peptides were based on the gene sequence of band 3 [13].

Enzyme Linked Immunoabsorbent Assays (ELISA). ELISA'S were performed in polystyrene microtiter plates [75]. Test peptides were dissolved in 0.2 M carbonate buffer, pH 9.6, at 10μg/ml and 100μl per well used to coat the plates. Antigens were coated on the plates by drying down the solution overnight at 37°C. Plates were washed 3 times with phosphate-buffered saline, pH 7.8, containing 0.05% Tween 20 (PBST) and then quenched with diluent (3% gelatin in PBST). The antisera, diluted in diluent, is reacted with antigen for 1 hr in the plate wells. The wells were then washed 4 times with diluent. Peroxidase conjugated goat anti-rabbit immunoglobulin, diluted 1/1000 in diluent, is incubated in the wells for 1 hr. After 5 washings with PBS, 0.03% 2,2'-axino-di- (3- ethylbenzthioazoline-sulfonic acid) in 0.1 M citrate buffer, pH 4.0, with 0.01% hydrogen peroxide was added. The reactions are read at 414 nm in a Titertek Multiscan.

Phosphorylation of Band 3. The phosphorylation reaction was performed as described [58,59] by aliquoting 100 μg brain homogenate and 5 mM HEPES buffer (pH 8.0) to a final volume of 50 μl and incubating this in 200 mM Tris-HCl (pH 7.6), Protein kinase A Inhibitor (100 μg/ml) (Sigma), for 12 minutes at 37°C with (gamma ^{32}P) ATP, 0.015 μCi/mM (37GBq/mmol)(ICN). At the end of the phosphorylation reaction the samples were diluted with a 1M Tris-HCl buffer (pH 6.8) containing 40% glycerol and 25 mM EDTA. Fifty micrograms (50 μg) in 100 μl of each sample were loaded into a 6-23% SDS-PAGE gel and allowed to run within 2 cm from the bottom of the gel. A Western blot was performed on PVDF membrane at 28 amps for 22 hours. Upon completion of transfer, the membrane was then washed 3 times for 15 each in a Tris Buffered Saline (TBS) containing 50 mM Trizma base, 3% fish gelatin, 0.1% Tween 20 then incubated for one hour in a polyclonal antibody to band 3 carboxy peptide diluted in TBS. The PVDF membrane was then washed 3 times for 5 minutes in TBS and incubated for one hour in 50 μl goat anti-rabbit alkaline phosphatase (Zymed), 0.1% MgZn, diluted in TBS. The membrane was then washed 3 x 5 minutes each in 50 mM Tris (pH 7.4), 0.1% Tween 20. Band 3 labeled antibodies were identified by incubating the membrane in 2.5% Diethanolamine, 0.3% 5-Bromo-4-chloro-3-indolyl-beta-D-galactoside, 0.7% p-Nitro Blue Tetrazolium Chloride, just until the color began to develop. Radioactively phosphorylated proteins were identified by exposing the membrane in a lightning plus cassette at -70 °C for 24 hrs.

EBV-Transformation of Peripheral Blood Lymphocytes (PBL)

Peripheral blood lymphocytes (PBL) were isolated by ficoll-Hypaque centrifugation using standard techniques. The lymphocytes were then incubated overnight with spent tissue culture supernatant from the marmoset cell line, B95-8, at a ratio of one part B95-8 supernatant to two parts RPMI 1640 supplemented with 10% fetal calf serum under standard cell culture conditions. Cells were then resuspended in fresh media and were further incubated for two weeks. The supernatant was then assayed for IgG and IgM directed against the band 3 peptides corresponding to residues 538-554 and residues 812-827 [23]. EBV transformed cells which produced positive results in band 3 ELISAs, were subcloned in 96-well plates at densities of 5 and 50 cells per well. Mitomycin C treated allogeneic PBL were added as a feeder cells. After two weeks the subcloned EBV transformed B cells were assayed again for reactivity against band 3 peptides. Cells from wells demonstrating positive reactivity were subcloned and assayed once again in the same manner as above and then expanded to 1 x 10^6 cells per ml.

RESULTS AND DISCUSSION

Band 3 Aging

Senescent cell antigen (SCA), an aging antigen, was discovered in 1975 [28]. It is a protein that appears on old cells and marks them for death. It acts as a specific signal for cellular termination by initiating the binding of IgG autoantibody and subsequent removal by phagocytes [1,2,10-12,19,22-48]. This appears to be a general physiologic process for removing senescent and damaged cells in mammals and other vertebrates [31]. Although the initial studies were done using human erythrocytes as a model, senescent cell antigen was discovered on cells in addition to erythrocytes in 1981 [31]. It occurs on all cells examined [1,2,31]. The aging antigen itself is generated by the degradation of an important structural and transport membrane molecule, protein band 3 [32]. Besides its role in the removal of senescent and damaged cells, senescent cell antigen also appears to be involved in the removal of erythrocytes in clinical hemolytic anemias, sickle cell anemia and the removal of malaria-infected erythrocytes. The aging and demise of band 3, which is synonymous with generation of senescent cell antigen, occurs in two distinct steps. Structurally, band 3 undergoes an as yet uncharacterized initial change during cellular aging that triggers a series of events terminating the life of the cell. We have recently developed antibodies against aged band 3 that recognize this change because they bind to a distinct region of band 3 in old but not middle-aged or young cells. Following the change in intact band 3 with aging, band 3 undergoes degradation, presumably catalyzed by an enzyme [67,75]. Preliminary experiments indicate that it is a calcium dependent membrane bound protease [67,75]. Cleavage of band 3 occurs in the transmembrane, anion transport region [32,34,53,62]. Fragments of band 3 are detected in membranes of old but not young cells by immunoblotting with antibodies to normal band 3. Following degradation, band 3 undergoes a change in tertiary structure becoming senescent cell antigen [75]. We have created a synthetic senescent cell antigen that immunologically mimics the one on senescent cells. A physiologic IgG autoantibody binds to our synthetic senescent cell antigen and initiates cellular removal.

Band 3 ages functionally as well as structurally during cellular aging. The following functional changes in band 3 occur as cells age, regardless of the age of the individual. These changes are: decreased anion transport activity (increased K_m; decreased V_{max}), decreased number of high affinity ankyrin binding sites, and binding of physiologic IgG autoantibodies in situ [28-34,53,62-73]. In addition, band 3 undergoes an as yet undefined change that results in binding of antibodies to aged band 3 [34]. These antibodies recognize band 3 that has aged prior to its formation of SCA. Degradation of band 3 generates SCA [31-34, 71-73].

Band 3 and Senescent Cell Antigen in Brain

We then defined band 3 changes during aging of the individual to determine if band 3 in terminally differentiated cells ages along with the individual and undergoes distinct aging changes. This turned out to be the case, and we defined the aging changes as described below. For example, cellular degradation of neural tissue results in the generation of senescent cell antigen as indicated by the labeling of material in brain lysosomes or peroxosomes by antibodies against a peptide of synthetic senescent cell antigen [25]. Anion transport both by band 3 and by non-band 3 transporter are impaired in brain tissue from old mice (Figure 2).

Examination of frozen brain sections from 10 year old and 96 year old individuals revealed labeling of fibrillary structures and processes with senescent cell antigen-band 3 antibodies in sections from old but not young brains [2].

The anion exchange, band 3 protein(s) in mammalian brain performs the same functions as that of erythroid band 3 [1,10,16,45]. These functions are anion transport, ankyrin binding, and generation of senescent cell antigen. Structural similarity of brain and erythroid band 3 is suggested by the reaction of antibodies to synthetic peptides of erythroid band 3 with brain band 3, the inhibition of anion transport by the same inhibitors, and an equal degree of inhibition of brain and erythrocyte anion transport by synthetic peptides of erythroid band 3. One of these segments, COOH, contains antigenic determinants of SCA. This supports the hypothesis that the immunological mechanism of maintaining homeostasis is a general physiologic process for removing senescent and damaged cells in mammals and other vertebrates.

Band 3 Structural and Functional Changes during Brain Aging

Physiological and functional parameters of band 3 were studied on aliquots of brains from mice of various ages as described below.

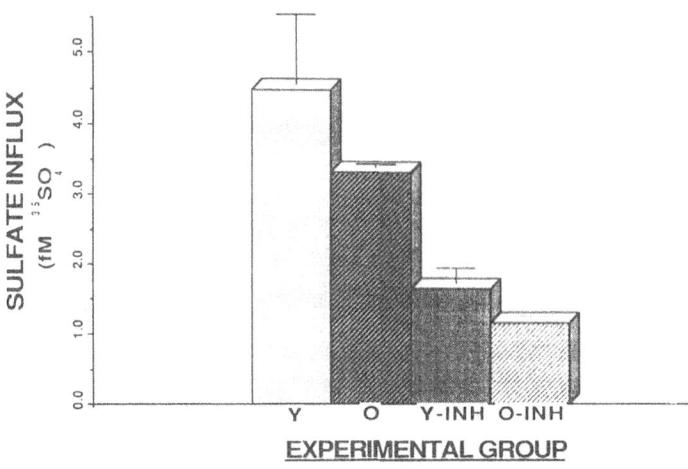

Figure 2. Anion transport (sulfate influx) by membrane vesicles prepared from young (3-4 month) and old (27 month) mice. The final concentration of protein was 0.1 mg per sample.

Anion Transport by Brain Frontal Cortex Decreases with Age. Results of anion transport studies indicate that transport by band 3 is impaired in brain tissue from old mice (Figure 2). Our data shows a consistent 22% difference between brains from young and 30 month-old, long-lived CBA mice. In addition to a decrease in transport by band 3 anion exchangers, non-band 3 related anion exchangers also exhibit a decrease in transport with age [24,55]. This is indicated by the fact that there is still a difference, although much smaller, in transport by young and old brains even when band 3 inhibitors are used to prevent transport by band 3.

Brain Band 3 Phosphorylation Changes with Age. Band 3 is a phosphorylated protein [76,77]. There are 10 tyrosines in the transmembrane, anion transport domain of band 3 and 7 in the cytoplasmic domain [13]. Membranes contain an associated tyrosine kinase which phosphorylates highly anionic peptide acceptor sites. Other membrane kinases phosphorylate band 3 on serine residues. The aging vulnerable site, 588-602, has a tyrosine adjacent to 2 serines, and is on an internal loop of band 3 in the cell cytoplasm according to our model (Figure 1). We determined that there was a change with age in the phosphorylation of brain band 3. Phosphorylation of band 3 from middle-aged (15-17 mo) and old (27-28 mos) mouse brains were compared by quantitating the amount ofα ^{32}P ATP binding to band 3 in the membranes of frontal cortex cells using sodium dodecyl sulfate- polyacrylamide gels [76,77]. Equal amounts of protein were loaded on gels as determined by Lowry's [78]. To determine which phosphorylated proteins were band 3 related, we incubated immunoblots (Western blots) with antibodies to band 3, and stained with alkaline phosphatase so that each lane was double labeled. We found that a 90-95 kD segment of band 3 from old brains (27 mo old mice with a mean life span of 30 mo) was more highly phosphorylated than that of middle-aged brains under experimental conditions based on quantitative scans of autoradiographs ($P \leq 0.003$; 55). A 25-28 kDa segment of band 3 is less phosphorylated in old brain, but the difference is not statistically significant ($P \leq 0.07$). This indicates that at least one post-translational modification of band 3 is an increase in phosphorylation sites. Studies have indicated that increased phosphorylation of a membrane protein can down-regulate its interaction with band 3 [79]. Other studies indicate that abnormal phosphorylation can interfere with the breakdown of molecules [80]. We are investigating whether increased phosphorylation of band 3 with age interferes with its interaction with other membrane proteins or with its degradation and removal, and whether increased phosphorylation of band 3 alters its functions such as anion transport. We did not determine whether the increased phosphorylation under experimental conditions is due to changes in the molecules configuration or a decreased in natural phosphorylation thus leaving more sites available for ^{32}P labeling.

Shifts in Band 3 and Its Breakdown Products with Age. We used antibodies band 3 and its peptides in an immunodot assay [81] to determine the amount of band 3, altered band 3, and band 3 breakdown products in young and old red cells and middle-aged (15-17 mo) and old mouse (27-28 mos) brain tissue. Controls in which no membranes or band 3 has been spotted were determined, and subtracted from experimental points. All samples to be compared were run at the same time. In our experience, these controls are less than 10% of experimental values. All determinations are performed in triplicate. Results showed that band 3 breakdown products increased with brain aging. Antibodies to band 3 breakdown products (bd3-brk) show a 5 fold increase in binding to old membranes ($P \leq 0.0001$). Antibodies to R 812-830, which carries SCA, show a 2 fold increased binding with age to brain membranes ($P \leq 0.01$). Antibodies to the cytoplasmic segment and to normal brain which were run as controls showed no difference.

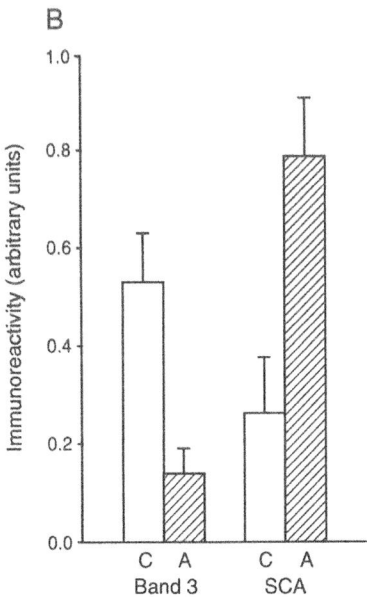

Determination of Band 3 Phosphorylation in Ad Brains

Phosphorylation studies were performed as described above. Brain membranes from AD patients were compared to normal age matched controls. The initial studies were done on frontal cortex and temporal lobe. The 25-28,000 kD segment appears to be more phosphorylated in control brains than in AD brain [55].

Studies on Peripheral Blood Cell Changes in Alzheimer's Disease

We have identified a defined patient population for which the inclusion criteria for our study is as follows. Patients must meet NINCDS-ADRDA criteria for clinical diagnosis of probable Alzheimer's disease [83]. These include: 1) dementia established by clinical examination and documented by the Mini-Mental State Examination, Blessed Dementia Scale, or some similar examination, and confirmed by neuropsychological tests; 2) deficits in two or more areas of cognition; 3) progressive worsening of memory and other cognitive functions; 4) non disturbance of consciousness; 5) onset between the ages of 40 and 90, most often after age 65; and 6) absence of systemic disorders or other brain diseases that in and of themselves could account for the progressive deficits in memory and cognition. In addition, the following neuroradiological inclusion criteria are employed by us as recommended: CT or MRI without evidence of focal lesion other than microangiopathic white matter changes frequently seen in the elderly. SPECT scan showing bilateral temporo-parietal hypoperfusion, highly characteristic of Alzheimer's disease. SPECT and MRI were performed on all participants in the study. This is fairly pure Alzheimer's disease without multi-infract dementia or other neurologic disease. Because of adherence to rigid criteria, only 8 individuals have been admitted to the study to date.

Our studies on peripheral blood cell changes in Alzheimer's disease have been very exciting. The results thus far on this defined patient population indicate a difference in glucose and anion transport, and in reactivity of cell membranes from individuals with Alzheimer's disease as opposed to their spouses and age-matched controls with select anti-band 3 peptide antibodies. Spouses were used to control for age and for environmental factors. We recognize that most of the Alzheimer's disease patients are male and the spouses are, of course, female. However, we do not observe differences in these assays between males and females as determined on both young and middle-aged donors, and on middle-aged and old mice. However, we compensated for sex differences by including age matched male controls as well. Glucose transport differences were observed in the V_{max} of glucose transport (AD, 214 pmol/10^8cells/min $\pm$ 35; control 269 $\pm$ 43; P $\leq$ 0.03), but not the K_m (AD, K_m 1.47 $\pm$ 0.41 mm; control, 1.81 $\pm$ 0.56; P $\leq$ 0.10). Anion transport differences are approaching significance for V_{max} and are expected to become significant as our sample size increases (V_{max} AD, 16.5 mmol/10^8cells/min $\pm$ 2.2; control, 20.4 $\pm$ 1.0; P $\leq$ 0.07); K_m (AD, K_m 3.46 $\pm$ 0.16 mm; control, 3.16 $\pm$ 0.18; P $\leq$ 0.05).

The "dot-blot"/ "immunodot assay was used to compare band 3, altered band 3, anti peptide antibodies, and band 3 breakdown products in red cells from individuals with Alzheimer's disease and age-matched controls. There was a significant difference between red cell membranes from patients from Alzheimer's disease and the spouse who was used as a control to the following antibodies to peptide segments: bd3-brk (P $\leq$ 0.01), AD-BD3 (P $\leq$ 0.04), and a glucose channel protein (P $\leq$ 0.05). Reactions with the other antipeptide and anti brain and erythrocyte antibodies was unchanged.

The immunoblot assay has turned out to be a sensitive reliable assay for detecting changes in Alzheimer's disease in peripheral blood cells and for detecting changes in the brain during aging. We are following these patients, repeating the tests every 6 months, and plan to obtain brain tissue at autopsy when they die.

The results of our studies, to date, indicate that brain band 3 undergoes the same age-related changes undergone by band 3 in other cell types including decreased anion transport, increased degradation of band 3, and generation of senescent cell antigen, increased phosphorylation of brain band 3 and a change in reactivity with antibodies to synthetic peptides of band 3. Studies on our defined patient population suggest that changes in brain band 3 are accompanied by changes in the peripheral blood. Autoantibodies to band 3 peptide 588-602 significantly increases with age (P ≤ 0.0001 by two-tailed t-test). This is one of the three key transport sequences of band 3 [25]. Increase in autoantibodies to this region are consistent with data indicating that this is a vulnerable region that is altered during aging.

In AD, preliminary results indicate that post-translational changes include decreased brain band 3 phosphorylation of a 25-28kD segment, increased degradation of band 3, alterations in band 3 recognized by antibodies, and decreased anion and glucose transport by blood cells. Serum autoantibodies were increased in AD patients compared to controls to band 3 peptide 822-839 by the two tailed t-test of P ≤ 0.06. Significance is expected to increase as sample size increases. This band 3 residue lies in an anion transport\binding region. This is consistent with alteration of this region in AD since it is recognized as antigenically different by the patient's immune system. Appearance/increase in antibodies to this region suggests that it may be altered in AD. This is consistent with the observed decrease in anion transport.

Naturally Occurring Serum Antibodies to Synthetic Peptides of Band 3, Brain Band 3, and SCA during Aging and Disease

Serum Autoantibodies to Brain Band 3, Synthetic Peptides of Band 3, and SCA during Aging and Disease. Since all human sera contain antibodies to senescent cell antigen, we quantitated naturally occurring autoantibodies to senescent cell antigen in human serum with ELISA during aging to determine whether these antibodies increased with age. At the same time, we also tested serum for antibodies to T cell receptor peptides. We selected T cell receptor peptides because we wondered if antibodies to the T cell receptor beta chain which is the antigen-specific receptor on thymus derived lymphocytes could contribute to the observed decrease in T cell activity with age and diseases [24,74]. We developed an ELISA that allows us to titer human serum against band 3 and T cell receptor peptides. We used this assay, thus far, to titer activity against band 3 peptides in serum from young and old individuals, and individuals with RA, SLE, and AD (below). Initially, we screened 20 peptides of band 3 with human and mouse serum to determine which ones were recognized by naturally occurring antibodies in mice and humans.

We showed that all sera contain low levels of antibodies to senescent cell antigen synthetic peptides residue 538-554 (pep-ANION 1) and 812-827 (pep-COOH). Absorption of sera with peptide affinity columns removed the peptide binding activity from the flow-through and increased it in the bound fraction that was eluted from the column with glycine-HCl, pH 2.3 [29,31]. Absorption with a cytoplasmic peptide did not affect anti pep-ANION 1 or anti pep-COOH activity in the eluate.

We titered antibodies to senescent cell antigen peptides pep-ANION 1 and COOH, an internal anion transport region peptide, ANION 2, and T-cell receptor sequences pNTERM (DAGVIQSPRHEVT-EMG) and pCDR1 (CKPISGHNSLFWYRQY). These sequences were chosen either as negative controls or test antigens on the basis of preliminary studies. Data were analyzed using the two-tailed Wilcoxon's test [74].

Studies of normal humans ranging in age from 20 to 90 suggests two major patterns for the IgM natural antibody response to synthetic peptides giving high responses. The first is that the level of IgM reactivity is high early in life and remains high throughout. The second pattern is one in which the reaction is high in younger individuals, but diminishes substantially in the latter decades of life. IgG autoantibodies to all 3 band 3 peptides tended to increase with age. Antibodies to ANION-2 and COOH are relatively low early in life but show substantial increases with increasing age. Antibody response to ANION-1 is relatively high in the third decade of life (geometric mean titer ≈ 250), decreases progressively to a geometric mean titer of under 40 during the sixth decade of life, and then rebounds to a geometric mean titer of ≥ 80 in the seventh and eighth decades. Although there is substantial variation in the IgG autoantibody responses, which most probably are strongly dependent upon T-cell help, the overall pattern is for these to be low early in life, and to increase with increasing age. We speculate that the antibody activities detected in the normal individuals carry out a physiologic regulatory role, as opposed to a destructive autoimmune dysfunction, because these molecules can be present in high levels throughout life.

The synthetic autoantigens that generate the highest titers of natural antibody are those that are either exposed on the surface of the cell (band 3 peptides) or are exposed in the predicted three-dimensional folding of molecule (T cell receptor β peptides).

We have performed ELISAs on serum from mice of various ages, reacting them with peptides of band 3. We studied the long-lived, CBA strain of mice which is resistant to autoimmune disease [84]. A significant difference was found in reactivity of sera of aged mice (27-28 mos) as compared to middle-aged mice (17 mo) to the peptide 588-602 (P ≤ 0.0001 by two-tailed t-test). No differences were found in reactivity to the peptide residues 538-554, and 812-827 in which prolines were substituted for histidine. Our data in other studies indicate that the histidine is a crucial amino acid. Peptide residue 588-602 is one of the three key transport sequences of band 3 [25].

Demonstration of B Cells Producing Antibody to SCA in Human Peripheral Blood. We were able to clone the B cells producing antibodies to senescent cell antigen from normal humans using EBV transformation. We transformed B cells from peripheral blood using Epstein-Barr virus. Some of these B cells secreted autoantibodies to senescent cell antigen. We have cloned B cell populations actively secreting autoantibodies that recognize SCA by limiting dilution analysis and obtained those cells secreting antibodies to SCA. Peripheral blood lymphocytes (PBL) were isolated by Ficoll-Hypaque centrifugation using standard techniques [30,34]. The lymphocytes were then incubated overnight with spent tissue culture supernatant from the marmoset cell line, B95-8, at a ratio of one part B95-8 supernatant to two parts RPMI 1640 supplemented with 10% fetal calf serum under standard cell culture conditions. Cells were then resuspended in fresh media and were further incubated for two weeks. The supernatant was then assayed for IgG and IgM directed against the band 3 peptides corresponding to residues 538-554 (anion peptide) and residues 812-827 (COOH peptide) [74]. Positive controls consisted of rabbit antibodies raised against pep-ANION 1 and COOH, which have strong titers at a 1:250 and a 1:3000 dilution, respectively (Figure 4). The antibody from EBV transformed clones reacted well at a dilution of 1:200 with pep-ANION 1, and 1:400 with pep-COOH. Thus, B cells producing antibody to SCA occur naturally in peripheral blood.

Long Term Effects of Radiation Exposure on Naturally Occurring Autoantibodies to Band 3 and SCA. We examined the effect of radiation on naturally occurring autoantibodies to band 3 synthetic peptides. The A-bomb survivors were in Hiroshima and had documented exposure. We obtained sera with the help of the Hiroshima Radiation Effects Research Foundation in 1980. Sera was stored frozen at -70°C until testing. The age range of the A-bomb survivors was 51-59 yrs. We titered serum antibodies to band 3 residues 538-554, 588-602, 822-830, and 812-830. A significant decrease was found in antibody titer to the peptide 588-602 (P ≤ 0.03) between those with documented radiation exposure and unex-

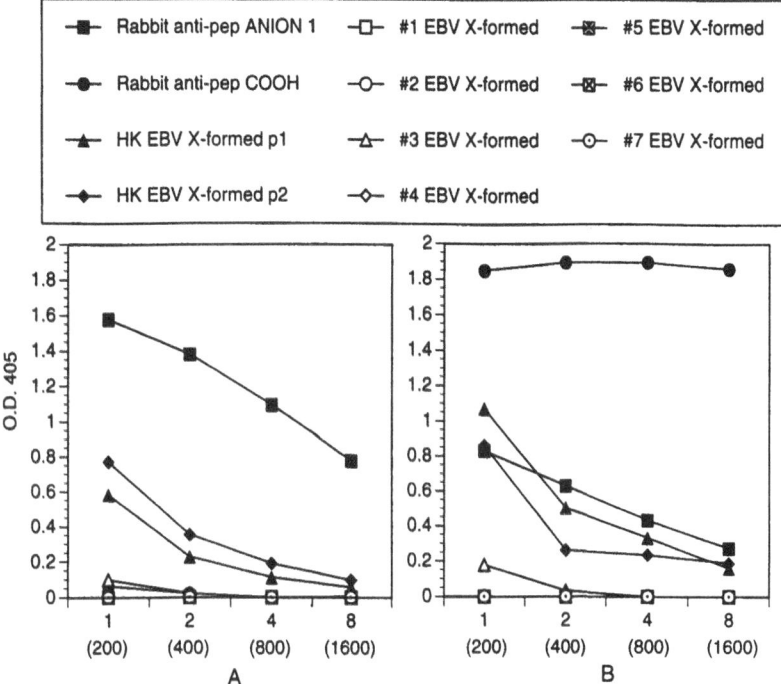

Figure 4. Reactivity of supernatants from EBV-transformed B cells against Band 3 peptides in ELISA. Two-fold serial dilutions of culture supernatants from EBV-transformed B- cells were added to peptide coated ELISA plates. ELISA was performed as described in methods section. High titer rabbit anti-sera were produced against peptides corresponding to residues 538-554 (J2 peptide) and residues 812-827 (J4 peptide). Numbers in parentheses along the X-axis correspond to dilutions of the high titer rabbit anti-peptide sera. "HK EBV X-formed p1 and p2" refers to passage number one and passage number 2 of the EBV transformed B cells from a normal donor. "#1 EBV X-formed through #7 EBV X-formed" refers to EBV transformed B cells from patients with "fast aging" band 3 disease.

posed, control Japanese (Figure 5). A decrease in antibodies was also found between controls and A-bomb survivors to peptide 822-830 (P ≤ 0.07), but it was not significant. When one considers that antibody titers to these two peptides tends to increase with age, the decrease in responsiveness in A-bomb survivors is even more striking. The implication of this finding is unclear. It is possible that the decrease in antibody titer is indicative of radiation-induced suppression of lymphogenesis or that band 3 is an potential tumor cell and autoantibodies are responsible for their removal.

Serum Autoantibodies to Brain Band 3, Synthetic Peptides of Band 3, and SCA Are Detected Between Normal Human Age-Matched Controls and Individuals with Rheumatoid Arthritis (RA), Systemic Lupus Erythematosus (SLE), and AD. Having demonstrated naturally occurring antibodies to senescent cell antigen peptides, we quantitated these naturally occurring autoantibodies to senescent cell antigen in human serum with ELISA in both generalized, systemic disease such as rheumatoid arthritis (RA) and systemic lupus erythematosus (SLE) and in a neurologic disease (AD). We examined antibodies to T cell receptor peptides in RA and SLE [24,74].

RA and SLE. The results of these studies show that RA patients exhibit an increase in IgM antibodies to pep-ANION 1 (a peptide of SCA) and CDR1, and an increase in IgG

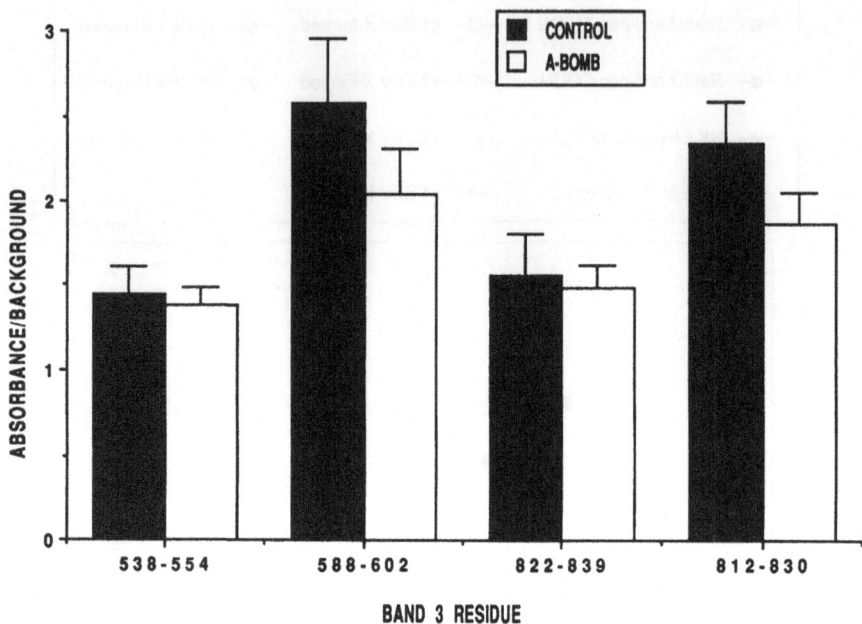

BAND 3 RESIDUE

Figure 5. ELISA titrations of sera from Hiroshima A-bomb survivors against band 3 synthetic peptides residues (R) 538-554, R 588-602, R822-839, and 812-830. Data is expressed as the mean absorbance divided by the background determined using an uncoated plate. Reactivity with R 588-602 was statistically significant (P ≤ 0.03). Controls were Japanese from near the sea who had no radiation exposure.

antibodies to peptides ANION 1 and COOH, the two peptides of SCA. SLE patients exhibit an increase in IgG to peptide COOH and to ssDNA. Thus, in tissue destructive diseases, IgG antibodies to SCA are elevated. We are extending our studies to determine whether antibodies to synthetic peptides can have diagnostic or predictive value in disease.

AD. Studies focused on the individuals and controls in our AD studies. These patients meet the criteria for clinical diagnosis of probable Alzheimer's disease. Results showed a difference between AD patients and controls to band 3 peptide 822-839 by the two tailed t-test of P ≤ 0.06. Significance is expected to increase as sample size increases. This band 3 residue lies in an anion transport\binding region [19,25]. It is one of the three key transport sequences of band 3 [25]. The effect of antibodies to this transport region on anion transport is being investigated to see if it contributes to the disease. We are currently testing AD serum against the T cell receptor peptides used above.

Mechanisms of Band 3 Aging

Identification and Localization of Aging Vulnerable Site on Aged Band 3. Our data indicate that the aging antigenic site(s) is different from the site in band 3 that is vulnerable to aging damage and initiates band 3 degradation. We have developed an antibody that recognizes subtle aging changes in the band 3 molecule prior to formation of senescent cell antigen [34,69,85]. Antibodies to aged band 3 bind to a distinct region of band 3 in immunoblots of membranes prepared from old cells but not middle-aged or young cells. In addition, this antibody binds to band 3 in cells and membranes from individuals with band 3 alterations that result in accelerated aging [34,57,85]. It does not bind to membranes from a band 3 mutation that has a normal or increased rbc lifespan [86].

Peptide binding experiments indicate that this antibody recognizes the band 3 segment which we refer to as ANION 2, residues 588-602 which is an internal segment (Figure 1) residing in one of 3 anion transport/binding regions [19,25]. These antibodies bind strongly to ANION 2 with trace binding to COOH, but not to ANION 1 or pep-CYTO in immunoblots. This suggests that the band 3 change that precedes generation of senescent cell antigen occurs in or near ANION2. Trace binding to COOH may indicate that region is involved.

These results are consistent with the physiological data demonstrating that old cells have impaired anion transport [25,34,53], and the biochemical and immunological data indicating that band 3 undergoes degradation with loss of a cytoplasmic segment during the aging process [19,34,75,76]. Peptides ANION 1 and COOH are in highly conserved regions of band 3 [19,50]. Our results are consistent with aging changes occurring in the region of ANION 2, residues 588-602. We found an increase in naturally occurring serum autoantibodies to band 3 residue 588-602 which is the key internal anion transport region [74]. In AD, there is an increase in naturally occurring serum autoantibodies to another transport region, namely, residue 822-839, which overlaps a senescent cell antigen peptide [25,26]. This increase is relative to age-matched controls, thus, this change is in addition to the age-related changes observed. The appearance /increase in autoantibodies to these regions with age or AD support the hypothesis that these regions are altered, and that the alteration is recognized immunologically.

Molecular Biology of Senescent Cell Antigen and Band 3

Mapping of Transport Function and Expression of Senescent Cell Antigen Onto a Structural Representation of the Band 3 Molecule. Based upon the translated sequence of cDNA, the human erythrocyte band 3 protein consists of 911 amino acids [13]. Protein sequences have also been derived for erythrocyte band 3 molecules of mouse and chicken as well as for molecules related to band 3 from lymphocytes, the cell line K562, and kidney. A schematic diagram of the band 3 model showing key peptides is shown in Figure 1. The N-terminal 393 amino acids are located within the cytoplasm of the cell where they associate with the cytoskeleton by binding to ankyrin. This so-called "cytoplasmic domain" can be isolated as a water-soluble fragment (NTr-41) by digestion of red cell ghosts with trypsin. The "membrane-associated domain" (TrC-55) consists of the C- terminal 518 amino acids. These separable fragments exhibit different behavior in solution: the NTr-41 is soluble in aqueous solution and the CTr-55 is soluble in diluted detergent. These properties of the protein fragments parallel the proposed states of aggregation of intact molecules within the cell plasma membrane where stable dimers and tetrameres are found.

Two crucial properties of the band 3 molecule namely the anion transport region and formation of the senescent cell antigen [25,26] map to the membrane- associated "domain". Application of hydrophilicity hydropathy algorithms to the derived protein sequence indicates that the membrane associated region consists of 11 hydrophobic membrane spanning helical regions and four major external and three major internal hydrophilic loops. The C-terminal 40 residues are located within the cytoplasm.

The locations of these predicted structures and the residues implicated in anion transport are given in Figure 1. Peptide technology was used to map the regions reactive with autoantibody to senescent cell antigen as summarized in Figure 1 [25-27,45,46]. Synthetic peptides were used to block either binding of senescent cell IgG autoantibody to intact red blood cells or the transport of sulfate. Peptides from the second (ANION 1) and fourth (COOH) major external loops also blocked the antibody. Furthermore, these peptides show a synergistic effect when admixed, thereby suggesting that these regions which are distant in amino acid sequence maybe closely associated in the 3-dimensional structure of the cell-bound band 3 molecule (or molecular complex). It is of interest that the autoantibodies also bind other peptides that do not block in the competition assay. This reflects the fact that aged cells binding the anti-senescent cell antigen autoantibody are removed from the circulation by phagocytes [28,29] which presumably process the antigen into peptides presented to T cells.

Peptides from external loop 04 (residues 812-830) and internal loop In-1 are the most effective inhibitors of anion transport [18,19,23-27,46]. The key residues tend to be positively charged lysyl (K) or arginyl (R) residues with some involvement of histidyl (H) and acidic glutamyl (E) or aspartyl amino acids. The involvement of lysines in external loop 04 is originally suggested by the capacity of these residues to bind the covalent inhibitor of anion-transport DIDS [87]. Comparison of band 3 sequences of different species strengthens the identification of residues Lys-539, 817, 829 and 851, Arg 589, 602,603, 808, 827, 832, 870 and 871 and His-703, 734, 819 and 834 as essential to anion transport because these are conserved among man, mouse and chicken (Figure 1). Comparisons among sequences of the N-terminal cytoplasmic regions do not provide clear indications of functional regions, however, because there are marked polymorphisms in length and no definite ankyrin-binding region shared with other ankyrin-binding proteins occurs. It is of further interest that searches of general computer databases by us and others showed considerable conservation of residues within the membrane-associated domains of band 3 of various species and band 3 related molecules but detected no strong homologies with other proteins.

Aging Antigenic Site. The aging antigenic site was identified using synthetic peptides and the IgG that binds to old cells [19,22-27]. Results indicate that crucial anion transport segments of the band 3 molecule carry the aging determinants and suggest that these may be the aging sites of the molecule.

Previous studies indicated that senescent cell antigen is a degradation product of band 3 that includes most of the ~35,000 Da carboxyl terminal segment and the ~17,000 Da anion transport region [32]. Both immunoblotting studies with IgG isolated from senescent cells and peptide mapping studies of senescent cell antigen indicated that senescent cell antigen lacks a ~40,000 molecular weight cytoplasmic segment which contains the amino terminus and, possibly, additional peptides of band 3 [19,32,34]. Peptide mapping studies and anion transport studies suggested that a cleavage of band 3 occurs in the anion transport region [22]. Furthermore, breakdown products of band 3 are observed in the oldest cell fractions but not in young or middle-aged cell fractions, and anion transport is impaired in old cells [32,34].

Based on structural, biochemical, and immunological data [31,32,34], cleavage of old band 3 was hypothesized to occur approximately a third of the way into the transmembrane anion transport region from the carboxyl terminus end. Therefore, peptides were synthesized from the anion transport domain of erythrocyte band 3. "Walking" of the anion transport segment was performed. "Walking" means the antigenic analysis of a series of synthetic overlapping peptides that encompass the entire polypeptide chain of the anion transport domain. The synthetic peptides are of uniform size and overlap their adjacent neighboring peptides by a predetermined number of residues to expect reasonable resolution of individual antigenic sites.

These studies focused on the 511 amino acid anion transport domain. Results indicate that this domain carries both the SCA antigenic determinants; whereas, the 40,000 Da cytoplasmic segment does not [32,34]. Therefore, investigation focused on segments within the transport domain that are most likely to be hydrophilic and therefore exposed based on hydrophobicity/hydrophilicity scales, amino acid composition, and location along the molecule based on our model described by Kay, *et al.* [18,19,23-37]. Peptides that corresponded to external regions predicted by the band 3 model presented earlier were synthesized. The peptides with residue number are:

CYTO	129-144	AGVANQLLDRFIFEDQ
	426-440	LLGEKTRNQMGVSEL
	515-531	FISRYTQEIFSFLISLI
	526-541	FLISLIFIYETFSKLI
	538-554	SKLIKIFQDHPLQKTYN
	549-566	LQKTYNYNVLMV-PKPQGP
	561-578	PKPQGPLPNTALLSLVLM
	573-591	LSLVLMAGTFFFAMMLRKF
ANION 2	588-602	LRKFKNSSYFPGKLR
	597-614	FPGKLRRVIGDFGVPISI
	609-626	GVPISILIMVLVDFFIQD
	620-637	VDFFIQDTYTQKLSVPD
GLYCOS	630-648	QKLSVPDGFKVSNSSARGW
	645-659	ARGWVIHPLGLRSEF
	684-704	ITTLIVSKPERKMVKGSGFHL
	752-769	IQEVKEQRISGLLVAVL
	776-793	MEPILSRIPLAVLFGIFL
	788-805	FGIFLYMGVTSLSGIQL
	800-818	LSGIQLFDRILLLFKPPKY
	813-818	FKPPKY
COOH	812-827	LFKPPKYHPDVPYVKR
	812-830	LFKPPKYHPDVPYVKRVKT
	818-827	YHPDVPYVKR
	822-839	VPYVKRVKTWRMHLFTGI
	869-879+881+883	LRRVLLPLIFRVL*
	887-903	DADDAKATFDEEEGRDE
	902-910	DEYDEVAMP

As a control, a peptide from the cytoplasmic segment of band 3 within the region of the putative ankyrin binding site was used [13] (CYTO, 139-159). Peptides were synthesized based on the sequence data from the paper by Tanner et al. [13]. Peptides were evaluated in a senescent cell IgG competitive inhibition assay. The specific physiologic autoantibody to the senescent cell antigen, senescent cell IgG (SCIgG) was isolated from old human red cells as described previously using affinity chromatography [29,31]. IgG eluted from senescent cells, rather than serum IgG, was used because normal serum contains antibodies to spectrin, actin, 2.1, etc. [73]. Competitive inhibition studies were performed using synthetic peptides to absorb the IgG isolated from senescent erythrocytes.

Results of these "walking" studies, summarized in Figure 1, indicate that senescent cell IgG recognizes antigenic determinants that lie within region 538-554 an anion transport site containing a cluster of lysines toward the carboxyl terminus, 812-827. More recent studies indicate that residues 812-830 is a more potent inhibitor of SCIgG binding than 812-827 (inhibition: 812-830, 45±6 (P ≤ 0.001); 812-827, 32±5 (P ≤ 0.01). The peptide "CYTO" from the cytoplasmic domain did not inhibit. Immunoblotting studies demonstrate binding of senescent cell IgG to peptides ANION 1 and COOH but not to CYTO, the peptide from the cytoplasmic segment of band 3 containing the putative ankyrin binding site [18,19,23-27].

---

*The actual sequence in band 3 is: LRRVLLPLIFRNVEL, but the N and E were not included in the peptide used for these studies.

Modification of Senescent Cell IgG Binding Segments to Identify Active Amino Acids.
Lysine has been implicated as an amino acid involved in anion transport based on DIDS
inhibition studies [87]. We suspected that lysine might be part of the senescent cell IgG
binding site. This was tested by substituting a neutral or positively charged amino acid for
the positively charged lysine or arginine in pep-COOH peptide. pep-COOH peptide was
selected because a) it is a highly conserved region of band 3 [18,19,23-27], b) it is a crucial
sequence in senescent cell antigen [18,19,23-37], and c) it is an senescent cell IgG binding
peptide. Substitution of either neutral glycines (pep-COOH-G/K) or positively charged
arginines (pep-COOH-R/K) the positively charged amino acid lysine for lysine in pep-
COOH still resulted in significant inhibition (P ≤ 0.05), but the inhibition by substituting
arginine for lysine (R/K) was significantly reduced compared to that of pep-COOH (Table 1)
[23]. Since both arginine and lysine are positively charged, the significant inhibition was
probably caused by the bulky guanidinium group on arginine. We can not determine at this
time whether all lysines are critical or whether antigenicity depends on a specific lysine or
all three of the lysines in the synthetic peptide pep-COOH.

We then synthesized pep-COOH related peptides in which glycines were substituted
for arginines (pep-COOH G/R) and separately, glycines were substituted for arginines and
lysines (pep-COOH G/KR). Significant inhibition occurred with both altered peptides (P ≤
0.01; Table 1), and was not significantly different than that of pep-COOH. Since both
substituted peptides caused inhibition, it seems that the presence of lysine or arginine is not
an absolute requirement for IgG binding. However, the size and/or configuration seems to
be important because substituting and arginine for a lysine alters antigenicity.

Because the amino acid sequence of human and chicken pep-COOH and pep-COOH-
N6 differ in several key amino acids, peptides were synthesized from the chicken sequence.
Chicken pep-COOH has methionine instead of lysine at a position corresponding to residue
814 of the human sequence, a lysine and glutamic acid instead of a aspartic acid and valine
at residue 821 and 822, and a threonine (T) instead of lysine (K) at residue 826. For example,
substitution of a methionine (M) for a lysine (K) is significant since the former is nonpolar

Table 1. Inhibition of senescent cell IgG binding by altered human and
by chicken pep-COOH synthetic peptides of band 3 protein

Synthetic Peptide		Inhibition
Residue	Sequence	(%) of IgG Binding
Human 538-554	SKLIKIFQDHPLQKTYN	46±2***
Chicken 538-554	AKLVTILQAHPLQQSYD	56±2***"**"
Human pep-COOH	LFKPPKYHPDVPYVKR	29±4***
Chicken pep-COOH	LLMPPKYHPKEPYVTR	28±4** (NS)
Human pep-COOH-N6	FKPPKY	43±4*** (*)
Chicken-pep-COOH-N6	LMPPKY	33±5**(*)[**]
Human pep-COOH-G/K	LFGPPGYHPDVPYVGR	22±5* (NS)
Human pep-COOH-R/K	LFRPPRYHPDVPYVRR	18±4* (*)
Human pep-COOH-G/R	LFKPPKYHPDVPYVKG	29±4**(NS)
Human pep-COOH-G/KR	LFGPPGYHPDVPYVGG	27±5**(NS)

Data are presented as the percentage inhibition ± standard deviation of senescent
cell IgG binding to human erythrocytes by 17μM peptide. *P ≤ 0.05; **P ≤ 0.01;
***P ≤ 0.001; NS, not significant compared to control without peptide; "**",
compared to human 538-554; (*P ≤ 0.05, NS) compared to human pep-COOH; [**]
compared to human pep-COOH-N6. There is no significant difference between
human pep-COOH-G/K and human pep-COOH-R/K. Peptide concentration is
17μm.

and the latter is a positively charged amino acid. Likewise, substitution of a negatively charged glutamic acid (E) for a nonpolar valine (V) or a positively charged lysine for a negatively charged aspartic acid (D) would be expected to alter tertiary configuration and binding properties. Thus, nature has performed "site-specific" mutagenesis for us.

Chicken pep-COOH inhibited senescent cell IgG binding which was not significantly less than that of human pep-COOH (Table 1) [23]. However, chicken pep-COOH-N6 produced significantly less inhibition than human pep-COOH -N6. Thus, the lysine at 814 in the human sequence for which methionine is substituted in the chicken sequence probably contributes to senescent cell IgG binding.

Both human and chicken residues 538-554 inhibit senescent cell IgG binding (P ≤ 0.001). Chicken 538-554 produced significantly greater inhibition than the human. This suggests that the second and third lysines may not be critical for senescent cell IgG binding since the chicken peptide has an uncharged threonine and a glutamine in these positions. Since chicken peptide COOH-N6 inhibited senescent cell IgG binding less than human COOH-N6, (Table 1). It suggests that the first lysine in peptide COOH and COOH-N-6 contributes to IgG binding. The two peptide changes probably compensate each other *in situ*. Differences between the human and chicken peptide are underlined:

```
Anion 1 human:   S K L I K I F Q D H P L Q K T Y N
Anion 1 chick:   A K L V T I L Q A H P L Q Q S Y D
COOH human:      L F K P P K Y H P D V P Y V K R
COOH chick:      L L M P P K Y H P K E P Y V T R
COOH-N6 human:   L K P P K Y
COOH-N6 chick:   L M P P K Y
```

Peptides inhibited IgG binding to red cells in a dose dependent manner. IgG binding regions of band 3 reside within residues 538-554, 593-601, 813-818, and 812-830 (P ≤ 0.001 compared to control without peptide). The inhibitory activity of the last peptide, 593-601 is puzzling because a larger peptide, residue 588-602 that includes it, does not inhibit IgG binding. On explanation for this is that the smaller peptide may assume a different configuration or the larger peptide may have groups that block binding. The next most active inhibitors were peptides 597-604, of which the 7 mer segment 588-594 is a most active peptide, 601-607, 630-648, subpeptides and overlapping peptides of the most active peptide 812-830, namely, 812-827, 818- 823, 818-827 and 822-827, 827-835 and 822-839, and carboxyl terminal peptides 869-879+881+883, 869-875, 879-884 and 902-910 (P ≤ 0.01). These are useful mainly for inferring active antigenic amino acids. Residue 813-818, also called pep-COOH -N6, is a more effective inhibitor of IgG binding than pep-COOH. Segments between 839 and 868 were not synthesized because they would be expected to be extremely hydrophobic. However, the data obtained from high transport band 3 which has a proline to leucine in position 868 and demonstrates 2-3 times higher anion transport support the data demonstrating that the carboxyl terminal stretch from approximately 812 to 910 represents a major transport region. It should be considered that band 3 transports anions both inside and outside.

The data suggest that the three dimensional configuration of a band 3 peptide or segment is more important than charge or actual amino acid sequence as a determinant of anion binding. For example, substituting glycines for lysines did not significantly change inhibition, but substituting arginine for lysines did. We suspect that the three dimensional structure of band 3 is a ring in which the two external loops on which peptide 538-554 and pep-COOH reside are in close spacial proximity forming senescent cell antigen as cells age.

Results, to date, indicate that (a) an anion transport region of band 3 that appears to be extracellular carries active antigenic determinants of an aging antigen; (b) this transport site is located toward the carboxyl terminal and overlaps or contains a stilbene disulfonate binding site, (c) a putative ankyrin binding region peptide is not involved in senescent cell antigen activity; and (d) synthetic peptides alone, without carbohydrate attached, abolish binding of senescent cell IgG to red cells. Therefore, carbohydrate moieties are not required for the antigenicity or recognition of senescent cell antigen.

Mapping of Aging Antigenic Sites in Relation to Anion Transport and Anion Transport Inhibitor Binding Site(s). Senescent cell antigen contains pep-ANION 1 and pep-COOH [18,19,23-27]. Data suggest that senescent cell antigen is related to the anion transport site because peptide mapping studies and anion transport studies indicate that a cleavage of band 3 occurs in the anion transport region [32]. Furthermore, breakdown products of band 3 are observed in the oldest cell fractions, but not in young or middle-aged cell fractions [32,34]. Fragments of the cytoplasmic segment of band 3 are detected with antipeptide antibodies in the cytoplasm of old red cells and of red cells with a band 3 alteration associated with accelerated aging [19,32,34]. Anion transport is impaired in old cells [22,33,45,47,62]) and band 3 mutations/alterations with damage to the transmembrane transport region result in accelerated aging and increased IgG binding [34]. To clarify the relationship between senescent cell antigen and anion transport, we mapped the anion binding/transport properties of segments of band 3 using equimolar amounts of synthetic peptide and sulfate in a competitive inhibition assay (Figure 1) [25]. Studies showed that the peptide was competing for the sulfate because increased amounts of sulfate overcame the inhibition [18].

Anion Binding/Transport Site. Peptides inhibited transport in a dose dependent manner [18,25]. Peptide residues 588-594 (a 7 amino acid peptide), 822-839, and 869-883 were the most active inhibitors of anion transport ($P \leq 0.001$ compared to control without peptide). The inhibitory activity of the last peptide, 869-883, could not be confirmed by testing adjacent peptides to the amino side since these regions are extremely hydrophobic. However, 6-7 amino acid peptides from this region produced inhibition of transport [25], but to a lesser degree than 869-883. The component residues are probably additive. Peptide 869-879+881+883 seems to have special anion binding/transport properties although both it and 869-883 produced significant inhibition ($P \leq 0.001$). Synergy was not observed.

Anion transport has been attributed to residues 538-554 or, more recently, to a carboxyl segment of band 3 including pep-COOH, by investigators based on indirect evidence [87,88]. Residue 538-554 includes two important amino acids. The lysine at 538 (558 in the mouse) is a covalent binding site for the anion transport inhibitor, DIDS, and the tyrosine at residue 553 is radioiodinated by extracellular lactoperoxidase [13]. However, our present results indicate that the residues 538-554 are not anion binding although segments carboxyl to them are. This agrees with the results of studies using mouse band 3 expressed in *Xenopus laevis* oocytes indicating that lysine 558 in the mouse sequence (539 on peptide 538-554 in the human sequence) is not involved in anion transport based on site directed mutagenesis [89,90]. There are two other lysines on peptide 538-554. Since the peptide does not inhibit anion transport, the other two lysines on peptide 538-554 at positions 542 and 551 do not function as anion binding sites. Since peptide 538-554 does not inhibit anion transport even though it has 3 lysines, anion transport binding must involve more than a mere positive charge. The configuration of the peptide is probably important. Results of the amino acid substitution studies support this interpretation.

Modification of Anion Transport/Binding Segments to Identify Active Amino Acids. Lysine has been implicated as an amino acid involved in anion transport based on DIDS inhibition studies [87]. Experiments were performed to test this hypothesis by substituting a neutral or positively charged amino acid for the positively charged lysine or arginine in pep-COOH peptide. pep-COOH peptide was selected because a) it is a highly conserved region of band 3 [18,19], b) it is a crucial sequence in senescent cell antigen, and c) it is an anion transport/ binding peptide. Substitution of either neutral glycines (pep-COOH-G/K) or positively charged arginines (pep-COOH-R/K) for lysine in pep-COOH still resulted in significant inhibition, but the inhibition was significantly reduced compared to that of pep-COOH. Since we changed all three of the lysines in the synthetic peptide pep-COOH, we can not determine at this time whether all lysines are critical or whether antigenicity depends on a specific lysine.

pep-COOH related peptides were synthesized in which glycines were substituted for arginines (pep-COOH G/R) and glycines were substituted for arginines and lysines (pep-COOH G/KR). Significant inhibition occurred with both altered peptides, but only pep-COOH G/KR was significantly different than pep-COOH. No significant difference was observed between pep-COOH G/R and pep-COOH. Since all of the substituted peptides caused inhibition, it seems that the presence of lysine or arginine is not an absolute requirement for anion binding. Furthermore, it appears that another amino acid, perhaps glycine, can participate in anion binding.

Because the amino acid sequence of human and chicken pep-COOH and pep-COOH-N6 differs in several key amino acids, peptides from the chicken sequence were synthesized. Chicken pep-COOH has an M instead of K at a position corresponding to residue 814 of the human sequence, a K and E instead of a D and V at residue 821 and 822, and a T instead of K at residue 826. For example, substitution of a methionine (M) for a lysine (K) is significant since the former is nonpolar and the latter is a positively charged amino acid. Likewise, substitution of a negatively charged glutamic acid (E) for a nonpolar valine (V) or a positively charged lysine for a negatively charged aspartic acid (D) would be expected to alter tertiary configuration and binding properties. Thus, nature has performed "site-specific" mutagenesis for us.

Chicken pep-COOH inhibited anion transport, but the inhibition was significantly less than that of human pep-COOH ($P \leq 0.05$). Since chicken pep-COOH-N6 inhibited to the same degree as human pep-COOH-N6, the lysine at position 814 in the human sequence for which methionine is substituted in the chicken sequence is probably not critical for anion transport. This suggests that the change to a K and E instead of a D and V may be a significant change. A glutamic acid is implicated in transport [87]. Neither human nor chicken pep-AN-ION 1, residues 538-554, inhibited transport.

Some investigators have suggested that lysines are not themselves part of the transport mechanism but are close to the transport site, and that arginine is involved in anion transport [91,92]. Residues 869-883 and 869-879+881+883, among the most potent inhibitors of anion transport, have arginines but no lysines. Other highly inhibitory peptides, residues 588-594 and 822-839, have lysines and arginines. Of the inhibitory peptides with $P \leq 0.01$, residues 804-811 and 830-835 have arginines and no lysines, and residues 549-566, 561-578, and 813-818 have lysines and no arginines. The substitution studies show that altered pep-COOH with glycines substituted for both lysine and arginine still inhibits anion transport although the percent inhibition is reduced. Histidine and glutamine have also been implicated in anion transport [87]. The peptide sequence recognized by SITS antiidiotypic antibodies overlaps a potent anion binding/ transport peptide, 822-839, which is consistent with data indicating inhibition of transport by stilbene disulphonates. This site could also be adjacent to the other 2 potent inhibitory peptides in three dimensional structure.

Pep-COOH, a peptide from the carboxyl terminus region, contains both hydrophobic and hydrophilic regions. The lysines found in this region comprise another binding site for the stilbene disulphonates based on data presented here and that of Jennings *et al.* [88]. Our data suggest that this region has anion binding/transport capability as well. Residues 812-827 (pep-COOH) and 813-818 (N6, the 6 amino acids on the amino side of pep-COOH) are inhibitors of anion transport. Pep-COOH (residues 812-827) is part of senescent cell antigen [19,23-27]. N6 is both an inhibitor of anion transport and of senescent cell IgG binding (IgG binding inhibition: 48±1% at 10μg) even though it is only six amino acids long. However, COOH-N6 does contain a proline-proline bend which may contribute to an anion pocket. It probably forms a loop in the membrane. These experiments suggest that at least part of a transport site is located on the same region of band 3 that generates senescent cell antigen.

Localization of SITS Binding Site on Band 3 Membrane Protein. Antiidiotype antibodies recognize the receptor of the ligand against which the idiotype is prepared [93,94]. This has provided an elegant method for preparing antibodies against membrane proteins without purifying them, and for localizing active ligand binding sites on membrane receptors and proteins [93,94].

SITS is an inhibitor of anion transport. Antiidiotypic antibodies to SITS react with band 3 and its breakdown products in erythrocyte membranes [93,94] and with band 3 peptide residues 788-805, 800-818, and strongest with 812-830. Residues 812-827, 3 amino acids smaller that the peptide giving the strongest reaction, reacted as well as the other 2 peptides, 788-805 and 800-818, but weaker than 812-830. Antiidiotypic antibodies did not react with residues 813-818 which is located at the strongly reactive amino end of the peptide which reacts the strongest. In contrast, the antibodies reacted with peptide 818-827 which is toward the carboxyl end of the peptide giving the strongest reaction. This is consistent with the band 3 model which we have presented predicting that these residues are on external loop 04 [19,23-27,50]. Reaction of antiidiotypic antibodies with breakdown products of band 3 suggests that these breakdown products carry transport sites. This would indicate that these segments are not derived from the cytoplasmic segment of band 3.

In summary, residues 812-830 contain both a SITS binding site and an anion binding/transport site. Since the antiidiotypic antibody does not react with residue 813-818, gives a weak reaction with residues 812-827, gives a strong reaction with residues 812-830 and reacts weakly with 818-827, a crucial epitope probably resides in the region of 828-830. Band 3 residues 788-805 and 800-818 may also be part of a SITS binding site. Data from chicken pep-COOH-N6 suggest that lysine 814 is not critical for anion transport. Data of Jennings *et al.* [88] indicate that one end of the dihydro derivative of DIDS, H2DIDS, reacts covalently with a lysine that is between 70 and 168 residues from the C terminus of band 3. This would be between residues 772 and 840 in the human sequence.

Results of this study, summarized in Figure 1, indicate that:(a) regions with residues 588-594, 822-839, and 869-883 being the most active transport regions (P ≤ 0.001);(b) residues 812-830, and, possibly, 788-805 and 800-818 are part of the stilbene disulphonate binding site;(c) residue 538-554, which has been reported to be a transport segment of band 3, does not bind anions; and (d) lysines themselves contribute to but are not required for anion binding and, thus, anion transport.

Synthetic peptide studies are consistent with the physiological data demonstrating that old erythrocytes have impaired anion transport [19,53,62], and the biochemical and immunological data indicating that band 3 undergoes degradation with loss of a cytoplasmic segment during the aging process [19,33, 34,47]. Localization of the active site of SCA will facilitate the next logical step, namely, definition of the molecular changes occurring during aging that initiate molecular as well as cellular degeneration. Peptides ANION 1 and COOH are in highly conserved regions of band 3 [19].

CONCLUDING REMARKS

Degradation of band 3 generates senescent cell antigen, an aging antigen that marks cells for removal by initiating the binding of IgG autoantibody and subsequent removal by phagocytes. It is generated on the transport domain of band 3. This appears to be a general physiologic process for removing senescent and damaged cells in mammals and other vertebrates. Although the initial studies were done using human erythrocytes as a model, senescent cell antigen has been found on all cells examined. The aging antigen itself is generated by the degradation of band 3. Besides its role in the removal of senescent and damaged cells, senescent cell antigen also appears to be involved in the removal of erythrocytes in clinical hemolytic anemias, and the removal of malaria-infected erythrocytes. Oxidation generates senescent cell antigen *in situ*.

Band 3 is a crucial structural and functional protein and is intimately involved in cellular aging. Band 3 and senescent cell antigen are present in the central nervous system, and differences have been described in band 3 between young and aging brain tissue, and in Alzheimer's disease. This suggests that band 3 and senescent cell antigen plays a role in neurological health and disease. The results of our studies, to date, indicate that band 3 from brain tissue undergoes the same age-related changes undergone by band 3 in other cell types including decreased anion transport, increased degradation of band 3, and generation of senescent cell antigen, increased phosphorylation of brain band 3 and a change in reactivity with antibodies to synthetic peptides of band 3. Studies on our defined patient population suggest that changes in brain band 3 are accompanied by changes in the peripheral blood as well. Autoantibodies to band 3 peptide 588-602 significantly increase with age ($P \leq 0.0001$ by two-tailed t-test). This is one of the three key transport regions of band 3 [25]. Increase in autoantibodies to this region are consistent with data indicating that this is a vulnerable region that is altered during aging.

In AD, preliminary results indicate that posttranslational changes include decreased brain band 3 phosphorylation of a 25-28kD segment, increased degradation of band 3, alterations in band 3 recognized by antibodies, and decreased anion and glucose transport by blood cells. Serum autoantibodies were increased in AD patients compared to controls to band 3 peptide 822-839 by the two tailed t-test of $P \leq 0.06$. Significance is expected to increase as sample size increases. This band 3 residue lies in an anion transport\binding region. This is consistent with alteration of this region in AD since it is recognized as antigenically different by the patient's immune system. Appearance/increase in antibodies to this region suggests that it may be altered in AD. This is consistent with the observed decrease in anion transport.

ACKNOWLEDGMENTS

This work was supported by NIH grants AG08444 and AG08574, a Veterans Administration Merit Review, the International Foundation for Biomedical Aging Research, the Arizona Disease Control Research Commission.

REFERENCES

1. Kay, M.M.B., Tracey, C.M., Goodman, J.R., Cone, J.C. and Bassel, P.S., 1983, Polypeptides immunologically related to erythrocyte band 3 are present in nucleated somatic cells. *Proc.Natl.Acad.Sci.USA* 80:6882-6886.

2. Kay, M.M.B., Bosman, G., Notter, M. and Coleman, P., 1988, Life and death of neurons: The role of senescent cell antigen. *Ann.N.Y.Acad.Sci.* 521: 155-169.
3. Demuth, D.R., Showe, L.C., Ballantine, M., *et al.*, 1986, Cloning and structural characterization of a human non-erythroid band 3-like protein. *EMBO J.* 5:1205-1214.
4. Kopito, R.R. and Lodish, H.F., 1985, Structure of the murine anion exchange protein. *J.Cell.Biochem.* 29:1-17.
5. Kudrycki, K.E. and Shull, G.E., 1989, Primary structure of the rat kidney band 3 anion exchange protein deduced from a cDNA. *J.Biol.Chem.* 264: 8185-8192.
6. Alper, S.L., Kopito, R.R., Libresco, S.M. and Lodish, H.F., 1988, Cloning and characterization of a murine band 3-related cDNA from kidney and from a lymphoid cell line. *J.Biol.Chem.* 263:17092-17099.
7. Hazen-Martin, D.J., Pasternack, G., Spicer, S.S. and Sens, D.A., 1986, Immunolocalization of band 3 protein in normal and cystic fibrosis skin. *J.Histochem.Cytochem.* 34:823-826.
8. Schuster, V.L., Bonsib, S.M. and Jennings, M.L., 1986, Two types of collecting duct mitochondria-rich (intercalated) cells: lectin and band 3 cytochemistry. *Am. J. Physiol.* 251:C347-C355.
9. Kellokumpu, S., Neff, L., Jamsa-Kellokumpu, S., Kopito, R. and Baron, R., 1988, A 115-kD polypeptide immunologically related to erythrocyte band 3 is present in Golgi membranes. *Science* 242:1308-1311.
10. Kay, M.M.B., Hughes, J.E. and Zagon, I., 1990, Aging of cell membrane molecules: Band 3 and senescent cell antigen in neural tissue. In: Molecular mechanisms of aging, edited by Beyreuther, K. and Schettler, G. Berlin: Springer-Verlag, p. 110-123.
11. Kay, M.M.B., Hughes, J., Zagon, I. and Lin, F., 1991, Brain membrane protein band 3 performs the same functions as erythrocyte band 3. *Proc. Natl. Acad. Sci. USA* 88:27780-27820.
12. Goodman J., Gamble D., and Kay, M.M.B., 1994, Distribution and function of multiple anion transporter proteins in brain tumor cell lines in relation to glucose transport. *Brain Res. Bull.* 33:411-417.
13. Tanner, M.J.A., Martin, P.G. and High, S., 1988, The complete amino acid sequence of the human erythrocyte membrane anion-transport protein deduced from the cDNA sequence. *Biochem. J.* 256:703-712.
14. Schofield, A.E., Martin, P.G., Spillett, D. and Tanner, M.J.A., 1994, The structure of the human red blood cell anion exchanger. *Blood* 84: 2000-2012.
15. Kudrycki, K.E., Newman, P.R. and Shull, G.E., 1990, cDNA cloning and tissue distribution of mRNAs for two proteins that are related to the band 3 Cl-/HCO3- exchanger. *J.Biol.Chem.* 265:462-471.
16. Steck, T.L., 1974, The organization of proteins in human red blood cell membranes. *J.Cell Biol.* 62:1-19.
17. Goodman, S.R. and Shiffer, K., 1983, The spectrin membrane skeleton of normal and abnormal human erythrocytes: a review. *Am.J.Physiol.* 244:C121-C141.
18. Kay, Marguerite M.B., Wyant, T., Poulin, J., and Johnson, R.C., 1994, Band 3 "high transport" mutation is expressed in human lymphocytes as well as erythrocytes. *Cell. and Mol. Biol.* in press.
19. Kay, M.M.B., Lin, F., Bosman, G., Marchalonis, J.J. and Schluter, S.F., 1991, Human erythrocyte aging: Cellular and Molecular biology. *Trans. Med. Revs.* 5:173-195.
20. Bennett, V., 1979, Immunoreactive forms of human erythrocyte ankyrin are present in diverse cells and tissues. *Nature, Ldn.* 281:597-599.
21. Goodman, S.R., Zagon, I.S. and Kulikowski, R.R., 1981, Identification of a spectrin-like protein in nonerythroid cells. *Proc. Natl. Acad. Sci. USA* 78:7570-7574.
22. Kay, M.M.B., 1992, Senescent cell antigen and band 3 in aging and disease. *Prog Cell Res.* 2:245-250.
23. Kay, M.M.B., 1992, Molecular mapping of human band 3 aging antigenic sites and active amino acids using synthetic peptides. *J. Prot. Chem.* 11:595-602.
24. Kay, M.M.B., 1994, Autoantibodies to band 3 during aging and disease and disease interventions. *Ann. N.Y. Acad. Sci.* 719: 419-447.
25. Kay, M.M.B., 1990, Molecular mapping of human band 3 anion transport regions using synthetic peptides. *FASEB J.* 5:109-115.
26. Kay, M.M.B., Marchalonis, J.J., Hughes, J., Watanabe, K. and Schluter, S.F., 1990, Definition of a physiologic aging auto-antigen using synthetic peptides of membrane protein band 3: Localization of the active antigenic sites. *Proc.Natl.Acad.Sci.USA* 87:5734-5738.
27. Kay, M.M.B. and Marchalonis, J., 1991, Synthetic aging antigen can be used to manipulate cellular lifespan. *Life Sci.* 48:1603-1608.
28. Kay, M.M.B., 1975, Mechanism of removal senescent cells by human macrophages *in situ. Proc.Natl.Acad.Sci.USA* 72:3521-3525.
29. Kay, M.M.B., 1978, Role of physiologic autoantibody in the removal of senescent human red cells. *J.Supramol.Struct.* 9:555-567.
30. Bennett, G.D. and Kay, M.M.B., 1981, Homeostatic removal of senescent murine erythrocytes by splenic macrophages. *Exp.Hematol.* 9:297-307.

31. Kay, M.M.B., 1981, Isolation of the phagocytosis-inducing IgG-binding antigen on senescent somatic cells. *Nature, Ldn* 289:491-494.
32. Kay, M.M.B., 1984, Localization of senescent cell antigen on band 3. *Proc.Natl.Acad.Sci. USA* 81:5753-5757.
33. Kay, M.M.B., Bosman, G.J.C.G.M., Shapiro, S.S., Bendich, A. and Bassel, P.S.. 1986, Oxidation as a possible mechanism of cellular aging: Vitamin E deficiency causes premature aging and IgG binding to erythrocytes. *Proc. Natl. Acad. Sci. USA* 83:2463-2467.
34. Kay, M.M.B., Flowers, N., Goodman, J. and Bosman, G.J.C.G.M., 1989, Alteration in membrane protein band 3 associated with accelerated erythrocyte aging. *Proc.Natl.Acad.Sci. USA* 86:5834-5838.
35. Hebbel, R.P. and Miller, W.J., 1984, Phagocytosis of sickle erythrocytes. Immunologic and oxidative determinants of hemolytic anemia. *Blood* 64:733-741.
36. Singer, J.A., Jennings, L.K., Jackson, C., Doctker, M.E., Morrison, M. and Walker, W.S., 1986, Erythrocyte homeostasis: Antibody-mediated recognition of the senescent state by macrophages. *Proc.Natl.Acad.Sci. USA* 83:5498-5501.
37. Glass, G.A., Gershon, H. and Gershon, D., 1983, The effect of donor and cell age on several characteristics of rat erythrocytes. *Exp.Hematol.* 11: 987-995.
38. Glass, G.A., Gershon, D. and Gershon, H., 1985. Some characteristics of the human erythrocyte as a function of donor and cell age. *Exp.Hematol.*13: 1122-1126.
39. Bartosz, G., Sosynski, M. and Kredziona, J., 1982, Aging of the erythrocyte. VI. Accelerated red cell membrane aging in Down's syndrome. *Cell Biol.Int.Rep.* 6:73-77.
40. Bartosz, G., Sosynski, M. and Wasilewski, A., 1982, Aging of the erythrocyte XVII. Binding of autologous immunoglobin. *Mech. Aging Dev* 20:223-232
41. Khansari, N. and Fudenberg, H.H., 1983, Immune elimination of autologous senescent erythrocytes by Kupffer cells *in vivo. Cell.Immunol.* 80: 426-430.
42. Lutz, H.U., Flepp, R. and Stringaro-Wipf, G., 1984, Naturally occurring autoantibodies to exoplasmic and cryptic regions of band 3 protein, the major integral membrane protein of human red blood cells. *J.Immunol.* 133:2160-2618.
43. Walker, W.S., Singer, J.A., Morrison, M. and Jackson, C.W., 1984, Preferential phagocytosis of *in vivo* aged murine red blood cells by a macrophage-like cell line. *Br.J.Haemat.* 58:259-266.
44. Muller, H. and Lutz, H.U., 1984, Binding of autologous IgG to human red blood cells before and after ATP-depletion. Selective exposure of binding sites (autoantigens) on spectrin-free vesicles. *Biochim.Biophys.Acta.* 729: 249-257.
45. Kay, M.M.B., 1991, Band 3 in aging and neurological disease. *Ann. N.Y. Acad.Sci.* 621:179-205.
46. Kay, M.M.B. and Lin, F., 1990, Molecular mapping of the active site of an aging antigen: Senescent cell antigen is located on an anion binding segment of band 3 membrane transport protein. *Geront.* 36:293-295.
47. Kay, M.M.B., 1989, Red cell aging: Senescent cell antigen, band 3, and band 3 mutations associated with cellular dysfunction. *Proc.Clin.Biol.Res.* 319:199-217.
48. Petz, L.D., Yam, P., Wilkinson, L., Garratty, G., Lubin, B. and Mentzer, W., 1984, Increased IgG molecules bound to the surface of red blood cells of patients with sickle cell anemia. *Blood* 64:301-304.
49. Friedman, M.J., Fukuda, M. and Laine, R.A., 1985, Evidence for a malarial parasite interaction site on the major transmembrane protein of the human erythrocyte. *Science* 228:75-77.
50. Kay, Marguerite M.B., Bosman, G., Johnson, R.C., Poulin, J., and Goodman, J., 1994, Molecular basis of human band 3 mutation associated with increased anion transport. *Expl Clin Immunogenet.*, in press.
51. Bruce, L., Kay, M. M.B., Lawrence, C. and Tanner, M,J.A., 1993, Band 3 HT, a human red cell variant associated with acanthocytosis and increased anion transport, carries the mutation pro-868 to leu in the membrane domain of band 3. *Biochem J.* 293:317-320.
52. Kay, M.M.B., Bosman, G., Gamble, D., Goodman, J., and Cover, C., 1994, Structural changes in band 3 during aging and in Alzheimer's disease: Electromicroscopic, phosphorylation, and peptide studies. Immunobiology of Proteins and Peptides, in press.
53. Bosman, G.J.C.G.M. and Kay, M.M.B., 1990, Alterations of band 3 transport protein by cellular aging and disease: Erythrocyte band 3 and glucose transporter share a functional relationship. *Biochem.Cell Biol.* 68:1419-1427.
54. Bosman, G., Bartholomeus, I., DeMan, C., VanKalmthout, P. and DeGrip, W.,1991, Alzheimer's disease: Indications for disturbed erythrocyte aging. *Neurobiol.Aging* 12:13-18.
55. Kay, M.M.B., Bosman, G., Gamble, D., Goodman, J., and Cover, C., 1994, Structural changes in band 3 during aging and in Alzheimer's disease: Electromicroscopic, phosphorylation, and peptide studies. *Brain Res Bull*, in press.
56. Poulin, J., Cover, C., Gamble, D., Wyant, T., and Kay, M.M.B., 1995 Band 3 functions during mouse aging: spleen and lymph node studies, submitted.

57. Kay, M.M.B., Goodman, J., Lawrence, C. and Bosman, G., 1990, Membrane channel protein abnormalities and autoantibodies in neurological disease. *Brain Res.Bull.* 24:105-111.
58. Saitoh, T. and Dobkins, K., 1986, Protein kinase C in human brain and its inhibition by calmodulin. *Brain Res.Bull.* 379:196-199.
59. Saitoh, T., Oswald, R., Wennogle, L.P. and Changeux, J.P., 1980, Conditions for the selective labeling of the 66,000 dalton chain of the acetylcholine receptor by the covalent non-competitive blocker 5-azido-[3H] trimethisoquin. *FEBS Lett.* 116:30-36.
60. Elgavish A., Smith, J.B., Pillion, D.J. and Meezan, E., 1985, Sulfate transport in human lung fibroblasts (IMR-90). *J. Cell. Physiol.* 125: 243-250.
61. Yam, P., Petz, L.D. and Spath, P., 1982, Detection of IgG sensitization of red cells with ^{125}I staphycoccal protein A. *Am.J. of Hematol.* 12:337-346.
62. Bosman, G.J.C.G.M. and Kay, M.M.B., 1988, Erythrocyte aging: A comparison of model systems for simulating cellular aging *in vitro*. *Blood Cells* 14:19-35.
63. Kay, M.M.B., 1993, Generation of senescent cell antigen on old cells initiates IgG binding to a neoantigen. *Cell. and Mol. Biol.* 39:131-153.
64. Kay, M.M. B., 1993, Vitamin E deficiency causes appearance of an aging antigen and accelerated cellular aging and removal in *Vitamin E: Biochemistry and Clinical Applications*, L. Packer and J. Fuchs (eds), Marcel Dekker, Inc., New York, pp. 287-296.
65. Kay, M.M.B., 1995, Regulatory autoantibody and cellular aging and removal. *Adv. Exp. Med. Biol.*, in press.
66. Kay, M.M.B., 1993, Cellular and molecular biology of an aging antigen. in *Immunobiology of Transfusion Medicine*, G. Garratty, (ed.), Marcel Dekker, Inc., New York, pp. 173-198.
67. Kay, M.M.B., 1991, Drosophila to bacteriophage to erythrocyte: The erythrocyte as a model for molecular and membrane aging of terminally differentiated cells. *Geront.* 37:5-32.
68. Kay, M.M.B., 1989, Molecular aging of membrane molecules and cellular removal. In: *Biomedical Advances in Aging*, edited by Goldstein, A. New York: Plenum Press.
69. Kay, M.M.B., 1990, Senescent cell antigen, band 3, and band 3 mutations in cellular aging. *Biomed.Biochim.Acta* 49:212-217.
70. Kay, M.M.B., 1986, Senescent cell antigen: A red cell aging antigen. In: *Red Cell Antigens and Antibodies*, edited by Garratty, G. Arlington, VA: American Association of Blood Banks, pp.35-82.
71. Kay, M.M.B., 1984, Band 3, the predominant transmembrane polypeptide, undergoes proteolytic degradation as cells age. *Monogr.Devel.Biol.* 17:245-253.
72. Kay, M.M.B., 1983, Antigenic changes associated with cellular aging. *Prog.Clin.Biol.Res.* 133:65-76.
73. Kay, M.M.B., Sorensen, K., Wong, P. and Bolton, P., 1982, Antigenicity, storage & aging: Physiologic autoantibodies to cell membrane and serum proteins and the senescent cell antigen. *Mol.Cell.Biochem.* 49:65-85.
74. Marchalonis, J.J., Schluter, S.F, Wilson, L., Yocum, D., Boyer, J., and Kay, M.M.B., 1993, Natural human autoantibodies to synthetic peptide autoantigens: correlations with age and autoimmune disease. *Geront.* 39:65-79.
75. Kay, M.M.B., 1985, Aging of cell membrane molecules leads to appearance of an aging antigen and removal of senescent cells. *Geront.* 31: 215-235.
76. Dekowski, S.A., Rybicki, A. and Drickamer, K., 1983, A tyrosine kinase associated with the red cell membrane phosphorylates band 3. *J.Biol.Chem.* 258:2750-2753.
77. Drakoulis Y., Vasseur, C., Piau,J.P., Wajcman, H., and Bursaux, E., 1991, Phosphorylation sites in human erythrocyte band 3 protein. *Biochemica et Biophysica Acta* 1061:253-266.
78. Lowry, O.R., Rosenbrough, N.J., Farr, A.L. and Randall, R.J. Protein measurement with the folin phenol reagent. *J.Biol.Chem.* **193**:265-275, 1951.
79. Cianci, C.D., Giorgi, M. and Morrow, J.S., 1988, Phosphorylation of ankyrin down-regulates its cooperative interaction with spectrin and protein 3. *J.Cell Biochem.* 37:301-315.
80. Biernat, J., Mandelkow, E.-M., Schröter, C., Lichtenberg-Kraag, B., Steiner, B., Berling, B., Meyer, H., Mercken, M., Vandermeeren, A., Goedert, M. and Mandelkow, E., 1992, The switch of tau protein to an Alzheimer-like state includes the phosphorylation of two serine-proline motifs upstream of the microtubule binding region. *EMBO J.* 11:1593-1597.
81. Jahn, R., Schiebler, W. and Greengard, P., 1977, A quantitative dot-immunobinding assay for proteins using nitrocellulose membrane filters. *Proc. Natl. Acad.Sci. USA* 81:1684.
82. Saitoh, T., Masliah, E., Baum, L., Sundsmo, M., Flanagan, L., Vikramkumar, R., and Kay, M.M.B., 1992, Degradation of proteins in the membrane-cytoskeleton complex in Alzheimer's Disease: Might amyloidogenic APP processing be just the tip of the iceberg? *Ann. N.Y. Acad. Sci.* 674:180-192.

83. McKhann, G. Drachman, D., Folstein, M., Katzman, R., Price, D., and Stadlan, E. M. 1984, Clinical diagnosis of Alzheimer's disease: report of the NINCDS-ADRDA Work Group under the auspices of Department of Health and Human Services Task Force on Alzheimer's Disease. Neurology 34:939-944.

84. McCay P.B., and King, M.M., 1980, Biochemical Function. I. Vitamin E: Its role as a biologic free radical scavenger and its relationship to the microsomal mixed-function oxidase system. In: Machlin LJ, ed. *Vitamin E, a comprehensive treatise.* New York: Marcel Dekker, pp. 298-317.

85. Kay, M.M.B., Goodman, J., Goodman, S. and Lawrence, C., 1990, Membrane protein band 3 alteration associated with neurologic disease and tissue reactive antibodies. *Clin.Exper.Immunol.* 7:181-199.

86. Kay, M.M.B., Bosman, G.J.C.G.M. and Lawrence, C., 1988, Functional topography of band 3: A specific structural alteration linked to functional aberrations in human erythrocytes. *Proc.Natl.Acad.Sci. USA* 85:492-496.

87. Jennings, M.L., 1989, Structure and function of the red blood cell anion transport protein. *Ann. Rev. Biophys. Chem.* 18:397-430.

88. Jennings M.L., Anderson, M.P., and Monaghan, R., 1986, Monoclonal antibodies against human erythrocyte band 3 protein. Localization of proteolytic cleavage sites and stilbenedisulfonate-binding lysine residues. *J. Biol. Chem.* 261:9002-9010.

89. Bartel, D., Hans, H. and Passow, H., 1989, Identification by site directed mutagenesis of lysine 558 as a covalent attachment site of DIDS in mouse erythroid band 3. *Biochem.Biophys.Acta.* 985:355-358.

90. Garcia, A.M. and Lodish, H.F., 1989, Lysine 539 of human band 3 is not essential for ion transport or inhibition of stilbene disulfonate. *J. Biol. Chem.* 264:19607-19613.

91. Bjerrum P.J., Wieth, J.O., and Minakami, S., 1983, Selective phenylglyoxalation of functionally essential arginyl residues in the erythrocyte anion transport protein. *J. Gen. Physiol.* 81:453-484.

92. Zaki, L., 1983, Anion transport in red blood cells and arginine specific reagents. (1) Effect of chloride and sulfate ions on phenylglyoxal sensitive sites in the red blood cell membrane. *Biochem.and Biophys.Resch. Comm.* 110:616-624.

93. Kay, M.M.B., 1990, Anti-idiotypic antibodies as membrane probes: Is the glucose transporter related to the anion transporter band 3? *Proc. Idiotypic Networks and Immune Regulation.*

94. Kay, M.M.B., 1985, Glucose transport protein is structurally and immunologically related to band 3 and senescent cell antigen. *Proc. Natl. Acad. Sci.* 82:1731-1735.

20

ESTABLISHMENT OF A MOUSE MODEL OF MYASTHENIA GRAVIS WHICH MIMICS HUMAN MYASTHENIA GRAVIS PATHOGENESIS FOR IMMUNE INTERVENTION

Premkumar Christadoss, Rashmi Kaul, Mohan Shenoy, and
Elizabeita Goluszko

The University of Texas Medical Branch
Department of Microbiology and Immunology
Galveston, Texas 77555-1019

INTRODUCTION

Myasthenia gravis (MG) is a classical antibody-mediated autoimmune disease of the neuromuscular junction. MG-like disease could be induced in animals by immunization with acetylcholine receptor (AChR) in complete Freund's adjuvant (CFA). We have established the mouse model of MG, experimental autoimmune myasthenia gravis (EAMG) and have characterized the clinical and immunopathological similarities with human MG. This paper reviews the immunopathological aspect of EAMG at the cellular and molecular level and its similarities with human MG, and our approach to treat the disease.

EAMG in susceptible mouse strains could be induced by multiple immunizations with AChR in CFA. Initial studies on congenic and recombinant strains suggested that H-2 genes control the cellular and humoral immune responses to AChR and the development of clinical EAMG (1-4). These studies established that strains with $H-2^b$ haplotype were highly susceptible to EAMG, those with $H-2^q$ were intermediate in susceptibility, and strains with $H-2^k$ and $H-2^p$ haplotypes were relatively resistant.

THE ROLE OF I-A MOLECULE AND ACHRα CHAIN PEPTIDE α146-162 IN EAMG PATHOGENESIS

The in vitro cellular and the in vivo humoral immune responses to AChR and the development of clinical disease were controlled by Ir genes within the H-2A subregion (2,4).

Immunobiology of Proteins and Peptides VIII
Edited by M. Z. Atassi and G. S. Bixler, Jr., Plenum Press, New York, 1995

Blockage of Ia antigens on the surface of antigen presenting cells or lymphoid cells by appropriate anti I-A antibodies prevented presentation of AChR to helper T cells for recognition and proliferation to occur (5,6). The I-Aβ chain mutation in the high-responder B6 strain gave rise to the congenic strain B6.CH-2^{bm12}(bm12). A gene conversion event that altered three amino acids at the C-terminal half of the first domain of I-Aβ^b of B6 mice (Ile-67 → Phe; Arg-70 → Glu; Thr-71 → Lys) resulted in low responsiveness of primed bm12 lymphocytes to AChR, compared to the high responsiveness observed in lymphocytes from parent B6 strain (5). Further, bm12 mice immunized with AChR in CFA failed to develop muscle weakness in contrast to EAMG-susceptible parent B6 mice. The bm12 mice also gave a significantly reduced cellular response to AChR, had low serum anti-AChR antibodies, low amounts of muscle AChR complexed with the antibody, and had lost very little muscle AChR. Therefore, a mutation in the bm12 I-Aβ chain at positions 67, 70, and 71 suppressed both the cellular and humoral immune responses to AChR and thus prevented the disease induction (7). Moreover, bm12 AChR immune lymphocytes responded well to only one (α111-126) of the three (α111-126, α146-162 and α182-198) dominant AChRα chain peptides (8). Thus, α146-162 and α182-198 were implicated as the possible pathogenic epitopes B6 mice. The pathogenicity of α146-162 was shown by successful prevention of EAMG by inducing neonatal tolerance to α146-162, and inducing EAMG by boosting with α146-162 in CFA after priming B6 mice with AChR in CFA (8,9). Thus the peptide α146-162 could be used as a tolerogen to suppress established EAMG. It is interesting to note that one of the dominant epitopes within AChR α chain was mapped within α146-162 in 3 of 5 patients with MG (10). Characterization of T cell receptor gene usage of pathogenic dominant AChR epitope (e.g. α146-162) specific T cells could reveal any restricted TCR gene usage in EAMG and MG.

IN VIVO EFFECT OF MONOCLONAL ANTI-I-A AND CD4 ANTIBODIES ON IMMUNE RESPONSE TO AChR

Treatment of SJL (H-2^s) mice with antibodies to I-A^s molecule before immunization with AChR in CFA suppressed antibody response to Torpedo AChR when compared to SJL mice treated either with non-crossreactive anti-I-A^b antibody or without treatment. Also in vivo treatment with anti-I-A antibody was effective in partially suppressing the disease incidence (11). Because the I-A molecule presents AChR epitopes to CD4$^+$ cells, the effect of in vivo depletion of CD4$^+$ cells on EAMG development was evaluated. The mAb (GK1.5) to CD4$^+$ T cells which depleted CD4$^+$ cells, not only prevented EAMG, but also suppressed established autoimmunity to AChR in B6 mice and induced prolonged clinical remission (12).

THE PROTECTIVE INFLUENCE OF IE MOLECULE ON THE DEVELOPMENT OF EAMG

Mice with H-2^b haplotype (e.g. B10 and B6) have a genomic deletion of the Eα gene, and therefore, fail to express cell surface I-E molecules. To test the hypothesis that this failure in the expression of the cell surface I-E molecule in B10 mice contributed to EAMG pathogenesis, I-Eα^k transgenic B10 mice expressing cell surface I-E molecule were screened for EAMG susceptibility (13). The expression of the I-E molecule on the cell surface of B10 mice significantly reduced the incidence of muscle weakness and suppressed anti-AChR antibody response. Thus autoimmunity to AChR could be modulated by the expression of the I-E molecule. The expression of IE molecule could have suppressed EAMG due to deletion of E reactive Vβ5, Vβ11 T cells, or by binding of IE molecule to epitope(s) of AChR which could have generated regulator/suppressor cells. This hypothesis is being tested.

THE EFFECT OF MHC CLASS II GENE DISRUPTION AND CD4+ T CELL DEFICIENCY ON EAMG

In order to demonstrate the direct genetic evidence for the involvement of MHC class II molecule and to assess the impact of lack of MHC class II gene expression and the resultant deficiency of CD4$^+$ cells on EAMG pathogenesis, MHC class II gene disrupted C57BL6 mutant (-/-) and EAMG-susceptible MHC class II wild-type C57BL6 (+/+) mice were evaluated for clinical and immunopathological manifestations of EAMG (14). The MHC class II and CD4$^+$ cell deficiency in the mutant (-/-) mice completely rendered them resistant to the development of muscle weakness. The reduction of bungarotoxin binding sites in the muscles and the increased level of serum autoantibodies to M-AChR of the wild-type (+/+) mice at various time points were significantly different to that of the mutant (-/-) mice. Thus, the deficiency of CD4$^+$ Th cells in the MHC class II mutant (-/-) mice resulted in a suppressed anti-AChR antibody production. These data unequivocally provided the first direct genetic evidence for the essential role of MHC class II molecule and the CD4$^+$ Th cells in the induction of EAMG, and also ruled out any pathogenic effector role for MHC class I-restricted CD8$^+$ T cells, γδ TCR-bearing cells, and NK cells, which are intact in the MHC class II mutant mice (14).

THE EFFECT OF MHC CLASS I AND CD8$^+$ T CELL DEFICIENCY ON EAMG PATHOGENESIS

There is no direct genetic evidence for and against an effector, or a regulator/suppressor role for MHC class I-restricted CD8+ T cells during the evolution and establishment of EAMG. We utilized β2 microglobulin (β2m)-deficient mice, lacking MHC class I and CD8+ T cells, to study the effector and regulator/suppressor role of the MHC class I-restricted CD8+ T cells in EAMG development (15). The MHC class I and CD8$^+$ T cell deficiency failed to prevent the induction of EAMG. The results of our study clearly indicated that class I and CD8$^+$ T cell-deficient β2m$^{-/-}$ mice not only developed EAMG, but also the incidence of the disease was higher in β2m$^{-/-}$ mice, compared to the β2m$^{+/-}$ mice. Therefore, class I-restricted CD8$^+$ T cells are not the effectors (in the destruction of AChR) during the evolution of EAMG in B10 mice (15). The increased incidence of clinical EAMG, observed in β2m$^{-/-}$ mice, could either be due to a suppressor/regulator role of class I restricted CD8$^+$ T cells, or to the effect of compensatory overexpressed class II-restricted CD4$^+$ T cells involved in helping the production of pathogenic anti-AChR antibodies by the B cells. In a recent study in vivo depletion of CD8+ T cells in B6 mice enhanced the incidence of EAMG, suggesting a regulator/suppressor role played by CD8+ T cells (Kaul et al, in preparation).

The murine EAMG model serves as an excellent model to test the therapeutic efficacy of specific agents for the treatment of MG. Table I illustrates the clinical and immunopathological similarities between MG and EAMG. Table II shows the various therapeutic strategy we plan to embark upon. As a first step, we selected interferon α, because it reduces MHC class II expression, and inhibits the Th1 responses. IFN-α (10^5 IU thrice weekly for five weeks) treatment started one week after the second immunization with AChR in CFA, when autoimmunity to AChR is well established, reduced the incidence of clinical EAMG

Table I. Clinical and immunopathogenic similarities between MG and EAMG

	MG	EAMG
Muscle weakness and temporary improvement with anti-cholinesterase	Yes	Yes
EMG[*]	decrement	decrement
MEPP[+]	reduction	reduction
Autoantigen	AChR[1]	AChR[1]
Genetic predisposition (Genetic Control)	MHC class II (HLA-DQβ chain polymorphism)[2]	MHC class II (IAβ chain 67-71 AA)[3,4]
Pathology	End plate AChR loss[5]	End plate AChR loss[6]
Cause of pathology	Anti-AChR antibody complement activation[1]	Anti-AChR antibody complement activation[7]
In vitro lymphocyte response	Class II restricted[7]	Class II restricted[3,4,8]
AChR-α subunit T cell epitopes	α146-162[9] α182-198[9]	α146-162[10] α182-198[10]
Anti-CD4 therapy	Benefit[11]	Benefit[12]

1. Lindstrom et al., 1988; 2. Bell et al., 1986; 3. Christadoss et al., 1985; 4. Kaul et al., 1994; 5. Fambrough et al., 1973; 6. Christadoss, 1988; 7. Hohlfeld et al., 1984; 8. Christadoss et al., 1982; 9. Oshima et al., 1990; 10. Shenoy et al., 1993; 11. Ahlberg et al., 1993; 12. Christadoss and Dauphinee, 1986.
[*]Electromyogram.
[+]Miniature end plate potential.
From Kaul, R., Shenoy, M., and Christadoss, P. Advances in Neuroimmunology (In Press), with permission.

Table II. Possible therapeutic strategy for EAMG

Agent	Possible Mechanism of Intervention
I. Non-specific	
(a) Interferon α	Reduction of MHC class II expression. Reduction in TH1 responses.
(b) Interferon β	Mechanisms similar to IFN-α. Shares the same receptor with IFN α.
(c) Non-depleting CD4 mAb	Inhibit activation of CD4. Long-term induction of tolerance.
II. Specific	
(a) Dominant AChR peptide (e.g.: α146-162)	Tolerance induction: Clonal deletion/anergy, suppressor cells.
(b) MHC class II AChR peptide conjugate	Tolerance induction by T cell anergy.
(c) MHC class II competitive peptide	Blockage of pathogenic AChR peptide binding to MHC class II.
(d) Oral AChR/peptide	Suppressor cells, TGFβ↑

by more than 50% in two separate experiments (p = 0.04 and 0.008). IFN-α-treatment also reduced the progression of EAMG. This preliminary report on the therapeutic efficacy of IFN-α has been submitted for publication (16).

ACKNOWLEDGMENTS

This work is supported by the MDA and Sealy Smith Endowment. Rashmi Kaul is a myasthenia gravis foundation Osserman postdoctoral fellow and Bo Wu is a J. W. McLaughlin postdoctoral fellow.

REFERENCES

1. Christadoss, P., Lennon, V.A., Lambert, E.H. and David, C.S. 1979. Genetic control of experimental autoimmune myasthenia gravis in mice. In. T and B Lymphocytes: Recognition and Function. Academic Press, New York, p.249.
2. Christadoss, P., Lennon, V.A. and David, C.S. 1979. Genetic control of experimental autoimmune myasthenia gravis in mice 1. Lymphocyte proliferative response to acetylcholine receptor is under H-2 linked Ir gene control. J. Immunol. 123:2540-2543.
3. Berman, P.W. and Patrick, J. 1980. Linkage between the frequency of muscle weakness and loci that regulate immune responsiveness in murine experimental myasthenia gravis. J. Exp. Med. 152:507-520.
4. Christadoss, P, Lennon, V.A., Krco, C.J., Lambert, E.H. and David, C.S. 1981. Genetic control of autoimmunity to acetylcholine receptors: role of Ia molecules. Ann. NY. Acad. Sci. 377:258-276.
5. Christadoss, P., Lennon, V.A., Krco, C.J. and David, C.S. 1982. Genetic control of experimental autoimmune MG in mice. III Ia molecules mediate cellular immune responsiveness to acetylcholine receptors. J. Immunol. 128:1141-1144.
6. Christadoss, P., Lindstrom, J. and Talal, N. (1983). Cellular immune response to acetylcholine receptors in murine experimental autoimmune myasthenia gravis. Inhibition with monoclonal anti-I-A antibodies. Cell Immunol. 81:1-8.
7. Christadoss, P., Lindstrom, J., Melvold, J.W., and Talal, N. 1985. I-A subregion mutation prevents murine experimental autoimmune myasthenia gravis. Immunogenetics 21:33-38.
8. Shenoy, M., Oshima, M., Atassi, M.A. and Christadoss, P. 1993. Suppression of experimental autoimmune myasthenia gravis by epitope-specific neonatal tolerence to synthetic region α146-162 of acetylcholine receptor. Clin. Immunol. Immunopath. 66:230-238.
9. Shenoy, M., Goluszko, E., and Christadoss, P. 1994. The pathogenic role of acetylcholine receptor α chain epitope within α146-162 in the development of experimental autoimmune myasthenia gravis in C57BL6 mice. Clin. Immunol. Immunopath. 73:338-343.
10. Oshima, M., Ashizawa, T., Pollack, M. and Atassi, M.Z. 1990. Autoimmune T cell recognition of human acetylcholine receptor: the site of T cell recognition in myasthenia gravis on the extracellular part of the α subunit. Eur. J. Immunol. 20:2563-2569.
11. Waldor, M.K., Sriram, S., McDevitt, H.O. and Steinman, L. 1983. In vivo therapy with monoclonal anti-I-A antibody suppresses immune response to acetylcholine receptor. Proc. Natl. Acad. Sci. USA 80:2713-2717.
12. Christadoss, P. and Dauphine, M.J. 1986. Immunotherapy for myasthenia gravis: A murine model. J. Immunol. 136:2437-2440.
13. Christadoss, P., David, C.S., Shenoy, M. and Keve, S. 1990. $E\alpha^k$ transgene in B10 mice suppresses the development of myasthenia gravis. Immunogenetics 31:241-244.
14. Kaul, R., Shenoy, M., Goluszko, E. and Christadoss, P. 1994. Major histocompatibility complex class II gene disruption prevents experimental autoimmune myasthenia gravis. J. Immunol. 152:3152-3157.
15. Shenoy, M., Kaul, R., Goluszko, E., David, C. and Christadoss, P. 1994. The effect of MHC class I and CD8 cell deficiency on experimental autoimmune myasthenia gravis. J. Immunol. 153:5330-5335.
16. Shenoy, M., Baron, S. and Christadoss, P. Interferon-α treatment suppresses the development of experimental autoimmune myasthenia gravis (submitted).

21

ANALYSIS OF MHC-SPECIFIC PEPTIDE MOTIFS

Applications in Immunotherapy

Douglas J. Loftus,[1] Ralph T. Kubo,[2] Kazuyasu Sakaguchi,[1] Esteban Celis,[2] Alessandro Sette,[2] and Ettore Appella[1*]

[1] Laboratory of Cell Biology
National Cancer Institute, National Institutes of Health
Bethesda, Maryland 20892
[2] Cytel
San Diego, California 92121

INTRODUCTION

Vertebrate immune responses hinge on T cell receptor (TCR)-mediated recognition of antigenic peptides bound to polymorphic class I and class II molecules of the Major Histocompatibility Complex (MHC). The polymorphism that exists at MHC class I and class II loci has no known parallel in the vertebrate genome. The desire to understand the origins and functional consequences of MHC polymorphism has engaged and challenged immunologists, molecular biologists, and population geneticists alike over the past 50 years. It is only recently that extensive structural and biochemical analyses have given way to "enlightenment" at the molecular level and provided fresh insights into MHC polymorphism and its functional concomitants. For both class I and class II molecules, structure and function are inextricably linked through the binding of peptides, and polymorphism is now understood in terms of its contribution to the peptide binding specificities ("motifs") associated with various MHC allomorphs.

The origins of MHC diversity and the selection forces that act to maintain polymorphism in a species are difficult to address experimentally. The link between polymorphism and antigenic peptide binding lends intuitive appeal to the prevalent notion that polymorphism is "needed" for optimal antigen presentation capability, though direct evidence to support this claim remains elusive. These issues are considered in several thoughtful reviews by Klein (Klein et al. 1990; Klein 1991; Klein et al. 1993). Here we broadly review the implications of diversity at the level of peptide binding to class I and class II molecules, and

[*] Address correspondence to: Dr. Ettore Appella, NCI/NIH, 37 Convent Dr. MSC 4255, Bethesda, MD 20892-4255. Tel: (301) 402-4177; Fax: (301) 496-7220.

Immunobiology of Proteins and Peptides VIII
Edited by M. Z. Atassi and G. S. Bixler, Jr., Plenum Press, New York, 1995

201

consider the extent to which knowledge of peptide binding motifs impacts peptide-based approaches to immunotherapy.

OVERVIEW

Antigens presented by class I molecules are recognized by the TCR of cytolytic effector cells (CTL) expressing the accessory molecule CD8, while class II molecules present antigens to $CD4^+$ T cells that act in a regulatory capacity (helper or Th cells). The basis for the presentation of antigenic peptides to T cells has been made clear by the solution of X-ray structures for both class I (Bjorkman et al. 1987; Madden et al. 1991; Fremont et al. 1992) and class II molecules (Brown et al. 1993; Jardetzky et al. 1994). The $\alpha 1$ and $\alpha 2$ domains of the class I heavy chain form a peptide binding groove; six sub-regions or "pockets" of this groove can be discerned and are designated by the letters A-F (Saper et al. 1991). Polymorphic residues lining these pockets determine the peptide-binding motif, or consensus sequence of peptides that a given class I allomorph preferentially binds. The ends of the groove are closed, which likely accounts for the restricted length (generally 8-10 residues) of class I-bound peptides. Similarly, a groove is formed by the association of the α and β chains of class II molecules, however, the ends of the groove are open and permit the accommodation of peptides ranging 10-34 residues in length.

The deduction of peptide motifs for MHC molecules has drawn upon both traditional and advanced bioanalytical techniques (Falk and Rotzschke 1993; Chicz and Urban 1994; Engelhard 1994). Consensus or individual sequences have been obtained for naturally-processed and/or synthetic peptides bound to a variety of class I and class II allomorphs, revealing both allomorph-specific and general features of peptides associated with MHC molecules. Class I-bound peptides show pronounced biases in the amino acid types preferred at 2-3 key positions of the peptide (termed anchor positions), while other positions exhibit various degrees of tolerance for diverse amino acids. Table I summarizes the anchor positions and consensus residues of class I HLA-associated peptides extracted from HLA-A, B, and C alleles (Engelhard et al. 1993; Falk and Rotzschke 1993; Kubo et al. 1994). The motifs listed derive from pool sequencing and a limited number of individual peptide sequences. Due to the low number of these individual sequences, the exclusive association of a particular amino acid with a given anchor position should be considered provisional. For example, pool sequencing of HLA-B27 suggests that peptides terminating with hydrophobic amino acids or tyrosine are more abundant than those terminating with lysine or arginine, which were found in the individual sequences of 7 out of 11 HLA-B27-associated self peptides (Falk and Rotzschke 1993; Jardetzky et al. 1991).

As indicated above, peptides associated with class II molecules are more variable in length compared to those associated with class I, and often comprise nested sets with extensions or truncations at either the NH_2- or COOH-terminal ends (Rudensky et al. 1991; Chicz et al. 1992). This size heterogeneity has precluded the definition of specific motifs by simple sequence alignment of naturally-derived peptides. However, by utilizing large peptide libraries, either synthetic (Sette et al. 1993; Sidney et al. 1994) or phage-displayed (Hammer et al. 1992; Hammer et al. 1993), motifs have been defined containing up to four anchor positions, with 2-4 different amino acids tolerated at each position. Table 2 summarizes the sequence alignment of allele-specific amino acid preferences for class II bound peptides.

A vast amount of detail has accrued regarding motifs for a variety of class I and class II alleles, and a comprehensive treatment of the combined findings would be cumbersome to relate, if not irritating to read. To further illustrate principles associated with peptide motifs, we limit specific discussion to the class I molecule HLA-A2.1, which has been well-characterized both biochemically and structurally, and generally consider class II

Table I. Structural characteristics and proposed motifs of peptides associated with human class I MHC molecules

Allomorph	Length	1	2	3	4	5	6	7	8	Carboxy terminus
HLA-A1	9-12		**T,S**	D,E						**Y**
HLA-A2.1	9-14		**L,***M,I*							**V,L,I,***A*
HLA-A3	9		**V,L,M**							**K,Y**
HLA-A11	9-12		**T,V**							**K**
HLA-A24	9		**Y**							**L,F**
HLA-Aw68	9-11		**V,***T*							**R,K**
HLA-B7	9	*A,R*	**P**	*R,K*						**L,I,***A,N,M*
HLA-B8	8-9			**K,***R*		**K,***R*				*L,I*
HLA-B27	9	*R,K*	**R**							**K,R**
HLA-Bw37	8-9		**D,***E*			**V,***I*				**F,M,I,L**
HLA-Cw3	8			*M,Y*	*P*		**F,Y**			**L**
HLA-Cw4	9		**Y,P**			*A,M*	*V,L*			**L,M,F**
HLA-Cw6	9					*I,F,M*	*V,I*			**L**
HLA-Cw7	8		**Y**			*F,M,Y*	*V,I*			**Y,L**

*Letters in bold indicate amino acids in single-letter code used as anchor residues, and found in most peptide sequences. Italicized letters indicate amino acids infrequently found and some interpreted as auxiliary anchor residues.

motifs, highlighting the similarities and differences that exist between class I and class II molecules.

PEPTIDE MOTIFS

Class I: HLA-A2.1

The natural peptides eluted from purified A2.1 molecules were initially analyzed by pool sequencing of unfractionated peptides by Rammensee and co-workers (Falk et al. 1991) and subsequently by combining microcapillary HPLC fractionation with electrospray ionization/tandem mass spectrometry (ESI/TMS; Hunt et al. 1992). In the latter study, sequence

Table II. Motif sequence biases for human class II epitopes

Allele	i	i+3	i+4	i+5	i+8	i+10
DR1	Y, F, W				L, Y, V	
DR2a/DR2b	I, L, V					H, K, R
DR3/DRw52	F, I, L, V	D, N, Q, T				
DR4	F, L, V				N, Q, S, T	
DR7	F, I, L, V, Y			N, S, T		
DR8	F, I, L, V, Y		H, K, R			
DR17[#]	*I, L ,V*		*D, E*			
DQ3.1[#]		*A, G, S, T*		*A, V, L, I*		

*Adapted from analysis of the percentage of identified residues in pool sequences following alignment (Chicz et al. 1993).
[#]Based on a set of single amino acid replacement analogs that still bind to either DR17 (Geluk et al. 1992) or to DQ3.1 (Sidney et al. 1994).

information for individual peptides was obtained by collision-activated dissociation. Both studies revealed the preference for aliphatic residues at anchor position 2 (P2; usually Leu or Ile) and at the carboxy terminus (Val, Leu or Ile) of the eluted peptides. Mass spectrometric analysis also permitted estimates of the heterogeneity and relative abundance of the peptides in the mixture; thousands of different peptides were present at levels of 0.01-0.1% of the total extract, corresponding roughly to 100-1000 complexes/cell. The majority of the peptides identified were 9-mers, although longer peptides (up to 14 residues) were observed as well. The combined data underscore the balance that is struck between degeneracy and specificity for peptide binding to class I molecules.

The structural basis for the A2.1 anchor preferences identified by sequencing can be discerned by examining the manner in which peptides bind in the class I groove. All X-ray crystallographic class I/peptide structures described to date (Fremont et al. 1992; Madden et al. 1992; Madden et al. 1993) indicate that peptides are bound with their amino and carboxy termini uniformly oriented, such that the peptide N terminus is bound in the A pocket, while the C-terminal side chain is accomodated in the F pocket. The P2 side chain is invariably bound in the B pocket of the groove. These observations account for the general P2/C-terminal motif pattern observed for most class I molecules, while allomorph-specific features of the B and F pockets determine the residue preferences at these positions (Corr et al. 1992; Guo et al. 1993).

Further refinement of the A2.1 motif has been accomplished through studies employing synthetic peptides, designed to identify sub-dominant peptide positions and their associated residue preferences. Using competitive peptide binding assays (Ruppert et al. 1993) or by measuring the ability of peptides to stabilize class I heavy chain/β_2-microglobulin association (Parker et al.1992 a,b), it is possible to index the "fitness" of a given amino acid or side chain type at a given position in the peptide. Additional positions at which side chain selectivity is observed have been termed secondary anchors (Ruppert et al. 1993); for A2.1, preferences for hydrophobic residues at P3, which binds in the D pocket, and for aromatic residues at P5, can be discerned. Among different class I allomorphs, secondary anchor positions and preferences differ, and further contribute to motif identity. The information gained from such studies has led to the development of algorithms (Parker et al. 1994) for predicting, among the array of potential epitopes found in a given protein antigen, which are likely to bind to class I with high affinity. A correlation between binding affinity and the immunogenicity of a given peptide sequence can be often be established (DiBrino et al. 1993; Sette et al. 1994a), however, this relationship has not been uniformly observed (Nijman et al. 1993; Wipke et al. 1993; see also under 'Immunotherapy' below).

The crystal structures for 5 homogeneous A2.1/peptide complexes (Madden et al. 1993) additionally have revealed that peptides bound to A2.1 adopt apparently sequence-dependent conformations, such that the extent to which peptide side chains at central positions (P4-P7) interact with class I side chains or are solvent exposed does not correlate well with the type of side chain found at a given position. Additionally, only marginal conformational alterations of the A2.1 molecule itself are observed among the 5 complexes. These observations have numerous implications.

For A2.1/peptide complexes, the relative contributions of nonanchor residues to overall binding affinity are likely to be context-dependent. This notion is also supported by a comprehensive peptide binding study (Kast et al. 1994). In this latter work, binding affinities for 5 HLA-A allomorphs were determined for all possible nonamers derived from the human papilloma virus proteins E6 and E7. The results obtained for A2.1 indicated that, generally, canonical anchor residues are likely to confer high affinity binding, however, the contributions to binding affinity by both anchor and non-anchor residues may depend on overall peptide sequence. The results further suggested that an expanded notion of motifs is necessary to account for tolerance of noncanonical anchor residues in peptides that bind to class I molecules.

The identification of an A2.1-presented peptide recognized by 5 independent, patient-derived CTL lines specific for melanoma cells (Cox et al. 1994) points to the immunologic relevance of peptides that deviate from canonical motifs. This peptide was identified by ESI/TMS from material eluted from purified A2.1 molecules expressed on a melanoma cell line, using previously established methods for isolating unknown epitopes (Henderson et al. 1993). The sequence obtained, YLEPGPVTA, derives from the melanocyte protein Pmel 17, and was bound by A2.1 with an affinity ~50-100 times lower than that of peptides deemed to be "good binders" (Ruppert et al. 1993). The low affinity for this peptide likely stems from the presence of the negatively charged Glu at P3, and Ala at P9, conditions which previously were identified as adversely affecting A2.1 binding affinity (Ruppert et al. 1993; Sette et al. 1994). Despite these features, this peptide sensitized cells for lysis by CTL at concentrations (in the picomolar range) 10^4-10^5 times lower than the concentration required for half-maximal inhibition of binding of a standard peptide (i.e., $IC_{50}/EC_{50} = 10^4$-10^5). In noting the apparent low abundance of the Pmel 17 peptide, as judged by MS, the authors of this study pointed out the bias inherent in most analyses of eluted material: low-affinity peptides are likely to be disproportionately lost through handling and purification prior to obtaining the eluate. Similarly, other melanoma-derived, A2.1-presented peptides recognized by patient-derived CTL have recently been described (Kawakami et al. 1994 a,b); these peptides have non-canonical Ala or Thr residues at P2 or Ala at P9.

The structural data of Madden et al. (1993) also suggest that the antigenic identity of A2.1/peptide complexes lies primarily in the conformational and stereochemical disposition of bound peptides. Furthermore, the absence of an orderly arrangement of side chains among various A2.1-bound peptides indicates the difficulty that lies in predicting, for a given epitope, which peptide residues are likely to form significant TCR contacts. While data relevant to these points are limited for other class I molecules, it has been suggested, at least for the murine class I molecule H-2K^b, that peptide-induced conformational changes in the K^b molecule may contribute to antigenic identity (Catipovic et al. 1992; Bluestone et al. 1993; Sherman et al. 1993). Additionally, it is possible that peptide binding to some class I molecules is marked by a more stable arrangement of peptide chains, such that certain peptide positions may be predisposed to contribute TCR contacts, as suggested by a small number of studies (Huet et al. 1990; Jameson et al. 1993; Kelly et al. 1993; Bowness et al. 1994). However, more data are needed to further assess how allomorph-specific motifs ultimately influence the antigenic identity of class I/peptide complexes.

The concept of allele-specific motifs recently has been extended to account for overlapping peptide specificities observed among related class I allomorphs. An A2 "supertype" has recently been described (del Guercio et al. 1995), which includes the alleles HLA-A2.1, A2.2, A2.5, A2.6, A68.2, and A69.1. These allomorphs exhibit similar selectivities for the canonical P2(Leu)/P9(Val) A2.1 motif, which permits a number peptides to be bound in common by these molecules, though with varying affinities. The similarity in ligand specificity among these alleles is correlated with their genetic similarity, as these alleles can also be seen to comprise a main branch of a phylogenetic tree drawn for the HLA-A locus (Klein et al. 1990). An HLA-B7 supertype has also been described (Sidney et al. 1995). These findings are likely to impact peptide-based vaccine development, given that a single, appropriately designed peptide might be used to extend vaccine coverage to a broader cross-section of individuals.

Class II Motifs

In contrast to class I molecules, binding motifs for class II molecules have been difficult to deduce from pooled sequence information alone. Additionally, binding motifs appear to be less stringent for class II molecules. Class II-associated peptides vary greatly in length and do not show amino and carboxy-terminal side chain biases; consequently, information for class II motifs is largely limited to amino acid preferences within a core region of 9-10 residues. Table 2 reveals a preference for hydrophobic residues at amino-proximal positions of the core region for most class II-bound peptides. Seven class II allele-specific binding motifs can be proposed by correlating sequence analyses with binding studies of synthetic analogues (Chicz et al. 1992; Chicz et al. 1993; Sette et al. 1993; Geluk et al. 1994; Sidney et al. 1994). However, it is apparent that the core regions do not share identical anchor positions or amino acid biases. The distance between the index position (i) and the anchor residue varies between alleles; this spacing may distinguish allele-specific peptide binding.

Self-peptides capable of binding to multiple DR alleles have been described (O'Sullivan et al. 1991; Hammer et al. 1993; Sinigaglia and Hammer 1994). The existence of these peptides suggests that allele-specific motifs for class II molecules are not rigid, or that there are broad peptide binding specificities among alleles which partially overlap one another. In this regard, the supertype concept, described above for some class I alleles, can be applied to class II molecules as well, and perhaps to a greater extent.

The crystal structure of a homogeneous class II HLA-DR1/peptide complex recently has been described (Stern et al. 1994). The refined structure indicates that the peptide binds in an extended conformation characteristic of the type II polyproline helix. This conformation directs several side chains to interact with HLA-DR1 while permitting extensive TCR interaction with the bound peptide. Pockets in the peptide binding site of the DR1 molecule accommodate five of the thirteen side chains of the bound peptide. The residues that line four of these pockets are highly polymorphic and are likely to determine the degree of peptide selectivity associated with a given class II allomorph. The crystal structure also shows that 12 out of 15 hydrogen bonds between the class II molecule and the bound peptide involve main-chain atoms of the peptide backbone alone. While hydrogen bonds also are formed between the class I molecule and the backbone of class I-bound peptides (Fremont et al. 1992; Matsumura et al. 1992), this type of interaction appears to play an even more pronounced role in stabilizing class II-bound peptides. Indeed, among class II allomorphs, residues that form hydrogen bonds with the peptide backbone are highly conserved. One of the major components of the class II-peptide binding interaction is therefore independent of peptide sequence, which partly accounts for the lack of stringency observed for class II peptide binding motifs.

IMPLICATIONS FOR IMMUNOTHERAPY

The characterization of allele-specific motifs has greatly advanced the prospects for the rational development and implementation of peptide-based immunotherapies. Numerous studies in the last several years have contributed to an appreciation of the significant role that cellular responses play in countering viral infection, and possibly tumor burden as well. However, traditional vaccine regimens, while effective in eliciting antibody responses, have proven ineffective in eliciting vigorous CTL responses. Coupled with the emergence of strategies for *in vivo* peptide delivery, a motif-driven approach to the design of peptide-based immunostimulants holds the promise for eliciting specific, effective cellular responses, tailored to the requirements imposed by MHC diversity. Given the issues raised here, "motif-driven" may more aptly describe efforts to elicit class I-directed responses, however the apparent low stringency of class II motifs might be exploited through the use of "universal immunogens" to elicit CD4$^+$ T cell responses (Panina-Bordignon et al. 1989). As described below, knowledge of motifs has led to an effective vaccination strategy combining the generation of both class I and class II-restricted responses.

One of the anticipated benefits of motif characterization has been the potential for prospectively identifying epitopes from the sequences of known, pathogen-derived proteins. As indicated previously, mixed results have been obtained by a number of groups investigating this approach. MHC binding affinity and the immunogenicity of a given peptide appear to be well correlated in many instances, but this correlation is not absolute. The failure to observe this concordance in all cases likely stems from the variable nature of antigen processing and T cell responsiveness, in that there is no guarantee that a synthetic peptide which binds well to a given MHC molecule will be generated intracellularly for assembly with an MHC molecule, or that the specificity for the epitope generated is adequately represented in the T cell repertoire. Nonetheless, the several instances in which epitopes have been successfully predicted provide a compelling demonstration of the utility of motifs. When combined with strategies for verifying the *in vivo* relevance of predicted epitopes, knowledge of motifs can greatly reduce the effort and cost of screening large numbers of peptides.

As a specific illustration of how epitope prediction may successfully be employed, a recent study (Celis et al. 1994) describes the identification of a melanoma-derived antigenic peptide. Peptides based on the HLA-A1 motif, derived from the known sequence of the melanoma antigen MAGE-3 (Gaugler et al. 1994), were synthesized and tested for binding to HLA-A1. Peptides exhibiting high affinity binding were then pulsed onto human peripheral blood mononuclear cells (PBMC) from healthy donors. The constitutive delivery to the cell surface of a small number of "empty" class I molecules permits loading of class I with exogenous peptides, resulting in the formation of stable cell surface complexes (Ljunggren et al. 1990: Ortiz-Navarette and Hammerling 1991). Peptide-pulsed PBMC were then used as antigen presenting cells for *in vitro* stimulation of responder PBMC that had been depleted of CD4[+] T cells. By this method, MAGE-3 -specific CTL were induced *in vitro* which were capable of lysing not only cells pulsed with a synthetic MAGE-3 -derived peptide, but lysed HLA-A1-positive melanoma cells as well.

This scheme for eliciting CTL reactivity has direct implications for optimizing tumor immunotherapy protocols employing tumor-infiltrating lymphocytes (TIL; Rosenberg et al. 1988), currently in limited use clinically for the treatment of malignant melanoma. TIL are obtained by culturing tumor explants *in vitro* in the presence of IL-2 to amplify melanoma-reactive CTL; these cells then are used to carry out adoptive immunotherapy. A recent study (Rivoltini et al. J. Immunol, in press) examined the effect of culturing A2.1-restricted TIL in the presence of cells pulsed with a peptide previously identified as an epitope for these CTL. A 20-100 fold enhancement of lytic acitivity, and a 500-fold reduction in the requirement for IL-2, were observed for these TIL compared to TIL subjected to the normal culture regimen. These results suggest that expanding the appropriate TIL subpopulation *ex vivo* through the use of defined epitopes could result in significantly improved adoptive tumor immunotherapy.

An effective application of peptides *in vivo* recently has been demonstrated in Phase I clinical trials of an anti-Hepatitis B virus (HBV) vaccine termed Theradigm™-HBV (Vitiello et al. 1995). The design of this vaccine has drawn upon numerous advances in the understanding of immune responses and represents a fully-developed, motif-driven approach to eliciting anti-viral reactivity in humans. The vaccine is based on the HBV core protein-derived peptide HBVc 18-27; this peptide previously was shown to bind to HLA-A2.1 with high affinity (Ruppert et al. 1993) and the HBV core antigen is known to provide an epitope for CTL isolated from HBV-infected patients (Penna et al. 1991). The immunogenicity of this peptide was enhanced by producing a modified, three-component vaccine construct. The HBVc CTL epitope is linked at its amino terminus to a tetanus toxoid peptide (TT 830-843), a peptide demonstrated to bind promiscuously to various class II HLA-DR allomorphs (Panina-Bordignon et al. 1989). This generic Th epitope serves to stimulate cytokine production, which is presumed to bolster the CTL response (Fayolle et al. 1991; Shirai et al. 1994). The TT peptide is in turn linked to a lipid moiety, a modification previously shown to enhance CTL responses to peptides *in vivo* (Deres et al. 1989) by mechanisms that remain unclear. An advantage of this design is that potentially toxic adjuvants are avoided; administration of this construct was associated with only mild untoward effects at the highest dose tested in the study. When administered to healthy human subjects, this vaccine elicited, in a dose-dependent manner, CTL capable of lysing transfected cells expressing the HBV core protein (Guilhot et al. 1992). Results from animal studes suggest that this type of vaccine is capable of inducing a long-lived (> 1 yr), anamnestic response. The combined results from both animal and human studies further suggested that all three elements of the vaccine design were key to its success.

Limited space has precluded discussion here of another peptide-associated activity, namely, the ability of MHC-bound peptides to act not only as positive regulators of the immune response, but also as antagonists, capable of inducing negative T cell signalling (reviewed in Sette et al. 1994b). MHC-presented peptides might therefore be viewed not only as antigens, but perhaps also as pharmacologic agents, with the ability to shape the immune response both quantitatively and qualitatively. Further understanding of the structure-activity relationships that exist among immunologic ligands and receptors is expected to contribute to the continued development of both prophylactic and therapeutic peptide-based immunomodulatory strategies.

SUMMARY

The structural features which underlie peptide binding to MHC molecules permit the binding of a diverse array of peptides. Polymorphic residues of class I, and to a lesser extent, class II molecules, determine the peptide selectivities associated with various allomorphs. The motifs which are described here and elsewhere in the literature mainly reflect peptide features which contribute to high affinity binding. While high affinity MHC binding is not an absolute prerequisite for the immunologic relevance of a peptide, motifs provide general guidelines for eliciting and characterizing cellular responses to epitopes presented by a given MHC allomorph or group of related allomorphs. The utility of motifs is underscored by emerging developments in the clinical application of peptides to elicit specific and effective cellular responses.

REFERENCES

Bjorkman, P. J., M. A. Saper, B. Samraoui, W. S. Bennett, J. L. Strominger and D. C. Wiley, *Nature* **329**, 506 (1987).

Bluestone, J. A., A. Kaliyaperumal, S. Jameson, S. Miller and R. Dick, *J Immunol* **151**, 3943 (1993).

Bowness, P., R. L. Allen and A. J. McMichael, *Eur J Immunol* **24**, 2357 (1994).

Brown, J. H., T. S. Jardetzky, J. C. Gorga, L. J. Stern, R. G. Urban, J. L. Strominger and D. C. Wiley, *Nature* **364**, 33 (1993).

Catipovic, B., J. Dalporto, M. Mage, T. E. Johansen and J. P. Schneck, *J. Exp. Med.* **176**, 1611 (1992).

Celis, E., V. Tsai, C. Crimi, R. DeMars, P. A. Wentworth, R. W. Chesnut, H. M. Grey, A. Sette and H. M. Serra, *Proc Natl Acad Sci U S A* **91**, 2105 (1994).

Chicz, R. M. and R. G. Urban, *Immunol Today* **15**, 155 (1994).

Chicz, R. M., R. G. Urban, J. C. Gorga, D. A. Vignali, W. S. Lane and J. L. Strominger, *J Exp Med* **178**, 27 (1993).

Chicz, R. M., R. G. Urban, W. S. Lane, J. C. Gorga, L. J. Stern, D. A. Vignali and J. L. Strominger, *Nature* **358**, 764(1992).

Corr, M., L. F. Boyd, S. R. Frankel, S. Kozlowski, E. A. Padlan and D. H. Margulies, *J. Exp. Med.* **176**, 1681 (1992).

Cox, A. L., J. Skipper, Y. Chen, R. A. Henderson, T. L. Darrow, J. Shabanowitz, V. H. Engelhard, D. F. Hunt and C. J. Slingluff, *Science* **264**, 716-9 (1994).

del Guercio, M. F., J. Sidney, G. Hermanson, C. Perez, H. M. Grey, R. T. Kubo and A. Sette, *J. Immunol* **154**, 685-693 (1995).

Deres, K., H. Schild, K. H. Weismuller, G. Jung and H.-G. Rammensee, *Nature* **342**, 561 (1989).

DiBrino, M., T. Tsuchida, R. V. Turner, K. C. Parker, J. E. Coligan and W. E. Biddison, *J Immunol* **151**, 5930 (1993).

Engelhard, V. H., *Ann. Rev. Imm.* **12**, 181 (1994).

Engelhard, V. H., E. Appella, D. C. Benjamin, W. M. Bodnar, A. L. Cox, Y. Chen, R. A. Henderson, E. L. Huczko, H. Michel, K. Sakaguchi, et al., *Chem Immunol* **57**, 39 (1993).

Falk, K. and O. Rotzschke, *Seminars in Immunology* **5**, 81 (1993).

Falk, K., O. Rotzschke, S. Stevanovic, G. Jung and H.-G. Rammensee, *Nature* **351**, 290 (1991).

Fayolle, C., E. Deriaud and C. LeClerc, *J. Immunol.* **147**, 4069 (1991).

Fremont, D. H., M. Matsumura, E. Stura, P. A. Peterson and I. A. Wilson, *Science* **257**, 919 (1992).

Gaugler, B., Van, den, Eynde, B, van, der, Bruggen, P, P. Romero, et al., *J Exp Med* **179**, 921 (1994).

Geluk, A., K. E. Van Meijgaarden, A. A. Janson, J. W. Drijfhout, R. H. Meloen, R. R. De Vries, T. H. Ottenhoff, *J Immunol* **149**, 2864 (1992).

Geluk, A., K. E. van Meijgaarden, S. Southwood, C. Oseroff, J. W. Drijfhout, R. de Vries, T. H. Ottenhoff and A. Sette, *J Immunol* **152**, 5742 (1994).

Guilhot, S., P. Fowler, G. Portillo, R. F. Margolskee, C. Ferrari, A. Bertoletti and F. V. Chisari, *J Virol* **66**, 2670 (1992).

Guo, H. C., T. S. Jardetzky, T. P. Garrett, W. S. Lane, J. L. Strominger and D. C. Wiley, *Nature* **360**, 364 (1992).

Guo, H. C., D. R. Madden, M. L. Silver, T. S. Jardetzky, J. C. Gorga, J. L. Strominger and D. C. Wiley, *Proc Natl Acad Sci U S A* **90**, 8053 (1993).

Hammer, J., B. Takacs and F. Sinigaglia, *J Exp Med* **176**, 1007 (1992).

Hammer, J., P. Valsasnini, K. Tolba, D. Bolin, J. Higelin, B. Takacs and F. Sinigaglia, *Cell* **74**, 197 (1993).

Henderson, R. A., A. L. Cox, K. Sakaguchi, E. Appella, J. Shabanowitz, D. F. Hunt and V. H. Engelhard, *Proc Natl Acad Sci U S A* **90**, 10275 (1993).

Huet, S., D. F. Nixon, J. B. Rothbard, A. Townsend, S. A. Ellis and A. J. McMichael, *Int Immunol* **2**, 311 (1990).

Hunt, D. F., R. A. Henderson, J. Shabanowitz, K. Sakaguchi, H. Michel, N. Sevilir, A. L. Cox, E. Appella and V. H. Engelhard, *Science* **255**, 1261 (1992).

Jameson, S. C., F. R. Carbone and M. J. Bevan, *J Exp Med* **177**, 1541 (1993).

Jardetzky, T. S., J. H. Brown, J. C. Gorga, L. J. Stern, R. G. Urban, Y. I. Chi, C. Stauffacher, J. L. Strominger and D. C. Wiley, *Nature* **368**, 711 (1994).

Jardetzky, T. S., W. S. Lane, R. A. Robinson, D. R. Madden and D. C. Wiley, *Nature* **353**, 326 (1991).

Kast, W. M., R. M. Brandt, J. Sidney, J. W. Drijfhout, R. T. Kubo, H. M. Grey, C. J. Melief and A. Sette, *J Immunol* **152**, 3904 (1994).

Kawakami, Y., S. Eliyahu, C. H. Delgado, P. F. Robbins, K. Sakaguchi, E. Appella, J. R. Yannelli, G. J. Adema, T. Miki and S. A. Rosenberg, *Proc Natl Acad Sci U S A* **91**, 6458 (1994)a.

Kawakami, Y., S. Eliyahu, K. Sakaguchi, P. F. Robbins, L. Rivoltini, J. R. Yannelli, E. Appella and S. A. Rosenberg, *J Exp Med* **180**, 347 (1994)b.

Kelly, J. M., S. J. Sterry, S. Cose, S. J. Turner, J. Fecondo, S. Rodda, P. J. Fink and F. R. Carbone, *Eur J Immunol* **23**, 3318 (1993).

Klein, J., *Hum Immunol* **30**, 247 (1991).

Klein, J., M. Kasahara, J. Gutknecht and F. Figueroa, *Chem Immunol* **49**, 35 (1990).

Klein, J., Y. Satta, C. O'hUigin and N. Takahata, *Annu Rev Immunol* **11**, 269 (1993).

Kubo, R. T., A. Sette, H. M. Grey, E. Appella, K. Sakaguchi, N. Z. Zhu, D. Arnott, N. Sherman, J. Shabanowitz, H. Michel, et al., *J Immunol* **152**, 3913 (1994).

Ljunggren, H. G., N. J. Stam, C. Ohlen, J. J. Neefjes, P. Hoglund, M. T. Heemels, J. Bastin, T. N. Schumacher, A. Townsend, K. Karre, et al., *Nature* **346**, 476 (1990).

Madden, D. R., D. N. Garboczi and D. C. Wiley, *Cell* **75**, 693 (1993).

Madden, D. R., J. C. Gorga, J. L. Strominger and D. C. Wiley, *Nature* **353**, 321 (1991).

Madden, D. R., J. C. Gorga, J. L. Strominger and D. C. Wiley, *Cell* **70**, 1035- (1992).

Matsumura, M., D. H. Fremont, P. A. Peterson and I. A. Wilson, *Science* **257**, 927 (1992).

Nijman, H. W., J. G. Houbiers, van, der, Burg, Sh, M. P. Vierboom, P. Kenemans, W. M. Kast and C. J. Melief, *J Immunother* **14**, 121 (1993).

O'Sullivan, D., J. Sidney, T. Arrhenius, M. F. del Guercio, M. Albertson, M. Wall, S. Southwood, M. Colon, F. C. Gaeta and A. Sette, *J. Immunol.* **147**, (1991).

Ortiz-Navarette, V. and G. J. Hammerling, *Proc. Natl. Acad. Sci.* **88**, 3594 (1991).

Panina-Bordignon, P., A. Tan, A. Termijtelen, S. Demotz, G. Corradin and A. Lanzavecchia, *Eur. J. Immunol.* **19**, 2237 (1989).

Parker, K. C., M. A. Bednarek, L. K. Hull, U. Utz, et al., J Immunol **149**, 3580 (1992)a.

Parker, K. C., M. L. Silver and D. C. Wiley, *Mol Immunol* **29**, 371(1992)b.

Parker, K. C., M. A. Bednarek and J. E. Coligan, J Immunol **152** 163 (1994).

Penna, A., F. V. Chisari, A. Bertoletti, G. Missale, P. Fowler, T. Giuberti, F. Fiaccadori and C. Ferrari, *J Exp Med* **174**, 1565 (1991).

Rivoltini, L., Y. Kawakami, K. Sakaguchi, S. Southwood, A. Sette, P F Robbins, F. M. Marincola, M. L. Salgaller,J. R. Yannelli, E. Appella and S. A. Rosenberg, *J. Immunol.* (in press).

Rosenberg, S. A., B. S. Packard, P. M. Aebersold, D. Solomon, S. L. Topalian, S. T. Toy, P. Simon, M. T. Lotze, J. C. Yuang, C. Seipp, et al., *N. Eng. J. Med.* **319**, 1676 (1988).

Rudensky, A. Y., P. Preston-Hurlburt, S. C. Hong, A. Barlow and C. A. Janeway, *Nature* **353**, (1991).

Ruppert, J., J. Sidney, E. Celis, R. T. Kubo, H. M. Grey and A. Sette, *Cell* **74**, 929 (1993).

Saper, M. A., P. J. Bjorkman and D. C. Wiley, *J Mol Biol* **219**, 277 (1991).

Sette, A., J. Sidney, G. M. F. del Guercio, S. Southwood, J. Ruppert, C. Dahlberg, H. M. Grey and R. T. Kubo, *Mol Immunol* **31**, 813(1994)a.

Sette, A., J. Alexander, J. Ruppert, K. Snoke, A. Franco, G. Ishioka and H. M. Grey, *Ann. Rev. Immunol.* **12**, 413 (1994)b.

Sette, A., J. Sidney, C. Oseroff, M. F. del Guercio, S. Southwood, T. Arrhenius, M. F. Powell, S. M. Colon, F. C. A. Gaeta and H. M. Grey, *J. Immunol.* **151**, 3163 (1993).

Sette, A., A. Vitiello, B. Reherman, P. Fowler, R. Nayersina, W. M. Kast, C. J. Melief, C. Oseroff, L. Yuan, J. Ruppert, et al., *J Immunol* **153**, 5586 (1994).

Sherman, L. A., S. Chattopadhyay, J. A. Biggs, R. Dick and J. A. Bluestone, *Proc Natl Acad Sci U S A* **90**, 6949 (1993).

Shirai, M., C. D. Pendleton, J. Ahlers, T. Takeshita, M. Newman and J. Berzofsky, *J. Immunol.* **152**, 549 (1994).

Sidney, J., M. F. del Guercio, S. Southwood, V. H. Engelhard, E. Appella, H.-G. Rammensee, K. Falk, O. Rotzschke, M. Takiguchi, R. Kubo, et al., *J. Immunol* **154**, 247 (1995).

Sidney, J., C. Oseroff, M. F. del Guercio, S. Southwood, J. I. Krieger, G. Ishioka, K. Sakaguchi, E. Appella and A. Sette, *J. Immunol* **152**, 4516 (1994).

Sinigaglia, F. and J. Hammer, *Curr Opin Immunol* **6**, 52 (1994).

Stern, L. J., J. H. Brown, T. S. Jardetzky, J. C. Gorga, R. G. Urban, J. L. Strominger and D. C. Wiley, *Nature* **368**, 215 (1994).

Vitiello, A., G. Ishioka, H. M. Grey, R. Rose, P. Farness, R. LaFond, L. Yuan, F. V. Chisari, J. Furze, R. Bartholomeuz, et al., *J. Clin. Invest.* **95**, 341 (1995).

Wipke, B. T., S. C. Jameson, M. J. Bevan and E. G. Pamer, *Eur J Immunol* **23**, 2005 (1993).

22

AUTOANTIBODIES AGAINST PEPTIDE-DEFINED EPITOPES OF T-CELL RECEPTORS IN RETROVIRALLY INFECTED HUMANS AND MICE

John J. Marchalonis,[1] Douglas F. Lake,[1] Samuel F. Schluter,[1]
Keivan Dehghanpisheh,[1] Ronald R. Watson,[2] Neil M. Ampel[3] and
John N. Galgiani[4]

[1] Microbiology and Immunology
University of Arizona, Tucson, Arizona
[2] Family and Community Medicine
University of Arizona, Tucson, Arizona
[3] Veterans Administration Medical Center
Tucson, Arizona
[4] Medicine, University of Arizona
Tucson, Arizona

ABSTRACT

Autoantibodies directed against peptide-defined epitopes of T-cell receptors occur in the serum of healthy humans with the levels and isotypic expression dependent upon physiological changes (aging, pregnancy) or upon the development of autoimmune disease. We carried out investigations of autoantibodies against Tcr peptide-defined epitopes in retroviral infections of humans (HIV-1) and mice (LP-BM5 strain of murine leukemia virus) to determine whether infection with these agents disrupted the regulation of the production of these antibodies. Retroviral infection in humans resulted in increased levels of autoantibody production against putative immunoregulatory regions of the Tcr β chain (Vβ CDR1 and Fr3), a result reflecting a disruption of regulation. In addition, antigenic mimicry was observed with a cross-reaction shared between the common portion of the V3 neutralizing loop of the HIV-1 gp120 molecule and the joining segment of T-cell receptors (Jβ). Infection of mice with the defective retrovirus resulted in the induction of antibodies directed particularly against Vβ CDR1 peptide-defined determinants. Analysis of the virally induced response to a set of 8 CDR1 peptide epitopes indicated a selectivity to the process. It was possible to partially reverse aberrant cytokine changes correlated with the onset of murine MAIDS by administration of T-cell receptor peptides in saline. These results show that retroviral infection in humans and mice has a profound dysfunctional effect on the regulation

Immunobiology of Proteins and Peptides VIII
Edited by M. Z. Atassi and G. S. Bixler, Jr., Plenum Press, New York, 1995

211

of autoantibodies to T-cell receptors. The function of these autoantibodies in the immunopathogenesis of acquired immunodeficiency remains to be determined.

INTRODUCTION

Recent studies support two emerging and related concepts regarding the pathogenesis of autoimmunity. The first is that infectious agents, particularly retroviruses, can initiate a series of events leading to autoimmune disease [1,2]. The second is that autoimmune reactions contribute to the immunopathogenesis in AIDS [3-5]. We have reported that healthy individuals possess low constitutive levels of autoantibodies directed against peptide-defined variable and constant domain epitopes of T-cell receptors [6], and that the levels of these antibodies as well as particular isotypes can vary with age [7,8], and with autoimmune disease [7-10]. The autoantibodies characterized differ from rheumatoid factor in specificity and in V_H and $V\kappa$ gene usage [11]. We proposed that the $V\beta$ epitopes described, CDR1 and Fr3, were most probably involved in immunoregulation [9], particularly inasmuch as Fr3 has been implicated in recognition of superantigens [11] and CDR1 has been suggested to function in the recognition of MHC in MHC/restricted antigen presentation [12]. In this study, we carried out experiments designed to determine whether retroviral infection in man (HIV-1) and in the C57Bl/6 mouse (LPBM-5 strain; the MAIDS model) affected the levels of IgG autoantibodies directed against the major peptide-defined autoepitopes. We present evidence for both species that retroviral infection substantially increases the titers of autoantibodies against these particular determinants.

MATERIALS AND METHODS

Peptides

Synthetic peptides were synthesized using Fmoc chemistry on an Applied Biosystems Peptide Synthesizer. Nested overlapping sets of synthetic peptides duplicating the covalent structures of human Mcg light chain [13,14], Tcr α chain [15], and Tcr β chain [16] were prepared and calibrated as previously described. Table 1 lists the synthetic peptide segments that were used extensively in the present study.

Table 1. Synthetic peptide segments forming epitopes analyzed in this study

	Synthetic sequence*	Description
pep–β3	C K P I S G H N S L F W Y R Q T	CDR1/Fr2 of Vβ8.1
pep–β8	K I Q P S E P R D S A V Y F C A	Fr3 of Vβ8.1
pep–β17	Q P L K E Q P A L N D S R Y C L	Cβ "loop"
pep–α3	N Y S S S V P P Y L F W Y V Q Y	CDR1/Fr2 of Vα1
pep–α8	P S A H M S D A A E Y F C A V S	Fr3 of Vα1
pep–α17	S A V A W S N K S D F A C A N A	Cα
pep–Mcg3	T G T S S D V G G Y N Y V S W Y	CDR1/Fr2 of VλVI
pep–Mcg8	S G L Q A E D E A D Y Y C S S Y	Fr3 of VλVI
pep–Mcg17	A A S S Y L S L T P E Q W K S H	Cλ

*single letter abbreviations

Immunoglobulins and Sera

Two types of purified pooled IgG from healthy individuals (IVIG) were used. Gammagard from Hyland Baxter (Costa, Mesa, CA) and Sandoglobulin (Sandoz Pharmaceuticals, (New Jersey) were used as normal controls in comparison with the plasma immunoglobulin preparations made from HIV-infected individuals (HIVIG). A European preparation was obtained from Merieux Institute (France). This IgG was prepared from the placentas of HIV infected African women. The other preparation of IVIG [17] was obtained from the AIDS Research and Reference Reagent Program sponsored by the National Institutes of Health (Bethesda, MD). These preparations were negative for HIV by polymerase chain reaction. By analysis of these preparations of SDS-polyacrylamide gel electrophoresis, both the IVIG and HIVIG preparations consisted predominantly of IgG immunoglobulin. Sera from healthy humans in the age range of 20-90 was obtained from Dr. David E. Yocum, Director of the Arizona Arthritis Center, and Dr. John Boyer, Chief of the Section of Gerontology in the Department of Medicine, University Medical Center. These samples were described previously. Ten sera of patients with osteoarthritis were obtained from the Arizona Arthritis Center. Ten sera of untreated individuals diagnosed with coccidiomycosis were obtained from the VA Medical Center. Sera from ten patients seropositive for HIV were obtained from the AIDS Clinic at the VA Medical Center at Tucson.

The Murine MAIDS Model

Female C57BL/6 mice, 3-5 weeks of age, were randomly assigned to two groups. One group was inoculated intraperitonealy once with 0.1 ml of the LP-BM5 strain of murine leukemia virus which was composed of an ecotropic titer of 4.5^{10} PFU/ml. This treatment induced disease with the same time course as previously described [18]. The control group was injected with the culture medium used for the LP-BM5. Immunization with peptides was carried out in the absence of adjuvant, as described below.

Affinity Chromatography

Peptide columns were prepared by coupling by free peptides to activated CH-Sepharose 4B (Pharmacia) under conditions described by the manufacturer. The affinity chromatography was carried out as previously described [6].

Enzyme Linked Immunosorbent Assay

ELISAs were performed in 96-well microtitrater plates as previously described for human IgG antibodies [6-8,14]. Conditions for assay of the murine IgG antibodies are described in detail elsewhere [19]. Horseradish peroxidase (HRP) conjugated goat antibodies to mouse IgG immunoglobulin (Jackson Immuno Research, West Grove, PA) was used as the detecting reagent. For inhibition ELISAs, serial dilutions of peptides were pre-incubated with anti-sera for one hour and then the mixture was added to the peptide-coated ELISA plates. Pig gelatin was used in the diluents and in the blocking buffers.

Cytokine Assays

Antigen capture assays for the cytokines IL-2, INFγ, IL-5 and IL-6 were carried out as previously described [18].

RESULTS

Dysregulation of Autoantibody Production in HIV infection

Table 2 presents the geometric mean titers of IgG autoantibodies binding to Vβ pep CDR1, Vβ pep Fr3 and the Cβ loop peptide (pep Cβ) in sera from healthy individuals in the age range 20-50 (27 individuals), 10 patients with osteoarthritis (a non autoimmune arthritis), 10 untreated patients infected with Coccidiomycosis ("Valley fever") and 10 individuals sero positive for HIV-1. The HIV positive individuals were seropositive but had not yet developed AIDS. The reaction with pig gelatin is used to measure background "stickiness" of the serum, and this reagent is also included in the dilution and blocking buffers.

As reported previously [6-9], normal healthy individuals have modest titers of antibodies against the 3 peptides. The same situation holds for the osteoarthritis patients, and the coccidiomycosis-infected individuals show increases in all of the autoantibody reactivities. By contrast, the HIV positive individuals show substantial increases in autoantibody levels in all specific categories. The magnitude of the effect is shown in Figure 1 which compares titration curves of an individual coccidiomycosis patient with those of an HIV positive individual. In addition to the Tcr β epitopes listed above, these titration curves illustrative reactivity to a peptide from the V3 loop [20] of HIV-1 gp120 and peptide defined epitopes corresponding to Vα CDR1 (pep α3), Vα Fr3 (pep α8) and Cα (pep α17). The coccidiomycosis infected individual has negative titers (less than 100) against the pig gelatin, the HIV-V3 peptide and the Tcrα peptides. It has modestly elevated titers against the Vβ CDR1 (β3), Fr3 (β8), and Cβ peptides (β17). The HIV positive individual has a low background titer (less than 100), but has extremely elevated titers against the HIV antigen and the Tcr β epitopes. The serum autoantibody level against the α peptides, particularly pep α8, are also elevated, but not to the degree of reactivity against the Tcrβ peptides. Thus, analysis of individual sera from HIV positive patients shows that elevated production of autoantibodies against putative regulatory sites of T-cell receptors is a feature of the disease. Although individual suffering from a fungal infection also show an elevation of some of these autoantibodies, the levels are no way comparable to those in the retroviral infection.

Antigenic Mimicry between HIV-1 and Human Tcrs

The preceding data indicate that retroviral infection in humans alters the mechanisms by which autoantibody levels against putative regulatory determinants of T-cell receptors are maintained with the results of a substantially increased levels of IgG autoantibodies to these determinants. The joining segment of Tcrβ, defined by the Jβ minigene, specifies a peptide that is highly conserved among Tcrs and immunoglobulin light chains [21]. Normally, autoantibodies to this peptide are not observed. In point of fact, it is extremely difficult to immunize rodents with this peptide [22], although rabbits produce high levels of specific antibodies against human Jβ-defined epitopes. We coupled the Jβ peptide to Sepharose and

Table 2. IgG autoantibody geometric mean titers against TCR
β chain peptide-defined autoantigens

Patients	Pig Gel	pep-CDR1	pep-Fr3	pep-Cβ
Healthy	< 50	58	278	172
Osteoarthritis	< 50	108	148	122
Coccidiomycosis-	< 50	637	1,067	1,043
HIV positive	< 50	10,357	6,672	1,373

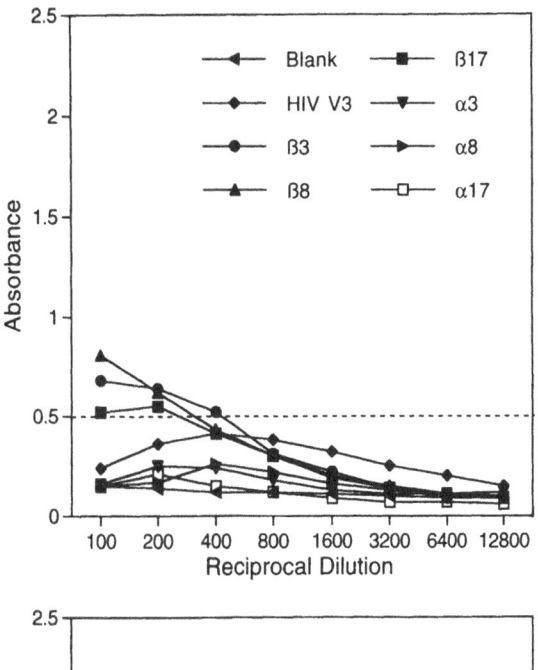

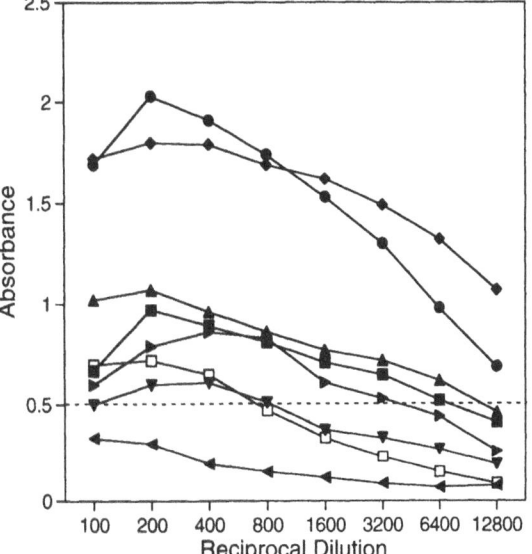

Figure 1. Titrations of IgG binding of serum from a typical coccidiomycosis patient (Upper panel) and an HIV positive individual to a panel of peptide antigens corresponding to HIV and Tcr epitopes. The peptides are defined in the text with those derived from Tcr α chain designated by α- and those from β chain by β-. Ordinate, Absorbance of the ELISA reaction; Abscissa, Reciprocal dilution of serum. The "titer" is defined as the reciprocal dilution at which the Absorbance is equal to 0.5 (dotted line).

affinity purified human IgG autoantibodies from gammaglobulin preparations of healthy individuals and HIV-infected individuals [23]. In addition, we constructed a solid phase immunoadsorbent from the sequence of the conserved segment of the V3 loop of HIV-1 viral gp120 (IIIB isolate). We found that antibodies against the determinants were not detectable in the intravenous immunoglobulin preparations obtained from normal individuals (Gammagard, and Sandoglobulin) but could be isolated from the HIVIG preparations (Merieux and the AIDS Research and Reference Standard from the National Institutes of Health). In addition to the cross-reaction between the V3 and Jβ peptides illustrated in Figure 2, the affinity purified anti-Jβ was able to react with an intact recombinant single chain molecule containing full length Vβ and Vα gene products [23] and also with intact recombinant HIV-1 gp120 (obtained from American Biotechnology Inc.). Both synthetic V3 loop peptide and

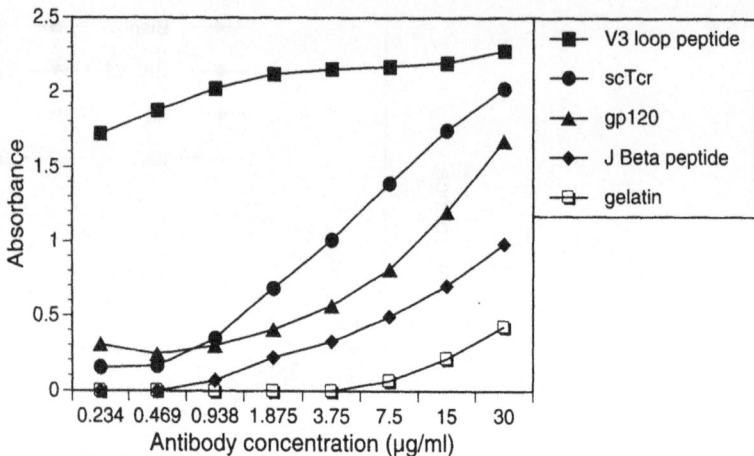

Figure 2. Reactivity of V3 loop-affinity purified HIVIG. HIVIG was affinity purified on a V3 loop column from gp120 (HIV$_{IIB}$) and teted for reactivity against scTcr, gp120, Jβ peptide, and porcine gelatin in ELISA. Peptides were coated onto the ELISA plates at 1 μg/well. Due to the enormouse expense of HIV-1 gp120, only 50ng was added to each well.

the Jβ peptide cross inhibited binding by either affinity purified anti-Jβ or anti HIV-V3 loop. Other peptides, however, showed no capacity to inhibit.

  The basis for this antigenic mimicry is illustrated in Table 3 which compares the sequence of the conserved region of the HIV-1 gp120 V3 loop with homologous regions of the joining segment of immunoglobulins and T-cell receptors. The features important to the cross-reaction are the conservation of the GXG motif and the large positively charged residue in close apposition to it. In an extended configuration, the 3-dimensional structures of these peptide regions would be extremely similar, and the similarity in conformation between the V3 region and the joining segment of immunoglobulin/Tcr chains [23] holds even in

Table 3. Comparative alignments of HIV-1 gp120 V3 loop peptides and TCR J-segments

```
            1 . . . 5 . . . .10 . . . .15 . . .

V3 (IIIB)ᵃ   R I H I Q R G P G * R A F Y T T K

NY 5         K S I R I Q R G P G * R A F V T I G

ZAIRE 6      T R Q S T P I G L G * Q A L Y T T R

Jβ (1.2)ᵃ    A N Y G Y T F G S G T R L T V V

Jβ2.7          S Y E Q Y F G P G T R L T V T

JαA          S A S K I I F G S G T R L S I R P

JαD          S W G K F E F G A G T Q V V V T P
```

ᵃSynthetic peptides used here.

Residues shared between any variant V3 - "common" sequences and the Tcr set are boxed.

comparison with the structure defined by Ghiara *et al.* [24] for the V3 loop complexed with the Fab of an antibody directed against it.

Autoantibodies to Tcr Peptide Defined Epitopes in LP-BM5 Mice

C57Bl/6 mice infected with the defective retrovirus mixture LP-BM5 show hyper-proliferation of B and T cells, hypergammaglobulinemia, splenomegaly and lymphade-nopathy followed by progressive immune deficiency [25]. We carried out immunochemical studies to determine whether this infection resulted in elevated production of IgG antibodies to T-cell receptor peptide-defined epitopes. Figure 3 compares the IgG autoantibody activity of pooled sera of five mice infected with the retrovirus and five control mice injected with saline. The assay was performed at 6 weeks following infection with the test carried out against the sets of overlapping 16-mer peptides duplicating the covalent sequence of the immunoglobulin domains of the human Tcr β chain YT35 and the human λ light chain Mcg. The retrovirally infected mice show substantially increased binding to the peptide epitopes of the β chain that are also recognized by autoantibodies present in the uninfected animals (β3, β8 and β17). An increase was also noted against some λ chain peptides (Peps Nos. 4, 8/9 and 17) but not against Mcg pep #3.

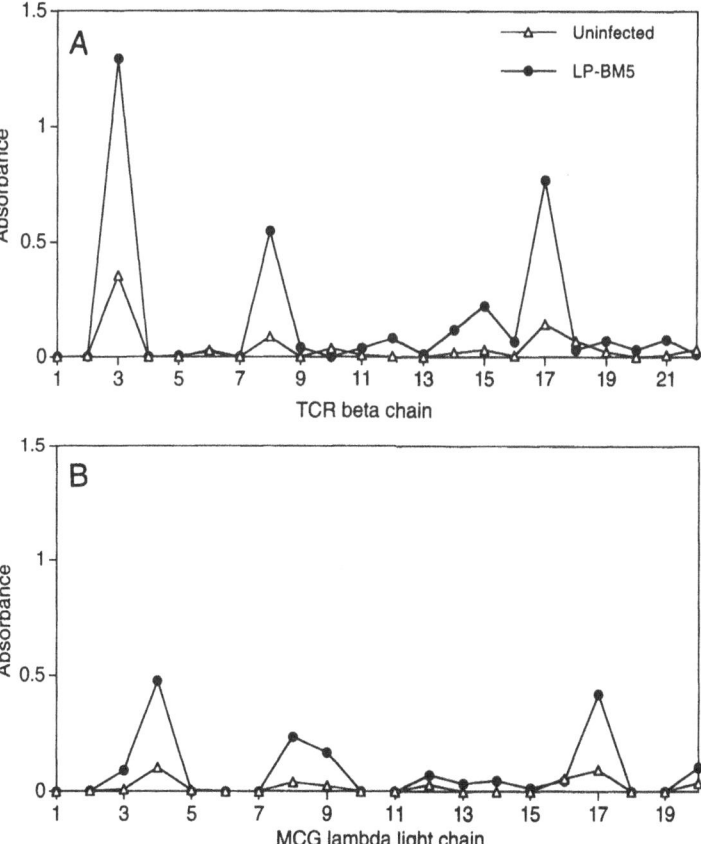

Figure 3. IgG antibody activity of pooled sera from LP-BM5 virus-infected C57Bl/6 to sets of synthetic overlapping peptides modelling the complete covalent structure of the immunoglobulin domains of the T-cell receptor β chain (A) and the human λ light chain Mcg (B). ELISA assays were carried out in quadruplicate at a serum dilution of 1:200 using serum pools, each corresponding to five individuals. The uninfected mice (Δ) were age and sex matched to the infected animals (●). The mice were bled six weeks following infection.

We found that the hypergammaglobulinemia in the infected animals was apparent at six weeks following infection, continued through twelve weeks and diminished to background levels by week fourteen. The autoantibodies tended to parallel this time course with the exception that antibody activity against certain Vβ peptide homologs, particularly those corresponding to the human Vβ 5.2 and murine Vβ8.1 sequences, continued to increase after the other activities had diminished. High IgG levels as a result of polyclonal activation do not directly correlate with the presence of autoantibodies in every instance. We have carried out a regression analysis between ELISA titer and total serum IgG content finding that two types of exceptions occur [19]. Some sera show high specific titer at low total IgG concentration. In addition, a number of sera show detectable bulk IgG, but do not have the predicted specific titers. In general, we found that of seven synthetic peptides corresponding to distinct human and murine T-cell receptor CDR1 segments, the levels of activity against five of these declined by four months but activity against the human Vβ5.2 and murine Vβ8.1 peptides continued to increase. To account for individual variation, we carried out a longitudinal study of individual uninfected or LP-BM5 infected young C57Bl/6 mice as shown in Table 4. The uninfected mice at two weeks post-infection had low levels of antibodies against the human Vβ8.1 CDR1 peptide, which is a homolog of murine Vβ11 CDR1. The levels of these autoantibodies gradually rose between two and six weeks post-infection and appeared to plateau at about ten weeks following infection. The observed level is that which would be expected for adult mice. With the LP-BM5 infection, however, a dramatic rise substantially greater than normal levels was observed by six weeks post-infection. These levels dropped significantly between ten and fourteen weeks post-infection. Interestingly in both studies, individual mice were observed with high levels of anti-Tcr autoantibodies during later stages of the infection. Here, this is observed at week fourteen and in other examples we have found high titers against the mouse Vβ8.1 peptide to continue as long as twenty weeks post-infection.

Table 4. Longitudinal comparison of IgG autoantibodies of individual mice following either sham infection or infection with LP-BM5

	Weeks following retroviral infection			
	2	6	10	14
Uninfected				
Individual mice				
1	0.59*	2.59	3.26	1.34
2	0.54	0.58	2.42	1.37
3	0.76	0.39	3.02	2.73
Mean ± SE	0.63±.07	1.19±.70	2.90±.25	1.81±.46
LP-BM5 Infected				
Individual mice				
1	0.47	2.86	9.61	3.49
2	1.00	4.91	6.66	6.21
3	0.49	2.79	6.17	7.64
4	1.79	8.20	9.74	3.91
5	2.09	6.62	7.36	1.04
6	2.86	6.74	6.86	4.40
Mean ± SE	1.45±.39	5.35±.91[a]	7.73±.63[b]	4.45±.93[a]

*Data expressed are Sum ELISA titers [19].
Note that the saline-injected animals show an increase in antibody activity that plateaus at the normal adult level by 10 weeks.
p values for comparison of mean ± SE of infected animals with the corresponding values for the infected mice: [a], p< .05; [b], p <.01.

Manipulation of Cytokine Levels in Infected Mice by Administration of Tcr Peptide

Although there is a clear correlation of increase in autoantibody levels to Tcr following viral infection in humans and mice, it is not known what the function of such antibodies would be in the development of the disease process. Evidence for possible antigen specific effects is suggested by the observation that the response in mice is to some degree selective in the sense that antibody levels continue to rise against some Vβ sequences, but not against the products of other genes. In addition, we have found that human and murine autoantibodies against Tcr variable region determinants stain appropriate human T-cells in immunocytofluoresence. In order to obtain sufficient murine antibodies to investigate possible effects such autoantibodies would have on immunoregulation *in vivo*, we have generated hybridomas by fusing spleens of LP-BM5-infected mice with the fusion partner P-3U1 8 at six weeks following infection. This approach should prove fruitful in the long run because it is exceedingly difficult to obtain enough purified autoantibody from mouse serum to obtain sufficient dosages for regulatory studies.

Another approach to investigating the effects of autoimmunity against T-cell receptors in the dysregulation of the immune system in MAIDS was to administer soluble peptides to infected and uninfected mice and determine the effects. The experimental group (eight mice) was infected with LP-BM5 and a control group was given saline at the same time point. Two infected groups were given respectively T-cell receptor CDR1 peptide (200 µg/mouse in saline) and a control group was given the corresponding CDR1 peptide from the human λ light chain Mcg. The peptide injections were given at day 10 and 14 following infection. It has been shown that one of the effects of LP-BM5 infection in C57Bl/6 mice is to diminish production of TH1 cytokines and enhance production of those cytokines elaborated by the TH2 cells [18,25]. Figure 4 presents the results of an experiment designed to determine whether the Tcr β or λ light chain peptides affected this process with the animals

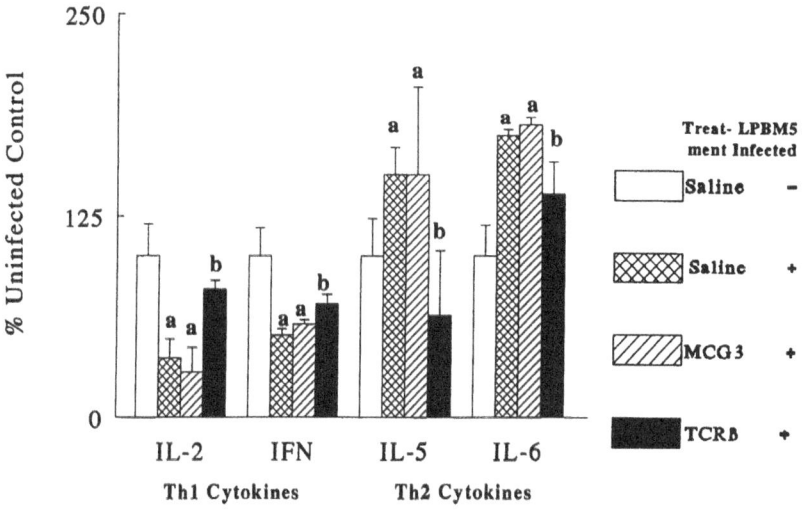

Figure 4. Effects Tcr Vβ peptide injected during retrovirus infection on production by T-helper 1 cells (Th1) of cytokines (IFN, IL-2) Th2 cells' cytokines (IL-5, IL-10), and T and B cell mitogenesis. Mice were allowed to develop severe immune deficiency (murine AIDS) at 16 weeks post infection. Cytokines were produced by splenocytes, mitogen-stimulated *in vitro*. Each sample was determined in triplicate. Values are mean ± SD, a:p<0.05 compared to saline-injected uninfected, normal mice (4). b:p<0.05 compared to saline-injected infected mice after 16 weeks of infection with development of murine AIDS.

being assayed at day 42 following infection. The retroviral infection significantly diminished the levels of the Th1 cytokines IL-2 and IFN (interferon γ) and decreased the levels of the Th2 cytokines IL-5 and IL-6. The λ light chain peptides had no effect on the cytokine levels. The Tcr β peptide, however, caused a significant increase in the TH1 cytokines and significant decreases in the TH2 derived cytokines. Thus, administration of soluble Tcr peptide following viral infection significantly reversed the aberrant effect on cytokine production usually observed in retrovirally infected mice.

DISCUSSION

There are two salient conclusions following from this study of the effects of retroviral infection on the production of IgG autoantibodies to T-cell receptors in humans and mice. The first is that in both HIV seropositive humans and in mice experimentally infected with the defective retrovirus LP-BM5, there is a profound dysregulation of the immune system that results in the production of high levels of autoantibodies to peptide-defined epitopes of T-cell receptors. These epitopes are the same ones that are recognized by antibodies constitutively present in the sera of healthy animals [6,19]. Since there is no detectable homology between these peptides and the sequences of the two retroviruses, it is reasonable to hypothesize that this overproduction results from a failure in regulation. There is, however, a specificity to the reaction which is particularly apparent in mice because the correlation between serum IgG levels and anti-Tcr autoantibodies shows a number of exceptions. Moreover, autoantibodies against CDR1-peptide defined regions of particular Vβ genes continue to increase after the hypergammaglobulinemia has subsided. In addition, the levels of increased autoantibody production against T-cell receptors in the murine model was substantially greater than that observed against unrelated antigens such as ovalbumin and bovine serum albumin. The precise mechanisms underlying this breakdown in regulation of autoantibody levels remain to be determined, but we found that administration of soluble T-cell receptor peptides could modify the aberrant cytokine production following LP-BM5 infection. Studies are in progress to establish the mechanism of this as well as its effect on the long term survival of mice in the MAIDS model.

Two potential mechanisms can be hypothesized. The first is based upon immunization against specific regulatory determinants on the products of individual Vβ genes; whereas the other, is predicated upon the possibility that peptides corresponding to the CDR1 and Fr3 segment of the Vβ chain [9,11,12] could interact with MHC molecules necessary for the presentation of peptide antigen.

The second salient discovery was that antigenic mimicry is found between the conserved portion of the V3 loop of HIV-1 gp120 and the highly conserved Fr4 segment of Tcrs which is specified by the joining gene segment. Our data show that antibodies specific for either the V3 loop peptide or the synthetic Jβ peptide can be isolated in detectable quantity by affinity purification from immunoglobulin preparations prepared from HIV-infected individuals. These antibodies show substantial cross-reaction among the two peptides and also have the capacity to bind to recombinant gp120 and recombinant single chain T-cell receptors. The studies we have carried out using the gammaglobulin from HIV infected individuals suggest that the virus is the actual immunizing antigen and that the recognition of the Jβ peptide segment is the result of a cross-reaction. We have previously shown that rabbit antibodies raised against the synthetic Jβ peptide react with the Tcr β chain in enzyme linked immunosorbent assay and Western blot analysis [21,22] and also with intact T-cells by immunocytofluorescence [26]. It is attractive to speculate that the autoantibodies generated by either a breakdown of regulation of normal control of autoimmunity or by antigenic mimicry play a role in the generation of acquired immunodeficiency. Experiments are in progress to test this hypothesis.

Our previous studies in humans were carried out using purified IgG immunoglobulin preparations. Here, we provide new information based upon titrations of autoantibodies in sera of healthy individuals, patients with osteoarthritis, untreated individuals diagnosed with coccidiomycosis, and healthy individuals seropositive for HIV. Our results confirm the low levels of constitutive autoantibodies occurring in normal sera against Tcr Vβ CDR1 and Fr3 epitopes and against an exposed loop peptide of the constant domain. The osteoarthritis patients were essentially equivalent to the normal individuals, which was the expected result because this disease does not have a strong component of autoimmunity. As might be predicted, the patients infected with the fungus, showed slightly elevated levels of autoantibodies. A similar phenomenon occurs with respect to levels of rheumatoid factor in various infections or purposeful immunizations [27]. The HIV-infected individuals had extremely high IgG titers against the characteristic Tcr β epitopes and also showed significantly elevated titers against the corresponding regions of the Tcrα chain. Thus, the presence of high levels of IgG autoantibodies against human Tcr determinants is a characteristic feature of HIV infection. In the case of the human patients, however, it was not possible to determine when the viral infection occurred. This deficiency could be experimentally approached in the murine MAIDS model where we determined precise time courses for the appearance of autoantibodies in individual mice following retroviral infection. We conclude that the production of autoantibodies to T-cell receptors is a consequence of retroviral infections in human and mice. These antibodies may play a significant role in the immunopathogenesis of acquired immune deficiency, and possible in autoimmune disease, as suggested by the fact that there is some selectivity in the response and antibodies of this nature have been found to have the capacity to bind directly to human T-cells.

ACKNOWLEDGMENTS

This work was supported in part by Grants #5-038 from Arizona Disease Control Research Commission and Grant # GM-45832 from the National Institutes of Health and a grant from Baxter Biotechnology/Hyland Divison to JJM and to a VA Merit Review #NO1-A1-5256 to JNG.

REFERENCES

1. Fauci, A.S., 1993, Multifactorial nature of human immunodeficiency virus disease: implications for therapy. *Science* 262:1011-1016
2. Valentine, F.T., 1990, Pathogenesis of the immunological deficiences caused by infection with the human immunodeficiency virus. *Seminars in Oncology* 17:321-334.
3. Clerici, M., Shearer, G., Hounsell, E.F., Jameson, B., Habeshaw, J. and Dalgleish, A.G., 1993, Alloactivated cytotoxic T cells recognize the carboxy terminal domain of human immunodeficiency virus-1 gp120 envelope glycoprotein. *Eur. J. Immunol.* 23:2022-2025.
4. Katz, D.H., 1993, AIDS: Primarily a viral or an autoimmune Disease? *AIDS Resch and Human Retroviruses.* 9:489-493.
5. Kion, T.A. and Hoffman, G.W., 1991, Anti-HIV and anti-MHC antibodies in alloimmune and autoimmune mice. *Science* 253:1138-1140.
6. Marchalonis, J.J., Kaymaz, H., Dedeoglu, F., Schluter, S.F., Yocum, D.E. and Edmundson, A.B., 1992, Human autoantibodies reactive with synthetic autoantigens from T-cell receptor β chain. *Proc. Natl. Acad. Sci. USA* 89:3325-3329.

7. Marchalonis, J.J., Schluter, S.F., Wilson, L., Yocum, D.E., Boyer, J.R. and Kay, M.M.B., 1993, Natural human antibodies to synthetic peptide autoantigens: correlations with age and autoimmune disease. *Geront.* 39:65-79.

8. Marchalonis, J.J., Kaymaz. H., Schluter, S.F., and Yocum, D.E., 1993, Human autoantibodies to a synthetic putative T-cell receptor β-chain regulatory idiotype: Expression in autoimmunity and aging. *Expt. & Clin. Immunogenet.* 10:1-15.

9. Marchalonis, J.J., Schluter, S.F., Wang, E., Dehghanpisheh, K., Lake, D., Yocum, D.E., Edmundson, A.B. and Winfield, J.B., 1994, Synthetic autoantigens of immunoglobulins and T-cell receptors: their recognition in aging, infection and autoimmunity. *Proc. Soc. Expt. Biol.* 207:129-147.

10. Dedeoglu, F., Kaymaz, H., Klein, G. and Marchalonis, J.J., 1993, Light and heavy chains specifying a human IgM κ autoantibody to a T-cell receptor Vβ-antigen. *Immunol. Letts.* 38:223-227.

11. Pullen A.M., Bill J., Kubo R.T., Marrack P., Kappler J.W., 1991, Analysis of the interaction site for the self-superantigen Mls-1ª on T-cell receptor Vβ. *J. Exp. Med.* 173:1183-1185.

12. Chothia, G., Boswell, D.R. and Lesk, A.M., 1988, The outline structure of the T-cell αβ receptor. *EMBO J.* 7:3745-3755.

13. Marchalonis, J.J., Dedeoglu. F., Kaymaz, H., Schluter, S.F., and Edmundson, A.B., 1992, Antigenic mapping of a human λ light chain: Correlation with 3-dimensional structure. *J. Prot. Chem.* 11:129-137.

14. Kaymaz, H. and Marchalonis, J.J., 1993, Autoreactive sites of human λ light chain mapped by comprehensive peptide synthesis. *J. Prot. Chem.* 12:659-666.

15. Marchalonis, J.J., Kaymaz, H., Schluter, S.F. and Yocum, D.E., 1994, Naturally occurring human autoantibodies to defined T-cell receptor and light chain peptides. in Immunobiology of Proteins and Peptides VII. (M.T. Atassi, ed.), Plenum Press, N.Y., pp 135-145.

16. Kaymaz, H., Dedeoglu, F., Schluter, S.F., Edmundson, A.B. and Marchalonis, J.J., 1993, Reactions of anti-immunoglobulin sera with synthetic T-cell receptor peptides: implications for the three-dimensional structure and function of the Tcr β chain. *Intl. Immunol.* 5:491-502.

17. Prince, A.M., Horowitz, B., Baker, L., Shulman, R.W., Ralph, H., Valinsky, J., Cundell, A., Brotman, B., Boehle, W., Rey, F., Piet, M., Teesink, H., Lelie, N., Tersmette, M., Miedema, F., Barbosa, S., Nemo, G., Nastala, C.L., Assan, J.S., Lee, D.R. and Eichberg, J.W., 1988, Failure of a human immunodeficiency virus (HIV) immune globulin to protect chimpanzees against experimental challenge with HIV. *Proc. Natl. Acad. Sci. USA* 85:6944-6948.

18. Huang, D.S., Wang, Y., Marchalonis, J.J., and Watson, R.R., 1994, The kinetics of cytokine secretion and proliferation by mesentric lymph node cells during the progression to murine AIDS, caused by LP-BM5 murine leukemia virus infection. *Regional Immunol.* 5:325-331.

19. Dehghanpisheh, K., Huang, D., Schluter, S.F., Watson, R.R. and Marchalonis, J.J., 1995, Production of IgG autoantibodies to T-cell receptors in mice infected with the retrovirus LP-BM5. *Intl. Immunol.* 7: in press.

20. La Rosa G.J., Davide J.P., Weinhold K., Waterburg J.A., Profy A.T., Lewis J.A., Langlois A.J., Dreesman G.R., Boswell R.N., Shadduck P., Holley L.H., Karplus M., Bolognesi D., Mathews T.J., Emini E.A., and Putney S.D., 1990, Conserved sequence and structural elements in the HIV-1 principal neutralizing determinant. *Science* 249:932-935.

21. Schluter, S.F. and Marchalonis, J.J., 1986, Antibodies to synthetic joining segment peptide of the T-cell receptor β-chain: serological cross-reaction between products of T-cell receptor genes, antigen binding T-cell receptors and immunoglobulins. *Proc. Natl. Acad. Sci. USA.* 83:1872-1876.

22. Marchalonis, J.J., Schluter, S.F., Hubbard, R.A., McCabe, C. and Allen, R.C., 1988, Immunoglobulin epitopes defined by synthetic peptides corresponding to joining- region sequence: Conservation of determinants and dependence upon the presence of an arginyl or lysyl residue for cross-reaction between light chains and T cell receptor chains. *Mol. Immunol.* 25:771-784.

23. Lake, D.F., Schluter, S.F., Wang, E., Bernstein, R.M., Edmundson, A.B. and Marchalonis, J.J., 1994, Autoantibodies to the α/β T-cell receptors in human immunodeficiency virus (HIV) infection: dysregulation and mimicry. *Proc. Natl. Acad. Sci. USA.* 91:10849-10853.

24. Ghiara, J.B., Stura, E.A., Stanfield, R.L., Profy, A.T., and Wilson, I.A., 1994, Crystal structure of the principal neutralization site of HIV-1. *Science* 264:82-85.

25. Jolicoeur P., 1991, Murine acquired immunodeficiency syndrome (MAIDS): an animal model to study the AIDS pathogenesis. *FASEB J.* 5:2398-2401.

26. Shankey, T.V., Schluter, S.F. and Marchalonis, J.J., 1989, Flow cytometric analysis of human lymphocytes using affinity purified antibody to T cell receptors synthetic J region. *Cell Immunol.* 118:526-531.

27. Carson, D.A., Chen, P.P., Fox, R.I., Kipps, T.J., Jirik, F., Goldfine, R.D., Silverman, G., Radoux, V. and Fong, S., 1987, Rheumatoid factor and immune networks. *Ann. Rev. Immunol.* 5:109-126.

23

CHARACTERIZATION OF AUTOANTIBODIES DIRECTED AGAINST T CELL RECEPTORS

Douglas F. Lake,[1] William J. Landsperger,[2] Ralph M. Bernstein,[1] Samuel F. Schluter,[1] and John J. Marchalonis[1]

[1] Microbiology and Immunology
University of Arizona, College of Medicine
Tucson, Arizona 85724
[2] Baxter Healthcare Corporation
Biotechnology Division
Duarte, California

ABSTRACT

Recently it has been observed that administration of intravenous immunoglobulin (IVIG) can have profound effects on a wide variety of diseases related to the dysregulation of the immune system. The mechanisms which explain these activities are poorly understood. Human IVIG and various Cohn plasma fractions contain autoantibodies directed against T cell receptors (Tcr). Previous studies have shown that IVIG contains autoantibodies against T cell receptor peptides. In order to further our understanding of autoantibody specificities, a single chain Tcr (scTcr) was constructed by recombinant DNA techniques from the variable α and variable β chains of the Jurkat cell line. Anti-Tcr autoantibodies were isolated from IVIG and Cohn fractions I + III using a scTcr affinity column. This scTcr affinity purified material reacted with the surfaces of T cells at 10ug/ml whereas non-purified IVIG did not. Sera from patients with rheumatoid arthritis (RA) as well as serum from patients with systemic lupus erythematosus (SLE) reacted with the scTcr at levels above that of normals.

INTRODUCTION

Intravenous Immunoglobulin (IVIG) has been used to treat a multitude of autoimmune diseases [1, 2, 3]. Since commercial preparations of IVIG are composed of purified IgG from thousands of individuals, the active components in IVIG responsible for the clinical benefit in the treatment of autoimmune diseases are unknown. There is some evidence that anti-idiotypic autoantibody networks play a role in ameliorating some autoimmune diseases [4]. For example, anti-idiotypic antibodies in IVIG have been shown to react with the

Immunobiology of Proteins and Peptides VIII
Edited by M. Z. Atassi and G. S. Bixler, Jr., Plenum Press, New York, 1995

223

combining site of anti-thyroglobulin autoantibodies, anti-neutrophil cytoplasmic antigen autoantibodies, anti-Factor VIII autoantibodies and anti-gp IIb IIIa autoantibodies [4-8]. In addition, *Myasthenia Gravis*, a classical autoimmune disease where autoantibodies exist against the acetyl choline receptor, can be treated IVIG [9]. Other evidence exists for Fc-dependent immunoregulation in IVIG treatment [10]. Fc-dependent immunoregulation might occur by Fc receptor blockade in which the Fc portion on IVIG nonspecifically blocks the binding of Fc receptors on phagocytic cells from interacting with pathogenic autoantibodies [11].

Our hypothesis is that the antibodies present in IVIG which are directed toward T cell receptors, and in particular, specific regions on the T cell receptor which we have defined by peptides from CDR1 (β3), Fr3 (β8) and a loop in the constant region (β17), are involved in immunoregulation [12-13]. We have previously shown that anti-T cell receptor (Tcr) peptide autoantibodies exist in IVIG [12]. Additionally, we have found that there are increased levels of anti-Tcr peptide autoantibodies in patients suffering from systemic lupus erythematosus (SLE) and rheumatoid arthritis (RA) [14, 15]. Here, we report the purification and partial characterization of anti-Tcr autoantibodies from different Cohn plasma fractions [16] of IVIG using a single chain T cell receptor (scTcr) affinity column.

METHODS

Intravenous Immunoglobulin

Gammagard (Baxter, Hyland Division), was used as a source of IVIG. Cohn fractions I + III were also obtained from Baxter.

Construction of Recombinant Single Chain T Cell Receptor

The single chain T cell receptor utilized the Vα and Vβ genes of the CD4+ T cell line, Jurkat. The scTcr was constructed, cloned, expressed and purified as described [17].

Coupling of scTcr to solid Support

One milligram of scTcr was dialyzed against 150mM NaHCO$_3$, pH 9.0 and coupled to a solid support by incubating with 0.25 grams of activated CH Sepharose 4B (Pharmacia) for four hours at 4°C. Unbound active groups on the Sepharose were blocked with 1M Tris-Cl, pH 7.9 at 25°C for one hour. An affinity column was then made by adding the scTcr-Sepharose to a 10mm x 10cm column.

Affinity Chromatography

The pH of both IVIG preparations (Gammagard and Cohn fractions I+III) were adjusted to 7.5 and loaded onto the scTcr column at 10mg/ml with a flow rate of 0.5ml/minute. The column was then washed free of unbound IgG with 10 bed volumes of Tris-buffered Saline (TBS). The bound, anti-Tcr antibodies were eluted with 150mM glycine, pH 2.0 in 2 ml fractions and neutralized to pH 7-8.0.

ELISA

ELISAs were performed as previously described [17, 18]. Briefly, peptides were coated onto Nunc Maxisorb ELISA plates at 1ug/well and scTcr was coated at 500ng/well.

Serial dilutions of affinity purified material was added to pre-coated, pre-blocked plates and incubated for one hour. The plate was washed with 0.5% Tween-20 in PBS and horse radish peroxidase conjugated goat anti-human IgG was added at an appropriate dilution. Finally, the plate was washed as before and ABTS substrate was added and the plate was quantitated at 405nm after one hour.

Flow Cytometry

CD4 positive cell lines, Jurkat, H9, and Molt-4 were employed in flow cytometry. Briefly, two million cells were removed from cell culture and washed once in PBS. They were then re-suspended in appropriate antibody solutions at 10 ug/ml in PBS and incubated on ice for one hour. The cells were then washed three times in cold PBS followed by the addition of an appropriate dilution of FITC coupled to either goat anti-human IgG or goat anti-rabbit IgG. The cells were incubated on ice for one hour in the dark and washed in PBS as before. They were then analyzed on a Becton Dickinson FACScan flow cytometer.

RESULTS

Reactivity of scTcr Affinity-Purified Antibodies Fractions I ± III

Specific antibodies were purified from both IVIG and Cohn fractions I + III on the scTcr affinity column. The eluates were tested for reactivity to the scTcr and peptides β3, β8 and β17 in ELISA to confirm purification of anti-Tcr autoantibodies. Figure 1 shows specific reactivity of affinity-purified anti-scTcr antibodies to Tcr peptides, β3, β8 and 17 as well as the recombinant scTcr itself. The sequence of the β17 peptide is not contained within the scTcr; it is composed of V-regions only. However, antibodies reactive with a determinant in an immunoglobulin-like domain located within the scTcr structure, may cross-react with the β17 peptide. Similar results were obtained with IVIG (data not shown).

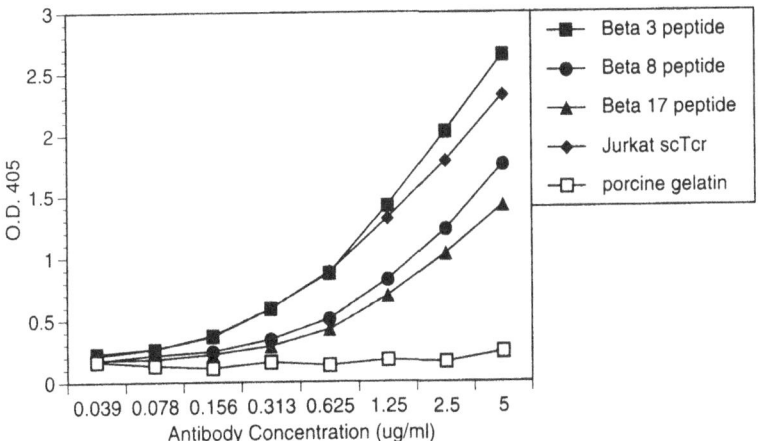

Figure 1. Reactivity of anti-scTcr affinity purified antibodies from Cohn fractions I + III. ELISA was performed as described in Methods. Affinity purified IVIG gave similar results in this assay.

2A

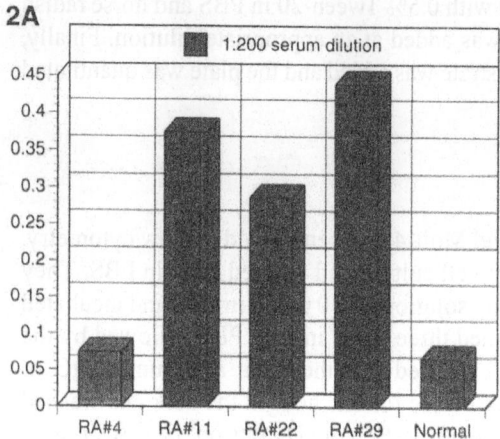

2B

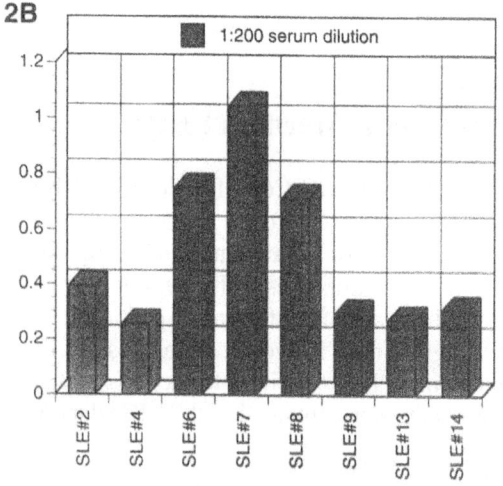

Figure 2. Reactivity of sera from patients with autoimmune diseases. Figure 2A demonstrates elevated antibodies to the scTcr in three of four seras from patients with RA. Figure 2B shows the sera from three of eight SLE patients reacting strongly with the scTcr. The remaining five SLE seras still showed reactivity against the scTcr above normal healthy sera.

Reactivity of Autoimmune Sera to scTcr

Sera from patients with autoimmune disease contain elevated levels of antibodies directed against T cell receptors. Using sera from RA (Fig. 2A) and SLE (Fig. 2B) patients in ELISA, increased reactivity over normal individuals to a recombinant scTcr was observed.

Flow Cytometry

Flow cytometric analysis was performed using IVIG and high titer rabbit anti-scTcr anti-serum. The results are shown in Figure 3. Figure 3A shows that non-purified IVIG at 10 ug/ml is not able to stain the surface of Jurkat T cells. However, when the scTcr-affinity purified IVIG was incubated with Jurkat cells at the same concentration, there was a shift from approximately 3 percent staining to 92 percent staining of cells with anti-scTcr (Fig. 3B). The shift seen for the affinity purified material was nearly identical to the shift seen for the high titer rabbit anti-scTcr at 95 percent positive (Fig. 3C). This indicates that almost all of the T cell receptors on the cell surface were bound by the scTcr-affinity purified IVIG.

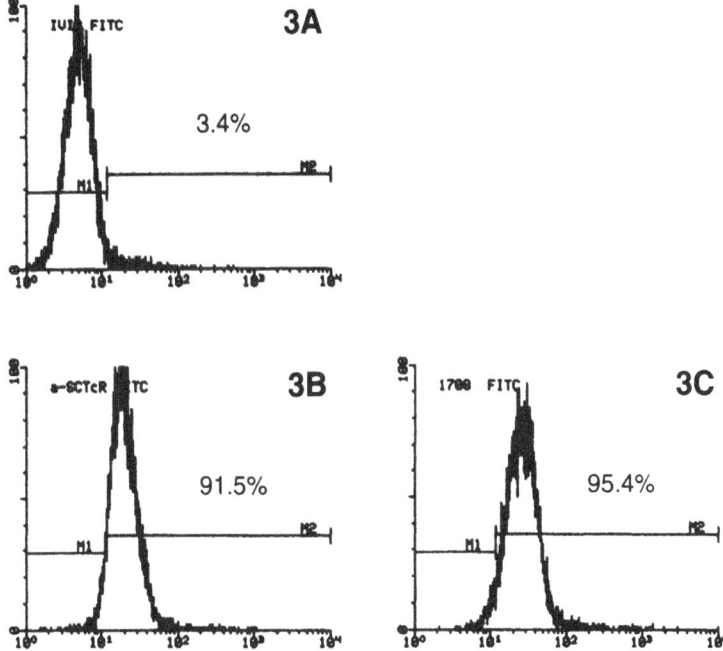

Figure 3. Flow cytometric analysis of affinity purified IVIG. Twenty-five milligrams of IVIG (Gammagard) was run through a scTcr column. Anti-Tcr IgG bound to the column was eluted with 150mM glycine, pH 2.0 and then neutralized to pH 7.5 with Tris-OH. The affinity purified material was then incubated with Jurkat cells as described in Methods.

Other CD4+ T cell lines tested were positive, but less positive than Jurkat due to the difference in Vα and Vβ family usage.

DISCUSSION

We have demonstrated the presence of anti-T cell receptor antibodies in commercial IVIG and in Cohn fractions from human plasma. The results showed that these antibodies can be purified from human serum using a recombinant scTcr and that the affinity purified antibodies react with the surfaces of T cells. Additionally non-purified sera from patients with rheumatoid arthritis and systemic lupus erythematosus had elevated levels of antibodies directed towards the scTcr.

We observed Tcr β17 peptide reactivity from Cohn fractions I + III scTcr affinity purification. This was interesting since the β17 peptide sequence is not contained within the scTcr sequence. This phenomenon might be explained by the presence of cross-reactive determinants in both the β peptide and in a domain in the three-dimensional configuration of the scTcr. Immunoglobulin domains have similar folds, so this particular cross reactivity to β17 peptide, may possess similar topography to a determinant in a domain in the scTcr. Studies are in progress to establish the basis of this cross reactivity.

We also observed a differential reactivity to the recombinant scTcr between patients with both RA and SLE. This might be attributed to stage of disease, but since we have no clinical information on the patients from the which the sera came, we cannot predict whether increased levels of anti-Tcr antibodies correlate with good or bad prognosis.

Cohn fractions I + III are normally discarded in the production process of commercial IVIG, so the finding of anti-Tcr antibodies in these fractions has attractive commercial potential. The yield of anti-Tcr antibodies from Cohn fractions I + III was 0.064%. This suggests that from one kilogram of IgG in the Cohn fractions, 640mg of pure anti-Tcr IgG could be obtained. Since our hypothesis is that the ability to purify anti-Tcr antibodies specific for individual Tcr V-region gene products from IVIG fractions has immunomodulatory properties, we can use these purified antibodies to perform functional studies. Some of the functional studies ongoing include studying the effects of anti-Tcr antibodies in lymphocyte mitogenic assays and in specific cytotoxicity assays.

One example of a clinical application for the use of these purified anti-Tcr autoantibodies is in Kawasaki's Disease where T cells bearing Vβ2 and Vβ8 are elevated. Since the current therapeutic dose of IVIG in Kawasaki's Disease is 2 g/kg [19,20], use of a specific more potent product would decrease the overall protein concentration given to an afflicted patient. This would also serve to decrease potential side effects of high dose IVIG administration, increase the half-life of the IVIG and dramatically increase the therapeutic index of the drug.

ACKNOWLEDGMENTS

This research was supported in part by a grant from Baxter Healthcare Corporation and a grant from National Institutes of Health #GM45832.

REFERENCES

1. Dwyer, J.M., 1994, Manipulating the immune system with immune globulin. *Lancet* 326:107-115.
2. Kaveri S.V., Dietrich, G., Hurez, V., and Kazatchkine, M.D., 1991, Intravenous immunoglobulins (IVIg) in the treatment of autoimmune diseases. *Clin. Exp. Immunol.* 86:192-198.
3. Schwartz, S.A., 1990, Intravenous immunoglobulin (IVIG) for the therapy of autoimmune disorders. *J. Clin. Immunol.* 10:81-89.
4. Sultan, Y., Kazatchkine, M.D., Maisonneuve, P., and Nydegger, U.E., 1984, Anti-idiotype suppression of autoantibodies to Factor VIII by high-dose intravenous immunoglobulin. *Lancet* 11:765-767.
5. Dietrich G., Kaveri S.V., and Kazatchkine, M.D., 1992, Modulation of autoimmunity by intravenous immune globulin through interaction with the function of the immune/idiotypic network. *Clin Immunol Immunopathol* 62:S73-S75.
6. Rossi F., Guilbert, B., Tonnelle, C., Ternynck, T., Fumoux, F., Avrameas, S., and Kazatchkine, M.D., 1990, Idiotypic interactions between normal human polyspecific IgG and natural IgM antibodies. *Eur. J. Immunol.* 20:2089.
7. Rossi F., Jayne, D.R., Lockwood, C.M. and Kazatchkine M.D., 1991, Anti-idiotypes against anti-neutrophil cytoplasmic antigen autoantibodies in normal human polyspecific Ig G for therapeutic use and in the remission serum of patients with systemic vasculitis. *Clin. Exp. Immunol.* 83:298-302
8. Berchtold, P., Dale, G.L., Tani, P. and McMillan, R., 1989, Inhibition of autoantibody binding to platelet glycoprotein IIb, IIIa by anti-idiotype antibodies in intravenous immunoglobulins. *Blood* 74:2414-2417.
9. Asura, E., 1989, Experience with intravenous immunoglobulin in *myasthenia gravis*. *Clin. Immunol. Immunolpath.* 53:5170-5175.
10. Fehr, J., Hofmann, V. and Kappeler, V., 1982, Transient reversal of thrombocytopenia in idiopathic thrombocytopenic purpura by high dose intravenous gamma globulin. V. *New Eng. J. Med.* 306:1254-1259.
11. Clarkson, S.B., Bussel, J.B., Kimberley, R.P., Valinsky, J.E., Nachman, R.L. and Unkless, J.C., 1986, Treatment of refractory immune thrombocytopenic purpura with an anti-FCγR antibody. *N. Eng. J. Med.* 314:1236-1240.
12. Marchalonis, J.J., Kaymaz, H., Dedeoglu, F., Schluter, S.F., Yocum, D.E. and Edmundson, A.B., 1992, Human autoantibodies reactive with synthetic autoantigens from T-cell receptor β chain. *Proc. Natl. Acad. Sci. USA*, 89:3325-3329.

13. Marchalonis, J.J., Schluter, S.F., Wang, E., Dehghanpisheh, K., Lake, D., Yocum, D.E., Edmundson, A.B. and Winfield, J.B., 1994, Synthetic autoantigens of immunoglobulins and T-cell receptors: their recognition in aging, infection and autoimmunity. *Proc. Soc. Expt. Biol.* 207:129-147.

14. Marchalonis, J.J., Schluter, S.F., Wilson, L., Yocum, D.E., Boyer, J.R. and Kay, M.M.B., 1993, Natural human antibodies to synthetic peptide autoantigens: correlations with age and autoimmune disease. *Geront.* 39:65-79.

15. Marchalonis, J.J., Kaymaz. H., Schluter, S.F., and Yocum, D.E., 1993. Human Autoantibodies to a synthetic Putative T-cell receptor β-Chain regulatory Idiotype: Expression in autoimmunity and aging. *Expt. & Clin. Immunogenet.* 10:1-15.

16. Cohn, E.J., Strong, L.E. *et al.* 1946, Preparation and properties of serum and plasma proteins, a system for the separation into fractions of the protein and lipoprotein components of biological tissues and fluids. *J. Am. Chem. Soc.* 68:459-475.

17. Lake, D.F., Bernstein, R.M., Hersh, E.M., Kaymaz, H., Schluter, S.F. and Marchalonis, J.J., 1994, Construction and serological characterization of a recombinant human single chain T-cell receptor (scTcr). *Biochem. & Biophys. Resch. Comm.* 201:1502-1509.

18. Marchalonis, J.J., Dedeoglu. F., Kaymaz, H., Schluter, S.F., and Edmundson, A.B., 1992, Antigenic mapping of a human λ light chain: Correlation with 3-dimensional structure. *J. Prot. Chem.* 11:129-137

19. Newburgh, J.W., Takashashi, M., Burns, J.C., Beiser, A.S., Chung, K.J., Duffy, C.E., Glode, M.P., Mason, W.H., Reddy, V., Sanders, S.P., Shulman, S.T., Wiggins, J.W., Hicks, R.V., Fulton, D.R., Lewis, A.B., Leung, D.Y.M., Colton, T., Rosen, F.S. and Melish, M.E., 1986, The treatment of Kawasaki syndrome with intravenous gamma globulin. *N. Eng. J. Med.* 315:341-347.

20. Rowley, A.H., and Shulman, S.T., 1988, What is the status of intravenous gamma globulin for Kawasaki syndrome? *Pediat. Infect. Dis. J.* 7:463-466.

24

AUTOREGULATION OF TCR V REGION EPITOPES IN AUTOIMMUNE DISEASE

Samuel F. Schluter,[1] Ena Wang,[1] John B. Winfield,[2] David E. Yocum,[3] and John J. Marchalonis[1]

[1] Microbiology and Immunology
College of Medicine, University of Arizona
Tucson, Arizona 85724
[2] Division of Rheumatology and Immunology
University of North Carolina
Chapel Hill, North Carolina 27599
[3] Division of Rheumatology
College of Medicine, University of Arizona
Tucson, Arizona 85724

ABSTRACT

Normal individuals possess low levels of autoantibodies specific for certain peptide defined regions of T-cell receptor (Tcr) variable regions, particularly CDR1 and Fr3. These regions are predicted to be exposed on the surface of the native molecule and, by analogy and comparison with immunoglobulins, correspond to public idiotype determinants. The anti-Tcr idiotype antibodies appear to be ubiquitous and we propose that they play a role in the regulation of T-cell function. To delineate the parameters of expression of these antibodies, we characterized anti-Tcr antibody activity in normal individuals, in those suffering from the autoimmune diseases rheumatoid arthritis (RA) and systemic lupus erythematosus (SLE), and in patients with non-autoimmune arthritis (osteoarthritis) as a disease control. There were significant increases in autoantibody levels in the autoimmune patients. There was also variation in isotype and the particular variable regions recognized. IgM autoantibodies directed against a few peptide defined determinants were elevated in RA, whereas SLE patient sera showed high levels of IgG binding to a broad spectrum of Tcr peptides.

INTRODUCTION

The antigen receptors of B cells and T cells are closely related members of the immunoglobulin family of molecules, sharing significant sequence homology and cross-reactive antigenic determinant. To characterize and map the antigenic determinants of Ig and

Immunobiology of Proteins and Peptides VIII
Edited by M. Z. Atassi and G. S. Bixler, Jr., Plenum Press, New York, 1995

231

Tcr, we applied the synthetic peptide approach [1] by generating sets of overlapping synthetic peptides duplicating the V(D)JC sequences of Tcrα, Tcrβ and Igλ light chains. Anti-Ig sera specific for human λ, human γ and human μ chains cross-reacted with particular segments of the human Tcrβ chain as defined by the synthetic peptides [2]. Significantly, all three anti-sera recognize the same regions of the Tcr β chain. These segments corresponded to peptide 3, peptide 8, peptide 11 and peptide 17. In association with Dr. Allen Edmundson of the Harrington Cancer Center [2,3] we have constructed a three-dimensional model of the αβ Tcr using the homologous Mcg λ structure as a template. From this model the peptide defined epitopes correspond to V region segments CDR1, Fr3, J-region and a loop in the constant region [3].

During these studies we used normal sera as controls and were surprised to find antibody reactivity to both Tcrβ [4] and Ig λ chains. In general, the autoantibodies reacted with the same regions recognized by the cross-reactive anti-Ig sera, namely, CDR1, Fr3, and the loop in the constant region. Normal people produce autoantibodies to a diverse array of self-antigens without any ill-effects. A well-known example are anti-idiotype antibodies which react with the binding site of another antibody. Anti-idiotype antibodies have been shown to play a role in the regulation of B cell activity. The CDR1 and Fr3 regions can be considered to represent idiotypic sites, suggesting that anti-Tcr autoantibodies may be analogous to anti-idiotype antibodies and have functional significance for the regulation of T-cell activity. In order to delineate the parameters of expression of anti-Tcr antibodies, we characterized the activity in normal individuals, in those suffering from the autoimmune diseases rheumatoid arthritis (RA) and systemic erythematosus lupus (SLE) and in patients with non-immune arthritis (osteoarthritis) as a disease control.

MATERIALS AND METHODS

Sera from thirty-two healthy individuals under the age of sixty years, fourteen patients suffering from rheumatoid arthritis, ten patients suffering from osteoarthritis, and seven patients suffering from systemic lupus erythematosus were obtained through the Division of Rheumatology, Department of Internal Medicine, College of Medicine, Arizona Health Sciences Center, Tucson.

Sixteen-mer peptides that overlapped by five residues were synthesized to model the immunoglobulin related domains of Tcrα chain, Tcrβ chain and immunoglobulin light chain. The peptides were synthesized by the University of Arizona Biotechnology Center, using an Applied Biosystems Peptide Synthesizer. Complete details of the Tcrβ chain and λ light chain peptides are described elsewhere [2,5]. The α chain peptides were derived from the gene sequence of human Tcrα chain PY14 [6].

Enzyme-linked immunosorbent assays (ELISA) were carried out as we have previously described [5].

RESULTS AND DISCUSSION

Profile of Autoantibody Reactivity in Autoimmune Disease

We have previously demonstrated that normal healthy people constitutively express low levels of autoantibody activity to Tcr [4]. The pattern of reactivity to the Tcr β chain is well defined. In general, three peptide defined autoantigenic regions are recognized corresponding to CDR1 (peptide β3), Fr3 (peptide β8), and a loop in the constant region (peptide β17)[4]. Several criteria indicate that these are specific antibodies and not just polyspecific natural antibodies. Firstly, the autoantibodies can be affinity purified on the appropriate peptide-sepharose immunoaffinity column. The immunoaffinity purified antibodies are specific, reacting only with the cognate peptide or with closely related peptides derived from the corresponding region of related Vβ [3,4]. Similarly, the antibodies can be inhibited by the correct free peptide [4]. Also we have isolated several human monoclonal antibodies, either as myeloma proteins or from EBV transformed B cells, that display very high specificity for CDR1 or Fr3 peptides [3,4].

Since we hypothesized that autoantibodies reactive with Tcrs are involved in regulation of the immune system, we predicted that the levels of these autoantibodies should be perturbed in autoimmune disease conditions. Figure 1 shows the profiles of autoantibody activity in representative serum from an RA patient and an SLE patient. The reactivity with overlapping peptides modelling the Ig domain of Tcrα chain, Tcrβ chain, and λ light chain was measured. The profile of reactivity to Tcr β chain peptides for the RA serum is very similar to normals in that there are three autoantigenic regions centering on peptide 3, peptide 8 and peptide 17. However, in contrast to normals, the levels are much higher, particularly of IgM antibodies. Autoantibody reactivity in the RA patient appears to be specific for the Tcrβ chain as the levels of reactivity to the Tcrα and λ chains are the same as those found in normals. The situation for the SLE patient is markedly different. There is a large general increase in IgG autoantibodies to all three proteins,with major regions of reactivity centered around the CDR1 and Fr3 peptides. However, there are also high levels of reactivity to many other regions, especially for the Tcrα and β chains.

Anti-Tcr Autoantibodies Are Elevated in Patients with RA and SLE Autoimmune Diseases

Our preliminary results suggested that there was an increase in autoantibody levels in patients with autoimmune disease. We performed a large scale study to test this hypothesis. Sera from fourteen RA patients, ten osteoarthritis patients, seven SLE patients and thirty-two normal people were tested for autoreactivity. Our initial results demonstrated that the CDR1, Fr3, and the constant region loop peptide were major autoantigenic sites and that reactivity to these was a good indication of autoantibody activity. Therefore, in this study we measured activity to these regions of the Tcr β chain. Ovalbumin was also used as an antigen since we have found that nearly all normal people have elevated levels of IgG antibodies to this ubiquitous environmental antigen and we used it as a measure of general immune competence. Patients with osteoarthritis were included as control since this is a degenerative joint disease that is not the result of autoimmune reactions. The results are shown in Figure 2 and are expressed for each group as the geometric mean titer. Statistical analysis of the data for the various groups compared to normals was performed using the Wilcoxon 2-sample test. The results for the osteoarthritis patients were the same as normals. However, the IgM autoantibody activity to the CDR1 peptide (P=0.016) and the constant region peptide (p=0.004) were significantly elevated in the RA patients. Although activity to Fr3 peptide showed a tendency to be higher, the result was not significant. IgM levels to CDR1 alone were elevated in SLE patients. In contrast, IgG autoantibodies were highly elevated to all three peptides in SLE patients but were indistinguishable from normal levels in the RA patients. Anti-ovalbumin IgG reactions were decreased in RA patients, a result consistent with the general anergy often observed in these patients. Thus, in general the results indicate that there is a prolonged IgM response against a limited set of peptides with a deficient secondary IgG response in RA patients. In contrast, there appears to be a strong secondary IgG response in SLE patients against a variety of peptides, with only moderate IgM levels.

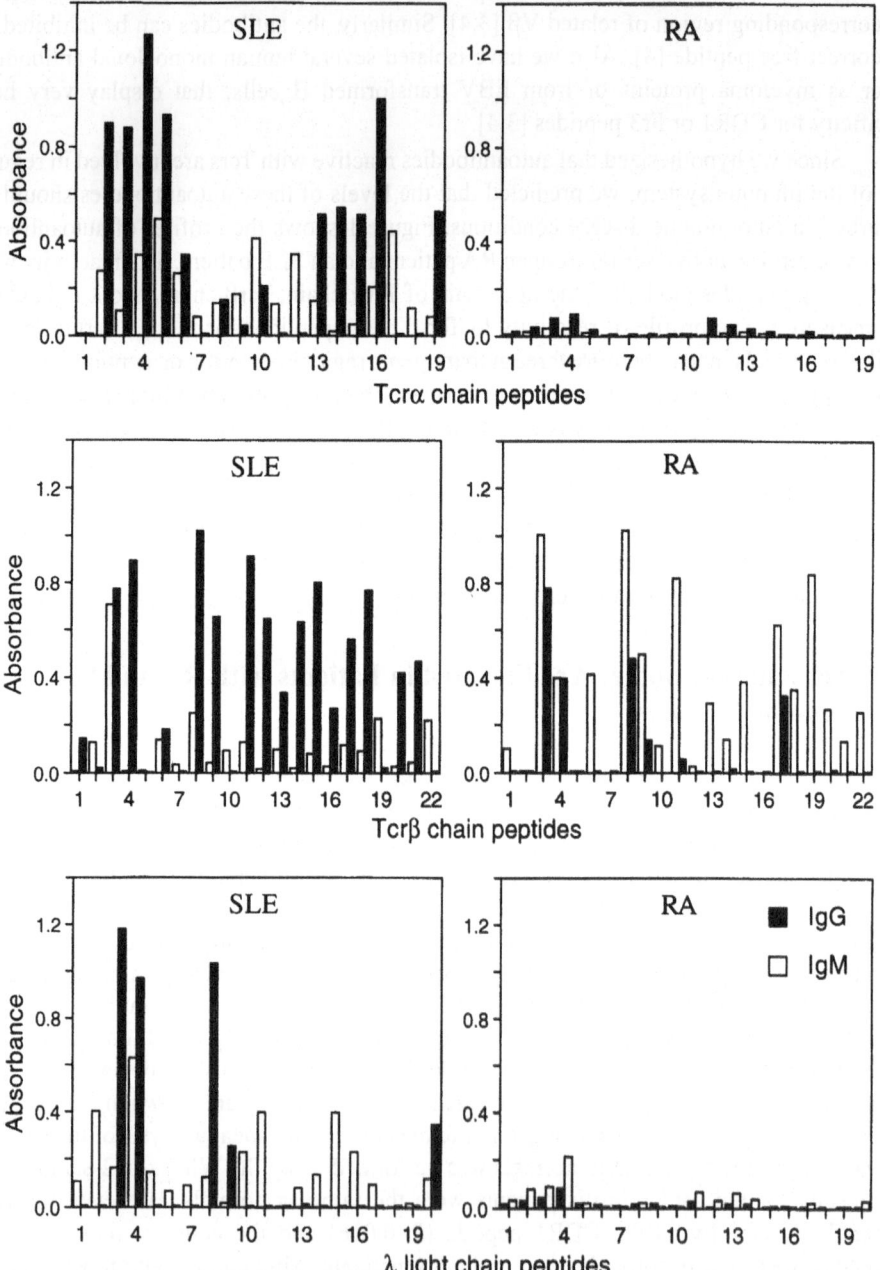

Figure 1. Profiles of anti-Tcr and anti-Ig autoantibody activity in serum from RA (Left panels) and SLE (Right panels) patients. Binding activity to sets of overlapping peptides duplicating the immunoglobulin domains of Tcrα(Top), Tcrβ (Middle), and λ light chain (bottom) were assayed in ELISA. IgM (□) and IgG (■) were assayed at a dilution of 1:200. The RA patient showed the highest activity of the patients studied and the SLE patient was from a group whose serum reacted with human T-cells.

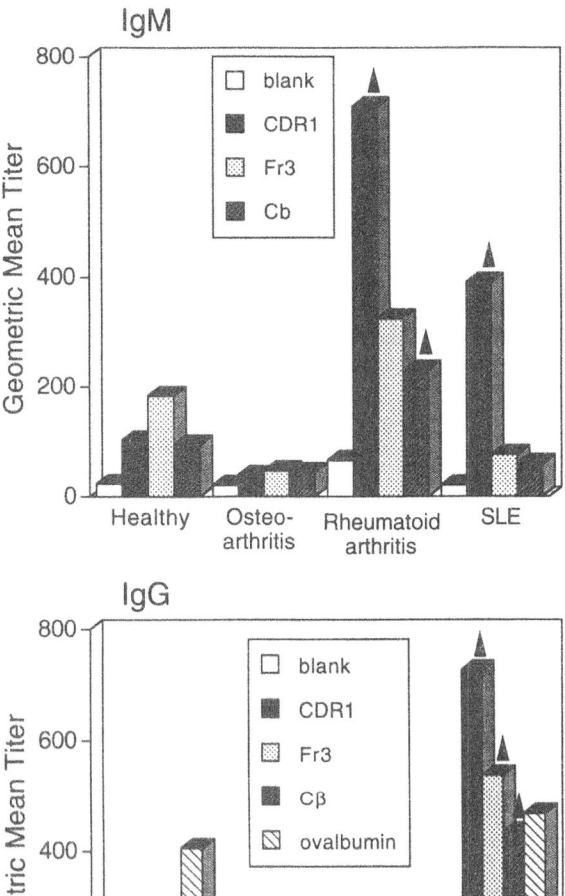

Figure 2. Statistical analysis and comparison of IgM and IgG autoantibody levels to Tcrβ chain peptides in autoimmune and control patients. IgM (top) and IgG (bottom) were measured for all individuals in each study group and the results are expressed as geometric mean titers. An up arrow (▲) indicates a statistically significantly higher value than for normals and a down arrow (▼) indicates a significantly lower value.

SUMMARY AND CONCLUSIONS

We have identified three major autoantigenic regions in Tcr chains corresponding to CDR1, Fr3, and a loop in the constant region. Autoantibodies to these regions are constitutively expressed at low levels in normal people but are significantly elevated in patients with autoimmune disease. Since CDR1 and Fr3 most likely are idiotypic regions on Tcr chains, we propose that the anti-idiotype autoantibodies play a regulatory role by binding to Tcrs and either activating or energizing the cells or blocking interaction with MHC-peptide complexes. Support for this contention is provided by the finding that antibodies to CDR1 peptide affinity purified from pooled human IgG bind to T cells bearing the cognate β chain as measured by FACS analysis [unpublished results].

ACKNOWLEDGMENTS

This research was supported in part by a Grant from Arizona Disease Control Commission #5-038 and from a grant from the National Cancer Institute #CA42049.

REFERENCES

1. Kazim, A.L., and Atassi, M.Z., 1980, A novel and comprehensive synthetic approach for the elucidation of protein antigenic stretchers. Determinations of the full antigenic profile of the α-chain of human haemoglobin. *Biochem. J.* 191:261-265.
2. Kaymaz, H., Dedeoglu, F., Schluter, S.F., Edmundson, A.B. and Marchalonis, J.J., 1993, Reactions of anti-immunoglobulin sera with synthetic T-cell receptor peptides: implications for the three-dimensional structure and function of the Tcr β chain. *Intl. Immunol.* 5:491-502.
3. Marchalonis, J.J., Schluter, S.F., Wang, E., Dehghanpisheh, K., Lake, D., Yocum, D.E., Edmundson, A.B. and Winfield, J.B., 1994, Synthetic autoantigens of immunoglobulins and T-cell receptors: their recognition in aging, infection and autoimmunity. *Proc. Soc. Expt. Biol.* 207:129-147.
4. Marchalonis, J.J., Kaymaz, H., Dedeoglu, F., Schluter, S.F., Yocum, D.E. and Edmundson, A.B., 1992, Human autoantibodies reactive with synthetic autoantigens from T-cell receptor β chain. *Proc. Natl. Acad. Sci. USA.* 89:3325-3329.
5. Marchalonis, J.J., Dedeoglu. F., Kaymaz, H., Schluter, S.F., and Edmundson, A.B., 1992, Antigenic mapping of a human λ light chain: Correlation with 3-dimensional structure. *J. Prot. Chem* 11:129-137.
6. Kabat, E.A., Wu, T.T., Perry, H.M., Gottesman, K.S. and Foeller, C., 1991, *Sequences of Proteins of Immunological Interest 5th Edition*, U.S. Department of Health and Human Services, Public Health Service, National Institutes of Health, Bethesda, MD.

25

INFLUENCE OF THE ANTERIOR CHAMBER OF THE EYE ON T-CELL PRODUCTION OF EXTRACELLULAR ANTIGEN-SPECIFIC PROTEINS

Robert E. Cone, C. Hadjikouti, Y. Wang, and J. O'Rourke

Vision-Immunology Center
Department of Pathology
University of Connecticut Health Center
Farmington, Connecticut 06030-3105

INTRODUCTION

Analysis of the immunobiological properties of the anterior chamber (AC) of the eye has shown that the "immunologic privilege" of this region is due to a complex interaction of factors which influence both the immune response *within* the AC, and the systemic immune response. The introduction of soluble or particulate antigens into the AC results in a modulation of the systemic immune response which is manifested by stimulation of IgM and IgG1 antibody production, and by suppression of delayed-type hypersensitivity and IgG2a antibody production. This so-called Anterior Chamber Associated Immune Deviation (ACAID, 1-6) response thus provides a major basis for the "immunologic privilege" of the AC by exerting a profound influence on systemic immunity, which, of course, includes the eye.

ACAID to soluble protein antigens is transmitted by antigen-presenting dendritic cells derived from the iris and ciliary body (7,8) that migrate from the eye, in concert with antigen, induce the activation of splenic suppressor T cells, which, in turn, effect a prolonged antigen-specific suppression of the systemic cell-mediated hypersensitivity response to the AC-injected antigen (7). Moreover, TGF-β present in the aqueous humor that fills the AC, can induce peritoneal exudate F4/80$^+$ macrophages to act similarly to the eye-derived dendritic cells in transmitting ACAID to naive mice (10), and is therefore probably responsible for the induction of ACAID-transmitting dendritic cells resident in the eye. However, the intracameral injection of particulate antigens (11), or a soluble antigen with a Th clone specific for the antigen (12) also results in the production in serum of an antigen-specific T cell-derived factor which induces the production of the splenic suppressor T cells that function in ACAID to particulate antigens (7,9). Like many antigen-specific extracellular immunoregulatory T cell "factors" (13-15) this serum ACAID-transferring factor bears T

Immunobiology of Proteins and Peptides VIII
Edited by M. Z. Atassi and G. S. Bixler, Jr., Plenum Press, New York, 1995

237

cell receptor for antigen (TcR) α and β chain constant region epitopes (12), and prevents the *in vivo* expression of delayed-type hypersensitivity (DTH) by sensitized T cells (16). Because this ACAID-transferring moiety is adsorbed to monoclonal antibodies specific for antigen-specific T cell proteins that induce suppressor T cells (TsIF) its role in ACAID may involve the amplification of suppressor T cells. The generation of the ACAID immunoregulatory T cell circuit and the production of soluble, antigen-specific immunoregulatory T cell proteins may thus be as complex as that already shown in many studies when various forms and/or doses of antigen are introduced intravenously (17).

While ACAID explains, in part, the local immunologic privilege of the anterior chamber in terms of its protection from damaging local immunologic reactions, its systemic effects also demonstrates the existence of a powerful ocular immunoregulatory principle which could eventually alter current immunotherapeutic procedures used in organ transplantation, autoimmunity and systemic immunization. Additionally, the observation that injection of particulate antigens into the anterior chamber results in the production of an antigen-specific T cell-derived immunoregulatory factor in serum underscores another important relationship: this factor may be related to extracellular antigen-specific T cell proteins that bind to nominal antigen (TABM) (18-25) since antibodies which identify TsIF have also been found bind TABM specifically (26).

We have shown previously that serum TABM produced in response to intraperitoneal immunization with soluble proteins and adjuvants (23,24) bear TcR Cα chain determinants, and limited amino acid sequence of monoclonal TABM suggests homology to TcR Vα (21). Thus, TABM are likely a soluble analogue of the TcR. Because TABM, like extracellular immunoregulatory T cell "factors" bear TcR Cα epitopes, we have investigated whether injection of soluble or particulate antigens into the anterior chamber influences the production in serum of TABM specific for the antigen used in ocular immunization. Our results suggest that AC injection of antigen into naive mice potentiates the production in serum of TABM specific for AC-injected antigen which occurs after systemic immunization, and induces TABM production in sensitized animals. These proteins may be the serum mediator of ACAID, and may therefore be a quantitative serologic indicator of the activation of immunoregulatory T cells.

Anterior Chamber Injection of Antigen into Naive Mice Potentiates the Production of Serum TABM after Sensitization

Because antigen-specific T cell "factors," which transfer ACAID, have been found in the serum of mice after intracameral injection of trinitrophenylated (TNP)-spleen cells (12,16). We adapted a TNP-ACAID model as described by Ferguson et al (16) to determine whether AC injection affects systemic TABM production. Approximately 2×10^5 BALB/c TNP-spleen cells were injected under microscopic control into the anterior chamber of the eye of anesthesized BALB/c mice. Two days after injection, the abdomens of the mice were painted with 100 μl 7% picryl chloride, and five days after application of picryl chloride, one ear of the mouse was painted with 15 μl of 7% picryl chloride. Twenty-four hours after challenge, swelling of the ear was measured with a micrometer, and the unchallenged ear served as a control. As shown in Figure 1, the DTH reaction of uninjected, sensitized mice was 16.9×10^{-3} cm, and that of mice receiving a subconjunctival (S.Con) injection of TNP-spleen cells before sensitization and challenge was 22×10^{-3} cm. In contrast, the DTH reaction of mice receiving an AC injection of TNP-spleen cells was reduced by 55% and 65% relative to uninjected mice, and mice receiving a subconjunctival injection, respectively. In other studies, AC injection reduced DTH as much as 80-95% depending on whether DTH was measured 1, 2 or 3 days after challenge. Sera were obtained two days after challenge (6 days after AC injection) and to detect TNP specific TABM in the sera, dilutions of the sera were added to TNP-BSA-coated microtiter tray wells. After

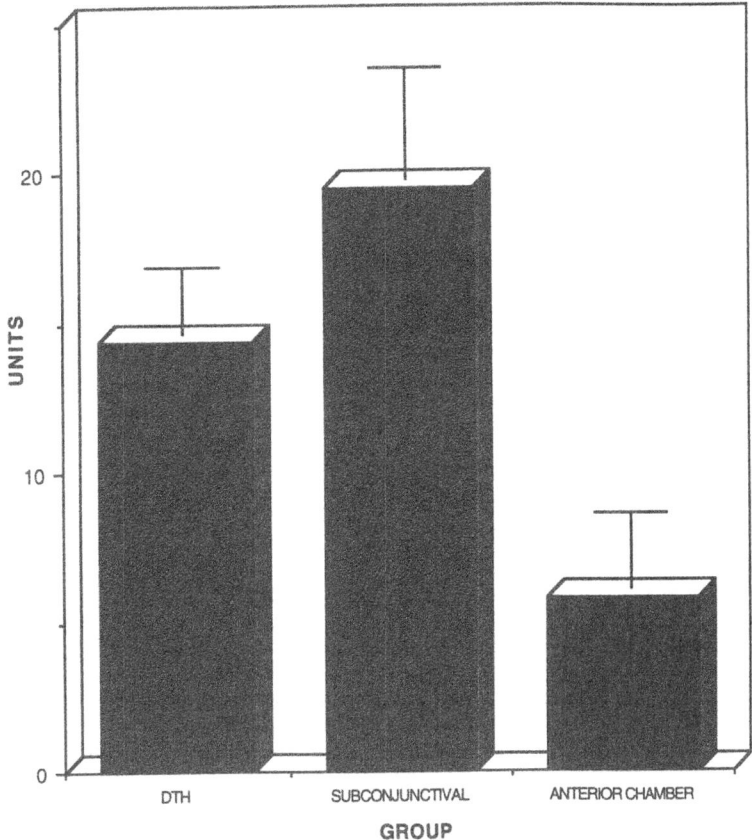

Figure 1. Intracameral injection of TNP-spleen cells before sensitization modulates DTH to TNP. BALB/c mice received an ocular injection of 5×10^5 TNP-spleen cells, and 24 hr later were sensitized by application of 7% picryl chloride to the abdomen. After five days, the mice were challenged by application of 15 µl 7% picryl chloride to one ear. The other ear served as a control. Ear swelling was measured with an micrometer 48 hr later. The thickness of the control ear (approximately 1 cm) was subtracted from the thickness of the challenged ear to determine units (cm x 10^{-3}). DTH = uninjected mice; AC = anterior chamber injection; Subconjunctival = subconjuctival injection.

incubation and washing, TABM bound to TNP were detected by the addition of rabbit anti-TABM antiserum, and then alkaline-phosphatase conjugated-anti-rabbit Ig and substrate. As shown in Figure 2, mice that were sensitized and challenged for DTH to TNP produced TNP-specific TABM (1:640). However, mice receiving an AC injection of TNP-spleen cells had a titer of 1:2560, while a subconjuctival injection of TNP-spleen cells before sensitization did not increase TNP-specific serum TABM above that obtained by sensitization only (1:640). In other experiments, sensitization and challenge for DTH only did not increase TNP-specific serum TABM above levels found in non-immune sera (data not shown).

As shown in Figure 3, serum TABM potentiated by AC-injection bound to TNP-BSA, but not BSA or azobenzenearsonate-BSA (data not shown). Binding was inhibited by soluble TNP-BSA, but not BSA. Moreover, sera from ovalbumin-sensitized mice receiving an AC injection of ovalbumin (27) contain an increase in ovalbumin-specific TABM, but no increase in TNP-specific TABM.

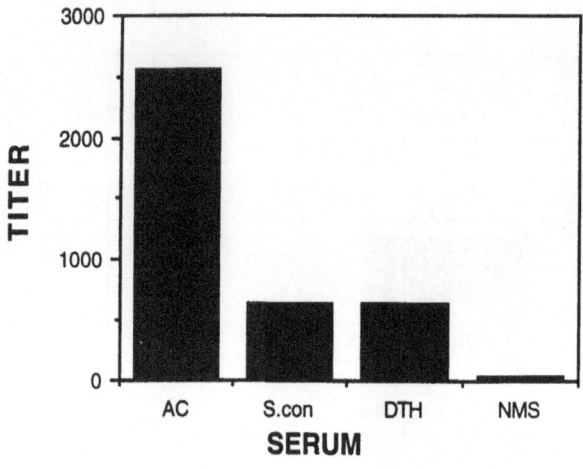

Figure 2. Intracameral injection of TNP-spleen cells before sensitization increases TNP-specific TABM in the sera of mice sensitized to TNP. BALB/c mice were treated as described in Figure 1, and blood was obtained 2 days after challenge. One hundred μl of sera diluted 1:20-1:5120 was added to microtiter trays coated with 1 μg/well TNP-BSA. After 1.5 hr incubation (37°C) incubation, and washing alkaline phosphatase-conjugated goat anti-rabbit Ig was added followed by p-nitrophenyl phosphate substrate. The titer is taken as the highest dilution giving an optical density of at least 0.2. Data represents results obtained with sera pooled from five mice. AC = sera from mice receiving an injection of TNP-spleen cells in the anterior chamber. S.con - sera mice receiving subconjuctival injection of TNP-spleen cells; DTH = sera from mice that did not receive an ocular injection of TNP-spleen cells; NMS = normal mouse serum.

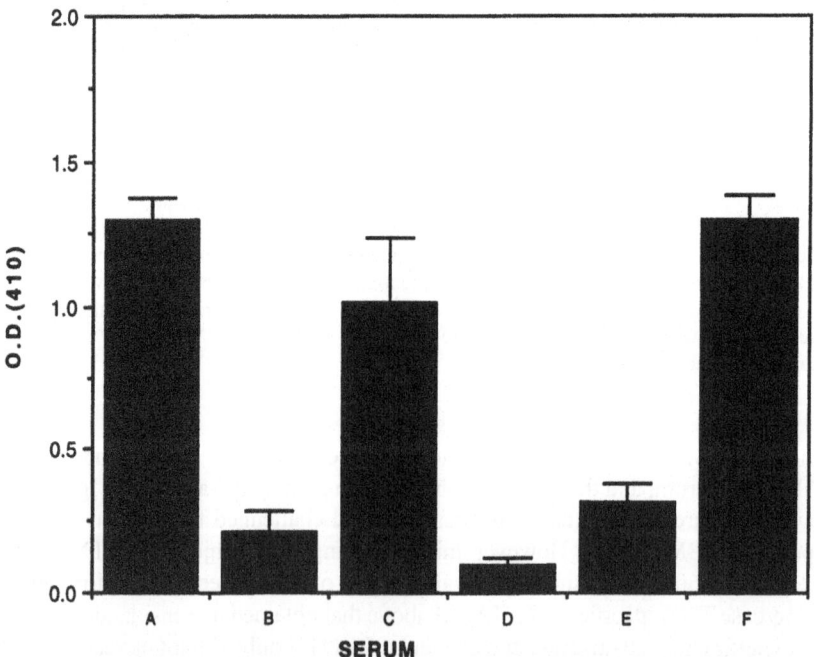

Figure 3. TABM increased by AC injection of TNP-spleen cells are specific for the injected antigen. BALB/c mice received an AC injection of 5 x 10^5 TNP-spleen cells, and were sensitized and challenged (Figure 1) with picryl chloride (A-D). Sera were obtained 2 days after challenge, and were assayed for binding to TNP-BSA or BSA. A) binding to TNP-BSA, B) binding to BSA, C) binding to TNP-BSA in the presence of 50 μg BSA, D) binding to TNP-BSA in the presence of 50 μg TNP-BSA, E) TABM in sera from ovalbumin (OVA)-sensitized mice obtained 48 hr after receiving an AC injection of OVA binding to TNP-BSA, F) TABM in sera from OVA-sensitized mice obtained 48 hr later after receiving an AC injection of OVA binding to OVA. Data represents the mean ± SE of 9 replicate results obtained from 5-10 mice.

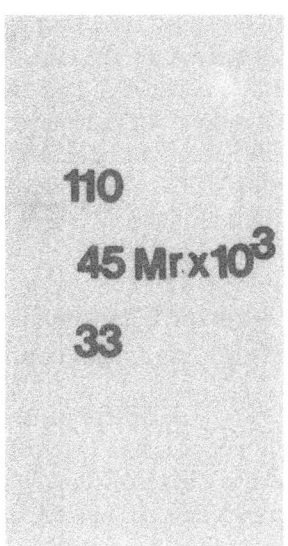

Figure 4. Immunoblot of serum TABM bound to and eluted from TNP-BSA sepharose. BALB/c mice received 5 x 10^5 syngeneic TNP-spleen cells AC and sera were obtained 2 days after sensitization and challenge with picryl chloride (Figure 1). One ml of a 1:10 dilution of the pooled serum was incubated for 3 hr with 300 μl TNP-BSA-sepharose beads and 200 μg BSA. After washing, the beads were eluted with one ml 2% SDS-urea, and the eluate reduced and resolved by SDS-PAGE. The resolved proteins were electroblotted onto an immobilon membrane and mixed with rabbit anti-TABM. Bound antibody was detected with glucose oxidase conjugated goat anti-rabbit Ig and diaminobenzidine substrate.

As shown in Figure 4, TNP-binding TABM in the sera of AC-injected mice bound to and eluted from TNP-BSA-sepharose affinity beads were resolved as Mr 110,000 polypeptides after reduction. These molecules were not detected in the sera of naive mice (data not shown).

The increase in serum TABM specific for the AC-injected antigen appeared to correlate with conditions that induce ACAID. Accordingly, the kinetics of the appearance of serum TABM after AC injection was investigated by obtaining serum 2-14 days after the ear was challenged with picryl chloride. As shown in Figure 5, 3 days after AC injection and/or

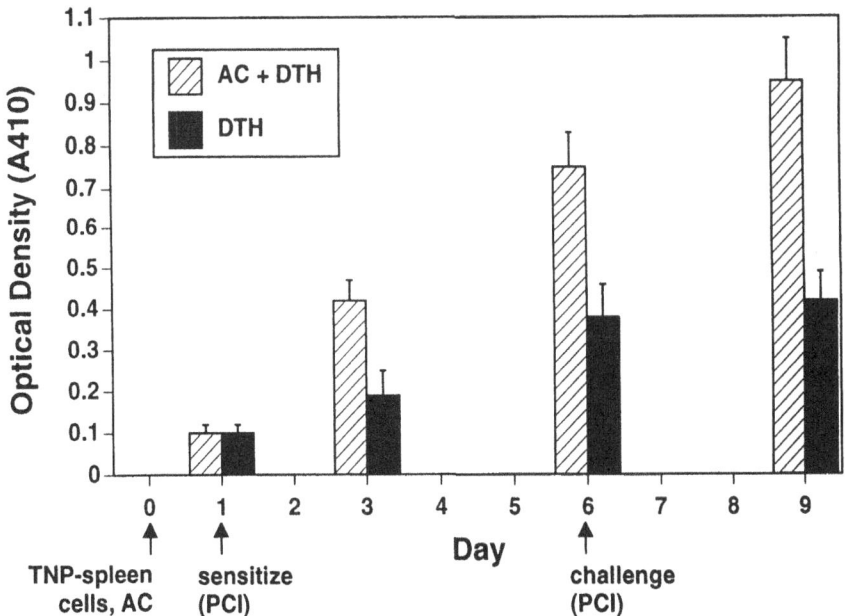

Figure 5. Kinetics of serum TABM in response following intracameral injection and sensitization. Five BALB/c mice received an AC injection of 5 x 10^5 TNP-spleen cells, and were then sensitized and challenged with picryl chloride (Figure 1). Mice were bled as indicated and tested by ELISA for TNP-specific TABM. Data represents the mean ± SE of 5 replicates of the optical density obtained with 100 μl of a 1:33 dilution of the pooled serum. The optical density of non-immune serum (0.15-0.3) has been subtracted.

sensitization, there was an increase in TNP-specific serum TABM, which peaked 6 days after intracameral injection (at the time of challenge for DTH). TNP-specific TABM were also detected in the sera of sensitized mice that did not receive an AC injection, however, TABM levels obtained 5 days after sensitization (6 days after AC injection) were similar to that obtained 3 days after mice received an AC injection (2 days after sensitization). In addition, TNP-specific TABM levels did not increase after mice sensitized only were challenged with picryl chloride while TNP-specific TABM levels increased further when AC-injected mice were challenged. Moreover, AC injection of TNP-spleen cells only, did not induce an increase in TNP-specific TABM in the serum.

Anterior Chamber Injection of Antigen into Sensitized Mice Induces Production of Serum TABM

The DTH response of mice previously sensitized to protein antigen can be suppressed by subsequent injection of the antigen into the anterior chamber (28). The studies described above suggested that, while AC injection alone of naive mice did not induce TABM production, AC injection primed the mice for TABM production elicited *after* sensitization. Therefore, we determined whether AC injection of *sensitized* mice elicits serum TABM production (27).

Mice immunized previously with ovalbumin (OVA) received an AC or S.Con injection of OVA, and the mice were bled 2 and 7 days later. AC injection of OVA resulted

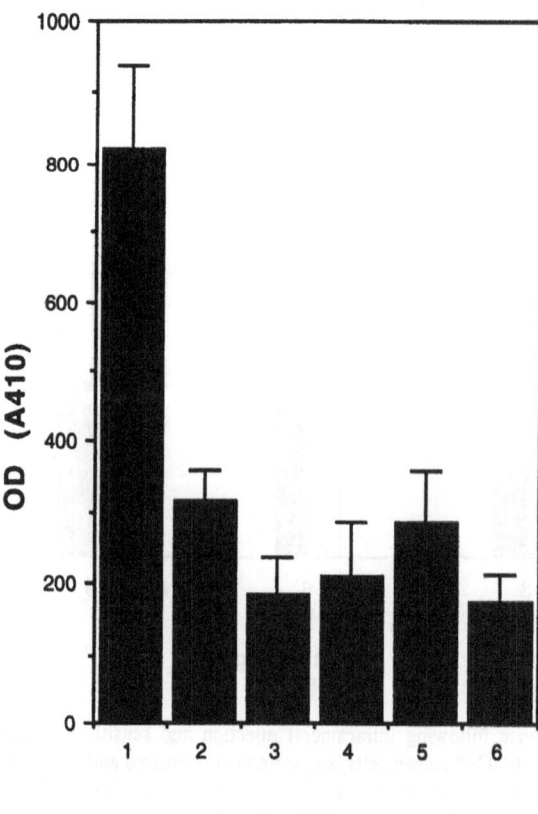

Figure 6. Intracameral injection of antigen into sensitized mice induces the production of serum TABM. Asplenic and sham-splenectomized BALB/c mice were sensitized to OVA by subcutaneous injection of OVA in complete Freund's adjuvant. After one week, the mice received AC or S.Con, 50 µg OVA and were bled 48 hr later. The sera were pooled and assayed (by ELISA) for OVA-specific TABM. Data represents the mean ± SE (15 replicates) of the O.D. obtained with 100 µl of a 1:160 dilution of serum minus the O.D. obtained with normal mouse serum. Group 1) splenectomized (SPX), OVA sensitized, AC injection of OVA, 2) SPX, OVA sensitized, S.Con injection of OVA; 3) SPX, OVA sensitized; 4) sham operated, OVA sensitized, AC injection of OVA; 5) sham operated, OVA sensitized, S.Con injection of OVA; 6) sham-operated, OVA sensitized.

in the production of OVA-specific TABM, while S.Con injection did not stimulate TABM production (Fig. 6). In addition, the sera containing an increase in OVA-specific TABM did *not* contain an increase in TABM binding TNP-BSA (Fig. 3). Moreover, we found that more TABM were found in the sera of splenectomized mice than non-splenectomized mice (Fig. 6). These observations are consistent with those of Ferguson (16) suggesting that the spleen may "filter" or suppress TABM production after AC injection.

Since AC injection of OVA into non-sensitized, splenectomized mice did not induce serum TABM, we reasoned that AC injection may activate the production of TABM by sensitized T cells or (as suggested by our studies with TNP-spleen cells), and may prime for TABM production. As shown in Figure 7, naive mice receiving an AC injection of BSA produced serum TABM when sensitized 7 days after AC injection, but did not produce serum TABM when sensitized 1 day after AC injection. These results are consistent with our hypothesis that AC injection "primes" for TABM production. After 7 days, there are sufficient numbers of T cells that produce TABM to the AC-injected antigen, and that challenge evokes a detectable TABM response in the serum. It is also apparent that TABM can be produced in response to AC injection of soluble (protein) antigen, as well as TNP-spleen cells.

Cyclophosphamide Inhibits ACAID and TABM Production

Cyclophosphamide has been shown to inhibit the activation or activity of immunoregulatory T cells in ACAID (9), and therefore, we determined whether cyclophos-

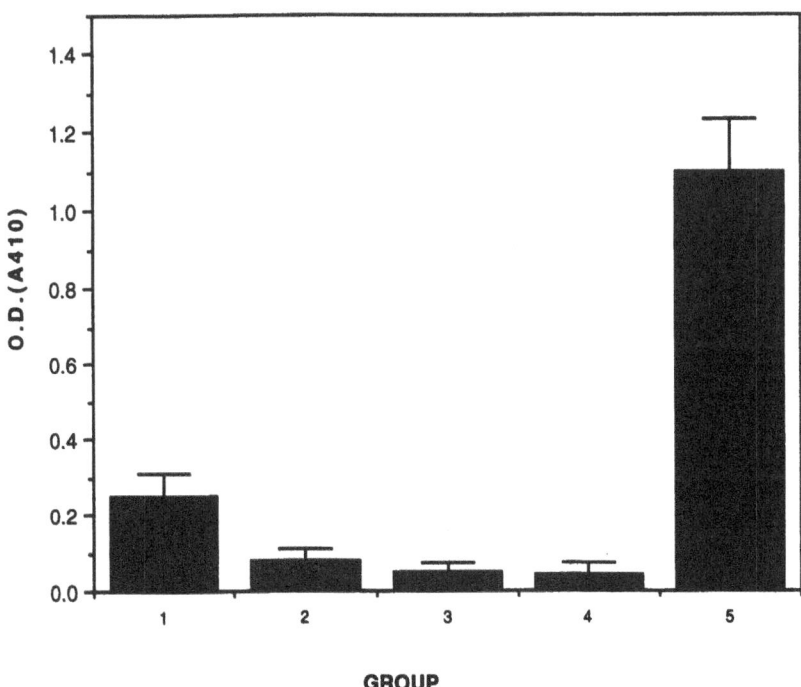

Figure 7. Intracameral injection of BSA primes for BSA-specific TABM production (27). Non-immune SPX and sham-operated mice received an AC or S.Con injection of BSA. Two days after injection, sera were obtained, pooled and assayed for TABM binding to BSA. Some mice received a subcutaneous injection of BSA 1 or 7 days after the AC injection. Data is the mean ± SE of 5 replicates of 100 µl of a 1:20 dilution of serum. Group 1) SPX, AC injection of BSA; 2) SPX, S.Con injection of BSA; 3) sham-operated + AC injection of BSA; 4) SPX, AC injection of BSA, BSA + Freund's Complete Adjuvant (CFA) 1 day after AC injection; 5) SPX + AC injection of BSA, BSA + CFA 7 *days* after AC injection.

phamide would directly interfere with ACAID. Moreover, we wished to determine whether cyclophosphamide would affect TABM production. BALB/c mice were injected with 200 mg/kg cyclophosphamide two days before receiving an AC injection. One day after receiving an AC injection, the mice were sensitized by applying picryl chloride to the abdomen. DTH was induced by applying picryl chloride to the ear 5 days after sensitization. Two days after challenge, ear swelling was measured and the mice were bled to obtain sera. As shown in Figure 8, the ear swelling of mice which received an AC injection was approximately 40% of control mice. In contrast, ear swelling of AC-injected mice that received cyclophosphamide was approximately 20% greater than sensitized mice which did not receive an AC injection. Moreover, non-AC injected mice that received cyclophosphamide had ear swelling almost 2-fold greater than mice which were not injected with cyclophosphamide. When TNP-specific TABM were measured (Fig. 8), there was a significant elevation of TABM in AC-injected mice which was eliminated by pretreatment with cyclophosphamide. The production of TNP-specific immunoglobulins was not affected, basal TABM levels in normal serum were not affected by cyclophosphamide (data not shown).

DISCUSSION

Previously, we have shown that the systemic injection of high doses of soluble protein antigens and adjuvants will induce a rise in serum of TABM specific for the antigen (23,24). That these serum proteins are T cell-derived is evidenced by (i) the production of TABM in immunized SCID mice only after reconstitution with T cells (23), (ii) the demonstration that T cells and T cell hybrid TABM purified to homogeneity share epitopes with serum TABM (24), and (iii) serum TABM bear TcR Cα and Cβ epitopes (23,24). Mice sensitized and challenged with picryl chloride produced moderate levels of TNP-specific TABM, which in some cases, was not apparent 4-5 days after sensitization. The production of TNP-specific TABM in sensitized mice is dependent on antigen dose and occurs whn the dose of antigen is supraptimal for the induction of DTH (J.W. Streilein, Y. Wang and R.E. Cone, unpublished observations). AC injection of low doses of soluble or particulate antigens (without adjuvants) alone did not induce a detectable rise in serum TABM, but enhanced the production of TABM after a second (sensitizing) exposure to antigen. At least one week after AC injection is required to detect the production of TABM suggesting that the TABM we detect are produced after antigen-induced expansion of TABM-producing cells. However, if the mice were sensitized previously, a single AC injection of antigen induced a rapid (2 days) rise in serum TABM in splenectomized mice, and therefore, TABM production was detected in the sera of splenectomized mice, these molecules are produced by T lymphocytes in the lymph nodes and circulation. Although some TABM may be produced in the spleen, clearly the spleen "filters" circulating TABM or depresses TABM production. Serum TABM bear TcR Cα and Cβ (23,24), and, like serum TsIF, are not detected in AC-injected, splenectomized mice. Since the TABM are detected with an antiserum raised against a T cell hybrid TABM with TsIF activity (20), it is probable that the serum TABM potentiated or induced by AC injection induce the activation of suppressor T cells.

The inhibition of TABM production (but not immunoglobulin production) and ACAID by cyclophosphamide suggests that TABM are produced by immunoregulatory T cells shown to be active in ACAID (1,3,4,9,29). Alternatively, cyclophosphamide-sensitive macrophages (30) might play a role in activating TABM-producing T cells. Moreover, these results provide further evidence that serum TABM may be a sensitive, quantitative indicator of immunoregulatory (T cell) activity. Positive and negative regulatory elements occur during an immune response, and the production of TABM in mice sensitized and challenged for DTH probably reflects a normal immunoregulatory T cell response. Injection of antigen

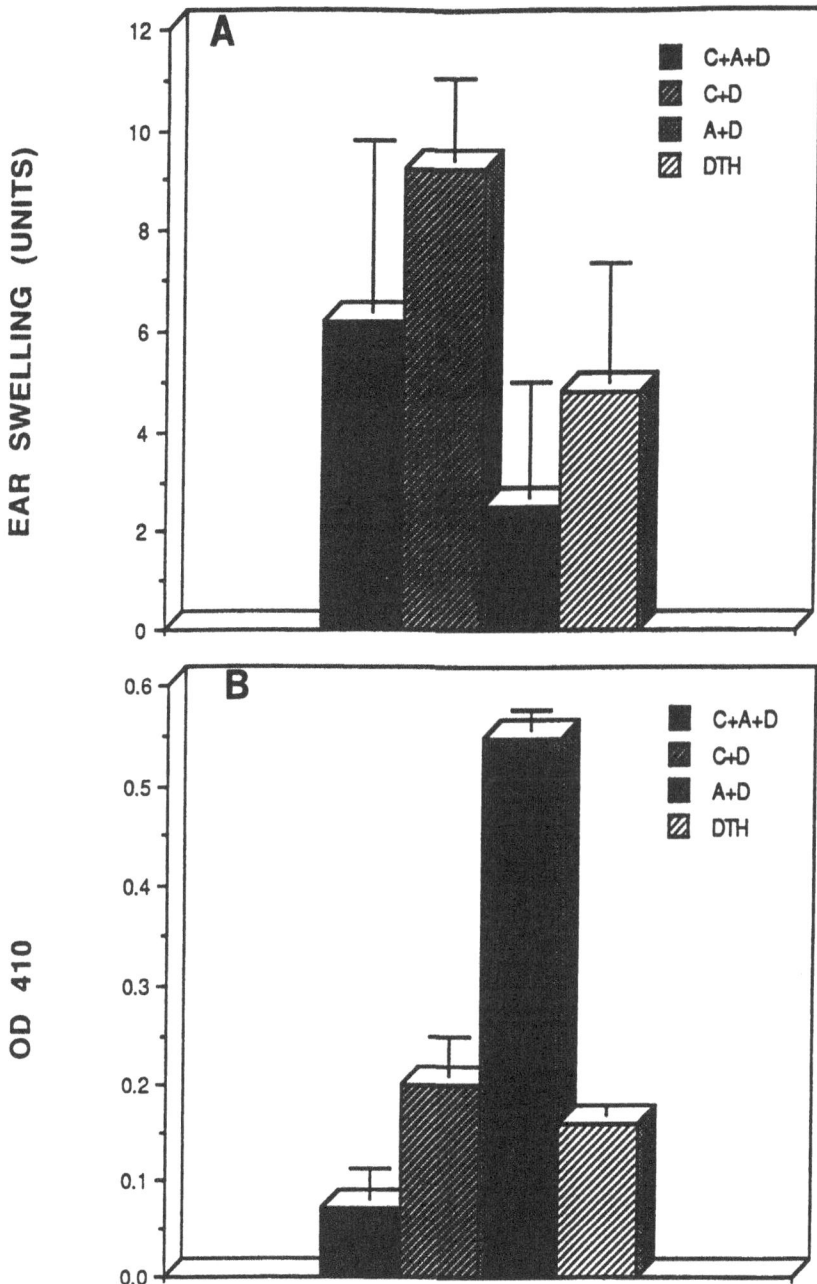

Figure 8. Cyclophosphamide inhibits ACAID and serum TABM production. Mice received 200 mg cyclo-phosphosphamide (CY)/g, and 24 hr later, received an AC injection of 5 x 10^5 TNP-spleen cells. The mice were sensitized and challenged (Figure 1) and bled 9 days after challenge. C + A + D = cyclophosphamide, anterior chamber, DTH C+ D = cyclophosphamide + DTH, A + D = anterior chamber + DTH, DTH = delayed hypersensitivity only. Data represents the mean ± SE of ear swelling (A) or (B) optical density of TABM in a 1:50 dilution of serum binding to TNP-BSA. Data is representative of the results of 2 separate experiments using 6-8 mice.

into the anterior chamber enhances negative immunoregulatory elements (for DTH) which may be reflected by increased TABM production above "normal" levels.

In relation to two models for the induction of ACAID, we believe that the role of TABM in ACAID (Fig. 9) may be explained by the sequence of events that lead to the systemic production of TABM that results from the transformation by TGF-β of F4/80+ macrophages which process antigen in the anterior chamber. These cells migrate from the anterior chamber to the peripheral blood, and then to the spleen where they induce the activation of suppressor T cells. Activation of suppressor T cells may be direct and/or by the activation of TsIF-producing T cells in lymph nodes, blood (and probably) spleen which then activate suppressor T cells. Preliminary evidence (R.E. Cone, S. Okamoto, J.W. Streilein, unpublished data) suggests that TGF-β-treated peritoneal exudate cells which transmit an ACAID-like "signal," induce the production of serum TABM. Thus, according to our model, the production of TABM is a distal step of ACAID resulting from the processing and presentation of antigen by eye-derived antigen-presenting cells.

Small increases in TABM production may be induced by systemic sensitization for contact sensitivity alone when large doses of antigen are used. However, intraperitoneal injection of large doses of antigen and adjuvant without sensitization induces large increases in serum TABM and immunoglobulin production (23,24). These observations suggest that serum TABM induced by AC-injection or systemic immunization might be derived from Th2 cells. However, Th1 cells probably also produce TABM which might be antigenically distinct from Th2 TABM. It is probable that TABM, like immunoglobulins are a "family" of different classes. The mode of action of TABM is not clear, but some evidence suggests that antigen-specific immunoregulatory "factors" may target cytokines to an appropriate target (31).

The induction of serum TABM by the introduction of small amounts of antigen into the anterior chamber of the eye indicates an additional relationship of the eye and the immune system. If TABM induced by AC injection of antigen participate in the suppression of DTH induced by AC injection, then these molecules may provide a sensitive quantitative and qualitative antigen-specific serologic indicator of immunoregulatory T cell activity. How-

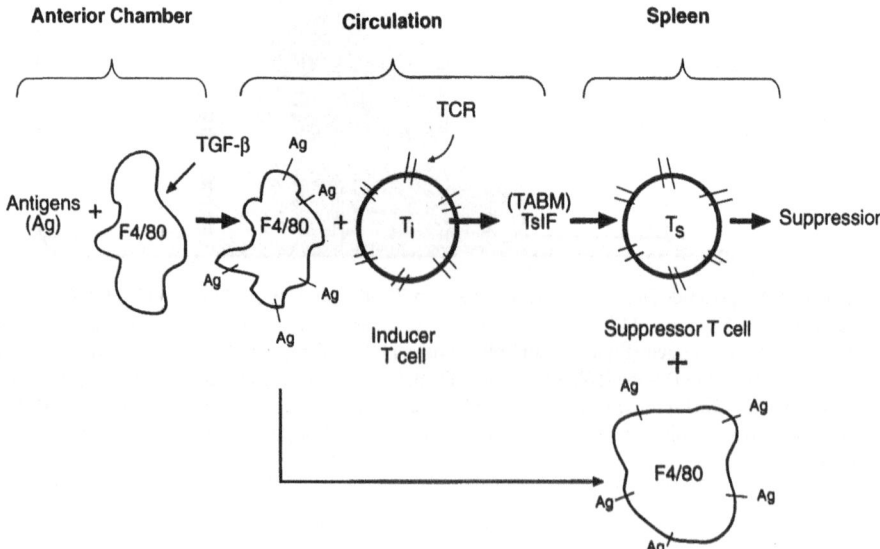

Figure 9. A model for cellular and humoral elements participating in the induction of ACAID.

ever, to be accepted into the immunological family, TABM need to be defined in molecular/genetic terms. Currently, our laboratory is making progress in this area.

ACKNOWLEDGMENTS

Supported by Connecticut Lions Research Foundation, Cyprus Leventis Foundation, and the University of Connecticut Health Center Research Advisory Committee.

REFERENCES

1. Streilein, J. W. Immune regulation and the eye: a dangerous compromise. *FASEB J.* 1:199 (1987).
2. Whittum, J.A., Niederkorn, J.Y., McCulley, J.P. and Streilein, J.W. The role of suppressor T cells in herpes simplex-induced immune deviation. *J. Virol.* 51:556 (1984).
3. Kaplan, H.J. and Streilein, J.W. Immune response to immunization via the anterior chamber of the eye I. F_1 lymphocyte-induced immune deviation. *J. Immunol.* 118:89 (1987).
4. Waldrep, J.C., Kaplan, H. T. Anterior chamber associated immune deviation induced by TNP-splenocytes (TNP-ACAID). I. Systemic tolerance mediated by suppressor T cells. *Invest. Ophthamol. Vis. Sci.* 24:1086 (1983).
5. Wilbanks, G.H. and Strelein, J.W. Distinctive humoral immune responses following anterior chamber and intravenous administration of soluble antigen. Evidence for active suppression of IgG2-secreting B lymphocytes. *Immunology* 71:566 (1990).
6. Strelein, J.W. and Niederkorn, J. Y. Induction of anterior chamber associated immune deviation requires an intact functional spleen. *J. Exp. Med.* 153:1058 (1981).
7. Williamson, J.S.P., Bradley, D. and Streilein, J. W. Immunoregulatory properties of bone marrow derived cells in the iris and ciliary body. *Immunology* 67:96 (1989).
8. Wilbanks, G.A. and Streilein, J.W. Studies on the induction of anterior chamber assoicated immune deviation (ACAID). Evidence that an antigen specific ACAID-inducing, cell-associated signal exists in the peripheral blood. *J. Immunol.* 146:1610 (1981).
9. Waldrep, J.C., Kaplan, H.J. Anterior chamber associated immune deviation by TNP-splenocytes (TNP-ACAID) I. Systemic tolerance indicated by suppressor T cells. *Invest. Ophthamol. Vis. Sci.* 24:1086 (1983).
10. Wilbanks, G.A., Mammolenti, M. and Streilein, J.W. Studies on the induction of anterior chamber-associated immune deviation (ACAID). III. Induction of ACAID depends on intraocular transforming growth factor β. *Eur. J. Immunol.* 22:165 (1992).
11. Ferguson, T.A., and Kaplan, H.J. The immune response and the eye I. The effects of monoclonal antibodies to T suppressor factors in anterior chamber-associated immune deviation (ACAID). *J. Immunol.* 139:346 (1987).
12. Kahm, M., Lima, S., Herndon, J.M., Kaplan, H.J. and Ferguson, T.A.. A role for soluble T cell receptor in HSV-ACAID. *Invest. Ophthalmol. Vis. Sci.* 34:903 (Abst. 1003) (1993).
13. Iwata, M., Katamwak, M., Kubo, R.T. and Ishizaka, K. Relationship between T cell receptors and antigen binding factors. J. Immunol. 143:3909 (1989).
14. Steel, J.K., Kawasaki, H., Kuchroo, V.K., Minami, M., Levy, J.G., and Dorf, M.E. Isolation of an antigen-specific T suppressor factor that suppresses the *in vivo* response of DBA/2 mice to ferredoxin. *J. Immunol.* 139:2629 (1987).
15. Fairchild, R.L., Kubo, R.T. and Moorhead, J.W. DNP specific class I MHC-restricted suppressor molecules bear determinants of the T cell receptor α and β chains. *J. Immunol.* 141:3342 (1988).
16. Ferguson, T., Hayashi, J.D. and Kaplan, H.J. The immune response on the eye III Anterior chamber-Associated Immune Deviation can be adoptively transferred by serum. *J. Immunol.* 143:82 (1989).
17. Ptak, W., Rosenstein, R.W, Ptak, M., and Gershon, R.K. Interactions between molecules (subfactors) relased by different T cell subsets that yield a complete factor with biological (suppressive) activity *Proc. Natl. Acad. Sci. (USA)* 79:2375 (1982).
18. Cone, R.E., Zheng, H., Chue, B., Beaman, K.D., Ferguson, T.A. and Green, D.R. T cell-derived antigen binding molecules (T-ABM). Molecular and functional properties. *Int. Rev. Immunol.* 3:205 (1988).
19. Cone, R.E., Weischedel, A-K, Urbanski, M. and Kristie, J. Specific antigen binding by proteins secreted by an antigen-specific T cell hybrid. *Mol. Immunol.* 29:689 (1992).

20. Chue , B., Ferguson, T.A., Beaman, K.D. and Rosenmann, S.J. An approach to the unification of suppressor T cell circuits. A simplified assay for the induction of suppression by T cell-derived antigen-binding molecules (TABM). *Cell. Immunol.* 118:30 (1989).

21. Cone, R.E. and Marchalonis, J.J. Partial amino acid sequence of monoclonal extracellular antigen-specific T cell proteins. *Immunol. Invest.* 22:541 (1993).

22. Cone, R.E. and Beaman, K.D. Antigen binding molecules of T cells-charge heterogeneity and structural ability. *Mol. Immunol.* 22:399 (1985).

23. Urbanski, M.J. and Cone, R.E. Appearance of T lymphocyte derived proteins specific for the immunizing antigen in serum during a humoral immune response. *J. Immunol.* 148:2840 (1992).

24. Urbanski, M. and Cone, R.E. Antigen-specific humoral T cell response II. Further characterization of the response and T cell immunoprotein. *Cell. Immunol.*153:131 (1994).

25. Beaman, K.D., Ruddle, N.H. Bothwell, A.L.M. and Cone, R.E. Messenger RNA for an antigen specific binding molecule from an antigen-specific T Cell hybrid. *Proc. Natl. Acad. Sci.* (USA) 81:1524 (1984).

26. Cone, R.E., Rosenstein, R.W., Janeway, C.A., Iverson, G.M., Murray, J.H., Cantor, H., Fresno, M., Mattingly, J.A., Cramer, M., Krawinkel, V., Wigzell, H., Binz, H., Frischnecht, H., Ptak, W. and Gershon, R.K. Affinity purified antigen specific products produced by T cells share epitopes recognized by heterdogous antisera raised against several different antigen specific products from T cells. *Cell. Immunol.* 82:232 (1983).

27. Cone, R.E., Wang, Y., O'Rourke, J., Kosiewicz, M.M., Okamoto, S. and Streilein, J.W. On the role of extracellular, antigen-specific T cell proteins in immune deviation elicited by intracameral soluble protein antigens. Sixth International Symposium on the Immunology and Immunopathology of the Eye. In press.

28. Kosiewicz, M.M., Miki, S., Okamoto, S., Ksander, B.R., Shimizu, T. and Streilein, J.W. ACAID can be induced in pre-primed mice. *Invest. Ophthalmol. Vis. Sci.* 35:1687 (Abst. 2004) (1994).

29. Wetzig, R., Foster, C.S. and Green, M.I. Ocular immune response 1. Priming of A/J mice in the anterior chamber with azobenzene arsonate-derivatized cells induces second-order like suppressor T cells. *J. Immunol.* 128:753 (1982).

30. Szcepanik, M., Bryniarski, K., Pryjmay, J. and Ptak, W. Distinct populations of antigen-presenting macrophages are required for induction of effector and regulatory cells in contact sensitivity response in mice. *J. Leuk. Biol.* 53:320 (1993).

31. Malley, A., Zaleny-Podey, M., and Murray, G. Third International Workshop on Immunosuppression, San Diego, CA (1993).

26

THE DEVELOPMENT AND USE OF T CELL RECEPTOR PEPTIDE VACCINES

Steven Brostoff

The Immune Response Corporation
5935 Darwin Court
Carlsbad, California 92008

INTRODUCTION

Autoreactivity is a normal feature of the immune system which must be able to distinguish self from non-self. However, when this autoreactivity results in pathological tissue damage, autoimmune diseases result. While autoimmune diseases can be associated with autoreactive T cells or autoantibodies, many of the major autoimmune diseases are considered to be T cell mediated. These include rheumatoid arthritis, multiple sclerosis, psoriasis and type I diabetes. In developing strategies to treat T cell mediated autoimmune disease conditions, one can consider the autoreactive T cells as pathogens. Thus, one can approach therapies for T cell mediated autoimmune disease by using strategies similar to those used to control other pathogens, such as virus or bacteria. These include strategies involving immunization with attenuated or subunit forms of these pathogens.

ANIMAL STUDIES

Early studies by Cohen and co-workers indicated, in an animal model of T cell mediated autoimmune disease, experimental autoimmune encephalomyelitis (EAE), that an attenuated form of the pathogenic T cell could be used to protect animals from EAE.[1] Although this approach is currently being tested in human clinical trials, its shortcoming is the need to use this approach as an autologous T cell immunization because of the outbred nature of the human population.

During development of the T cell repertoire, various T cell receptor gene elements are rearranged to form the α and β chains of the intact T cell receptor. Studies examining T cell receptor gene use in the EAE model in the late 1980s[2-5] indicated there was a remarkable conservation in the T cell receptor gene elements found on T cells causing EAE. In both rats and mice, these encephalitogenic T cells almost exclusively used Vβ8.2 and either Vα2 or Vα4 rearranged to very closely related Jα31 or Jα39. These studies suggested to us another approach for downregulation of pathogenic T cells. We employed a strategy of using

Immunobiology of Proteins and Peptides VIII
Edited by M. Z. Atassi and G. S. Bixler, Jr., Plenum Press, New York, 1995

249

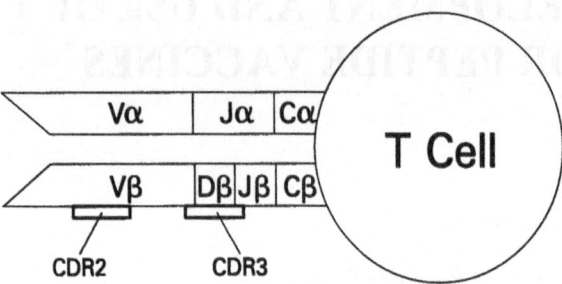

Figure 1. T Cell Receptor Structure. T cell receptor peptides were synthesized using sequences from complementarity-determining region (CDR) 2 and CDR3.

synthetic peptides comprising amino acid sequences from these gene elements, found to be conserved on encephalitogenic T cells, as subunit vaccines to treat EAE. Vβ peptides from both complementarity-determining region (CDR) 2 and CDR3, as well as a Jα peptide, were used.[6,7] Others reported the use of CDR2 peptides for both prevention[8] and treatment.[9] Figure 1 is a representation of the α and β chains of the T cell receptor with the location of the various gene elements and CDR2 and CDR3 regions indicated. A composite of the results of several experiments with CDR2 and CDR3 peptides are shown in Figure 2. In these experiments, Lewis rats were injected with T cell receptor peptides in complete Freund's adjuvant and were challenged 42 days later with myelin basic protein in complete Freund's adjuvant. The disease course was graded on a 3 point scale and the animals were followed until they recovered. Lewis rats generally spontaneously recover from disease within a week after the most severe clinical signs (usually paralysis, grade 3) have occurred. A significant impact on the disease was demonstrated. It is of significance that subunit vaccines comprised of T cell receptor peptides would not have to be given in an autologous fashion, and thus are more suitable for human therapy than attenuated whole T cell vaccines.

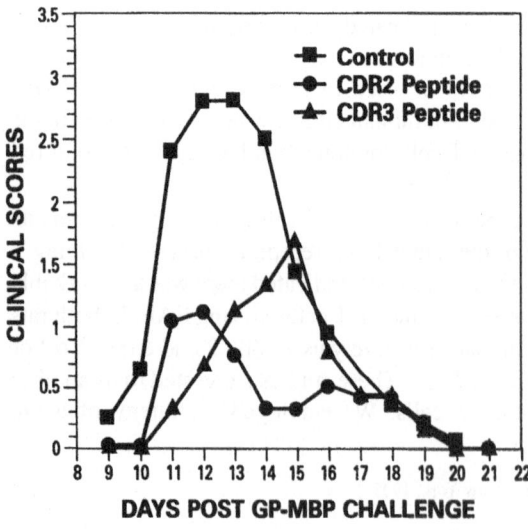

Figure 2. Prevention of EAE in the Lewis Rat by TCR CDR2 and CDR3 Peptides. Animals were graded on a 3 point scale as follows: 1) loss of motor control of tail, 2) hind leg paresis, 3) hind leg paralysis. Peptides were administered in complete Freund's adjuvant and challenged 42 days later with myelin basic protein in complete Freund's adjuvant. Control animals received an irrelevant peptide in complete Freund's adjuvant.

The animal studies using T cell receptor peptides to treat EAE suggest that this approach can be used to treat T cell mediated autoimmune disease, as well as B cell mediated disease to the extent that T cell help is needed and, in fact, any T cell mediated pathology involving oligoclonal T cell populations, such as T cell lymphoma.

RHEUMATOID ARTHRITIS

In order to apply T cell receptor peptide immunization therapy to human autoimmune disease, restriction in T cell receptor gene use must be a feature of the disease. Therefore, our basic strategy has been to try to identify T cells involved in human autoimmune disease and analyze their T cell receptor gene use. In rheumatoid arthritis, we obtained synovial tissue from patients undergoing joint replacement surgery and, using magnetic beads coated with antibody, we isolated IL-2R+ T cells infiltrating these tissues.[10] Previous animal studies indicated that most of the T cells infiltrating target tissues in EAE were not involved in the disease. Rather it was the small percent of activated (IL-2R+) T cells that correlated with disease activity.[11] This observation led us to focus on the small population of infiltrating IL-2R+ T cells for analysis of the T cells involved in rheumatoid arthritis. Our analysis of infiltrating T cells in rheumatoid arthritis indicated an over representation of three Vβ gene families: Vβ17, Vβ14 and Vβ3.[10] CDR3 sequence analysis of infiltrating IL-2R+ T cells bearing these Vβs indicated a clonal expansion *in situ* had occurred since many of these T cells were either clonal or oligoclonal. Of significant interest is the fact that these three Vβ gene families are quite homologous to each other, especially in the CDR4 region thought to bind superantigen.

PROPOSED SUPERANTIGEN INVOLVEMENT IN THE PATHOGENESIS OF AUTOIMMUNE DISEASE

Superantigens are products of bacteria or virus that stimulate T cells in a Vβ specific fashion by forming a bridge between the Vβ chain of the T cell receptor and the MHC molecule outside of the peptide groove associated with the stimulation of T cells by nominal antigen.[12] The close relationship of the three Vβ gene families found in rheumatoid arthritis, especially their close homology in the putative superantigen binding region, support a role for superantigen in the pathogenesis of autoimmune disease. In one scenario, following a viral or bacterial infection in which superantigens were released, there would be a Vβ specific activation of T cells. Included in the activated population would be T cells that are potentially autoreactive. These activated, autoreactive T cells would then migrate to the tissue containing their specific target antigen and continue to proliferate in response to the autoantigen, resulting in tissue damage. The clonal nature of the infiltrating T cells containing these Vβ gene families in rheumatoid arthritis support the continued expansion of these T cells in response to antigen *in situ*, rather than superantigen, since, in the latter case, a polyclonal, rather than a clonal, population of T cells bearing these restricted Vβs would be expected. Figure 3 illustrates this proposed role for superantigen in autoimmune disease.

PSORIASIS

In addition to rheumatoid arthritis, we have investigated the T cell receptor gene use in psoriasis.[13] In psoriasis, we have examined epidermal shave biopsies from multiple patients and have obtained repeat biopsies in four of these. We have compared the T cell Vβ gene repertoire in

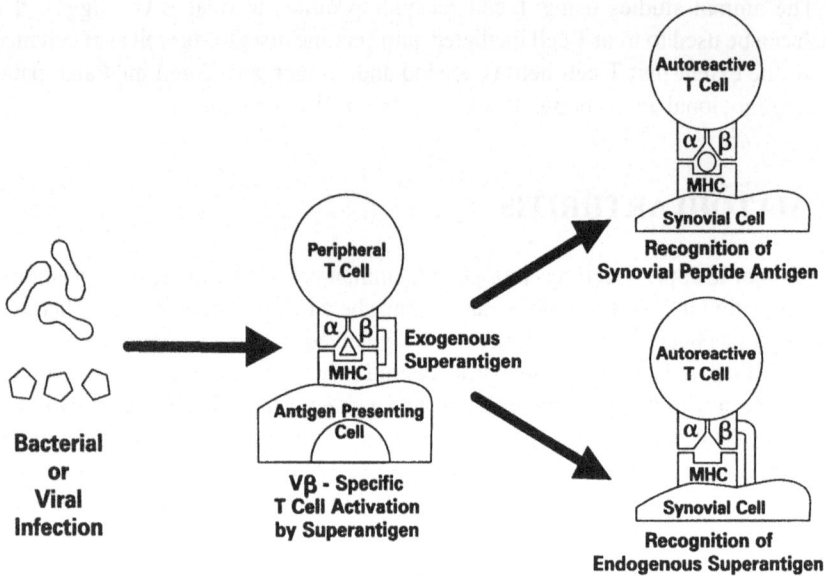

Figure 3. Proposed Involvement of Superantigen in Rheumatoid Arthritis. The clonal nature of the infiltrate suggests that after Vβ specific activation, recognition of a synovial peptide antigen, rather than endogenous superantigen, would occur.

these biopsies with those found in peripheral blood leukocytes (PBL) in order to identify T cells that were over represented. Since psoriasis is considered to be an MHC class I associated disease, we fractionated lymphocytes into CD4+ and CD8+ subsets and analyzed the CD4+,CD25+ and CD8+,CD25+ populations separately for Vβ gene use. The criteria we used for over representation of a given Vβ gene family was its presence at greater than 10% of the total Vβ genes and twice the value found in the PBL or an absolute value of greater than 30%. In the normal T cell repertoire, no Vβ gene family is found consistently at a level greater than 10%.

We recently reported results from seven patients.[13] In eight of nine biopsies from five of these patients (four repeat biopsies), Vβ13.1 T cells were over represented in the CD8+ population. In the ninth biopsy, Vβ13.1 was represented at an absolute value above 10%, but was not twice the level found in the PBL. In the four patients who underwent repeat biopsies, three also had an over representation of Vβ3 in each of the six biopsies in the CD8+ population. No such over representation of these two Vβ gene families was found in the CD4+ population from the skin biopsies of these patients. When CDR3 sequence analysis was performed on the T cell receptors from these patients, the Vβ13.1 and Vβ3 T cells were found to have only one or two dominant clones. The same dominant clones were found in three of these patients, in the same or different lesions, up to eight months apart, strongly suggesting their involvement in the disease process. A similar level of clonality was not found in any Vβ gene family in CD4+ T cells from the epidermal shave biopsies or in CD4+ or CD8+ subfractions of the PBL.

It is interesting to note that the two patients in which neither Vβ3 nor Vβ13.1 were over represented in lesions, Vβ17 was over represented and these T cells were dominated by one (one patient) or two (one patient) clones. However, in subsequent biopsies of eight additional patients, no over representation of Vβ17 has appeared, although Vβ3 and/or Vβ13.1 continue to be consistently elevated.[14]

The elevation and clonality of Vβ3 and Vβ13.1, especially the presence in an unresolved lesion sampled at two times, 3.5 months apart, give strong support for the involvement of these T cells in psoriasis and suggest that targeting of these T cells in psoriasis with a T cell receptor peptide vaccine may prove beneficial.

It is interesting to note that in recent reports,[15,16] two laboratories have described a superantigen from *Yersimia Pseudotuberculosis* that stimulate both Vβ3 and Vβ13.1 T cells. The stimulation of the specific Vβs found elevated and clonal in psoriasis by a common superantigen suggests the possibility that superantigens may be involved in the pathogenesis of psoriasis, as well as rheumatoid arthritis, and may be a common feature in the pathogenesis of other autoimmune diseases.

MULTIPLE SCLEROSIS

In multiple sclerosis, the target tissue is not readily available from living patients. An alternate strategy for attempting to identify the relevant T cells in this disease is to examine activated cerebrospinal fluid (CSF) T cells from multiple sclerosis patients. Because of the low numbers of T cells available, the CSF T cells must be grown in the presence of IL-2 to expand the population sufficiently for analysis of T cell receptor gene use. Presumably, the IL-2R+ T cells will expand preferentially in cultures containing IL-2 and the subsequent analysis will be meaningful. We have initiated clinical trials in pre-screened patients in which the same Vβ T cells have been found in repeat CSF samples.[17] In this trial, patients, in which Vβ6 T cells were found to be present at significant levels in cultures of T cells from more than one spinal tap,[17] were immunized with a Vβ6.2/6.5 T cell receptor peptide. This trial is currently ongoing. Since this approach, if successful, will require somewhat of an individualized therapy, development of T cell receptor peptide vaccines for treatment of multiple sclerosis is more challenging than either rheumatoid arthritis or psoriasis.

CLINICAL TRIALS WITH TCR PEPTIDES IN RHEUMATOID ARTHRITIS

The observation of restricted Vβ gene use in rheumatoid arthritis led us to initiate clinical trials using T cell receptor peptide vaccines. Following our observation that Vß17 was over utilized in rheumatoid arthritis, we initiated a phase I open-label safety and dose-ranging study.[18] For this study, we used a 17 amino acid sequence derived from the CDR2 region of the human Vß17 TCR. Fifteen moderate to severe rheumatoid arthritis patients received an intramuscular injection of one of four doses of the peptide in incomplete Freund's adjuvant, followed by a booster injection of the same dose four weeks later. The doses given were 10 µg (three patients), 30 µg (three patients), 100 µg (five patients) and 300 µg (four patients). Patients were followed for 48 weeks post immunization to monitor for safety. Nonblinded assessment of disease, measured by number of swollen & tender joints, showed trends towards improvement over the 48-week follow-up period. In addition to safety measurements, a variety of immunogenicity measurements were used to measure the affect, if any, of the immunization on the immune system and the target Vß17 T cell population. No antibody to the immunizing peptide was observed in the patients during the 48-week follow-up. Significant T cell proliferation (SI≥3) in response to the immunizing Vß17 peptide was observed at six weeks or later post-immunization in 1/3 patients in both the 10 and 30 µg dose groups, and in 2/5 and 2/4 of patients in the 100 and 300 µg groups, respectively. The Vß17+ T cell population remained constant in the range of 3-6% of total T cells in these patients. However, percentages of Vß17 T cells in blood that were IL-2R+ decreased (≥20%) in 3/5 patients in the 100 µg group after initial measurement at week 2 and in 3/4 patients in the 300 µg group after immunization. This decrease remained in 5/6 patients through 48 weeks. No measurable effect on the ability of the T cells from these patients to be stimulated by PHA was noted during the course of the study.

A second trial using a Vβ14 TCR peptide was also recently completed, as well. The observations with both trials indicate that TCR peptide vaccination is safe and well tolerated. No significant adverse events attributable to the treatment were experienced. In the Vβ17 trial, a statistical analysis combining the patients in all dose groups and comparing the mean pain and swelling scores prior to treatment and at week 48 indicate a statistically significant reduction of pain and swelling. Further analysis of this technology in controlled clinical trials in rheumatoid arthritis and psoriasis will be needed to provide definitive information on the efficacy of this approach for treating autoimmune disease.

REFERENCES

1. Ben-Nun A, Wekerle H, Cohen IR. Vaccination against autoimmune encephalomyelitis with T-lymphocyte line cells reactive against myelin basic protein. Nature 1981; 292:60-61.
2. Acha-Orbea H, Mitchell DJ, Timmermann L, Wraith DC, Tausch GS, Waldor MK, Zamvil SS, McDevitt HO, Steinman L. Limited heterogeneity of T cell receptors from lymphocytes mediating autoimmune encephalomyelitis allows specific immune intervention. Cell 1988; 54:263-273.
3. Urban JL, Kumar V, Kono DH, Gomez C, Horvath SJ, Clayton J, Ando DG, Sercarz EE, Hood L. Restricted use of T cell receptor V genes in murine autoimmune encephalomyelitis raises possibilities for antibody therapy. Cell 1988; 54:577-592.
4. Burns FR, Li X, Shen N, Offner H, Chou YK, Vandenbark AA, Herber-Katz E. Both rat and mouse T cell receptors specific for the encephalitogenic determinants of myelin basic protein use similar Vα or Vβ chain genes even though the major histocompatibility complex and encephalitogenic determinants being recognized are different. J Exp Med 1989; 169:27-39.
5. Chluba J, Steeg C, Becker A, Wekerle H, Epplen JT. T cell receptor ß chain usage in myelin basic protein-specific rat T lymphocytes. Eur J Immunol 1989; 19:279-284.
6. Howell MD, Winters ST, Olee T, Powell HC, Carlo DJ, Brostoff SW. Vaccination against experimental allergic encephalomyelitis with T cell receptor peptides. Science 1989; 246:668-670.
7. Brostoff SW, Howell MD. Immunoregulation of autoimmune disease by vaccination with T cell receptor peptides. Ann NY Acad Sci 1991; 636:71-78.
8. Vandenbark AA, Hashim G, Offner H. Immunization with a synthetic T-cell receptor V-region peptide protects against experimental autoimmune encephalomyelitis. Nature 1989; 341:541-544.
9. Offner H, Hashim GA, Vandenbark AA. T cell receptor peptide therapy triggers autoregulation of experimental encephalomyelitis. Science 1991; 251:430-432.
10. Howell MD, Diveley JP, Lundeen KA, Esty A, Winters ST, Carlo DJ, Brostoff SW. Limited T cell receptor β-chain heterogeneity among IL-2R+ synovial T cells suggests a role for superantigen in rheumatoid arthritis. PNAS 1991; 88:10921-10925.
11. Sedgwick J, Brostoff S, Mason D. Experimental allergic encephalomyelitis in the absence of a classical delayed-type hypersensitivity reaction. J Exp Med 1987; 165:1058-1075.
12. Marrack P, Kappler J. The staphylococcal enterotoxins and their relatives. Science 1990; 248:705.
13. Chang JCC, Smith LR, Froning KJ, Schwabe BJ, Laxer JA, Caralli LL, Kurland HH, Karasek MA, Wilkinson DI, Carlo DJ, Brostoff SW. CD8+ T cells in psoriatic lesions preferentially use T-cell receptor Vβ3 and/or Vβ13.1 genes. PNAS 1994; 91:9282-9286.
14. Chang JCC, Smith LR, Froning KJ, Schwabe BJ, Kurland HH, Karasek MA, Wilkinson DI, Carlo DJ, Brostoff SW. Unpublished data.
15. Abe J, Takeda T, Watanabe Y, Nakao H, Kobayashi N, Leung DYM, Kohsaka T. Evidence for superantigen production by Yersinia pseudotuberculosis. J Imunol 1993; 151:4183-4188.
16. Uchiyama T, Miyoshi-Akiyama T, Kato H, Fujimaki W, Imanishi K, Yan X-J. Superantigenic properties of a novel mitogenic substance produced by Yersinia pseudotuberculosis isolated from patients manifesting acute systemic symptoms. J Immunol 1993; 151:4407-4413.
17. Gold D, Wilson D, Smith R. Unpublished data.
18. Moreland LW, Heck LW Jr, Koopman WJ, Saway PA, Adamson TC, Fronek Z, O'Connor RD, Morgan EE, Diveley EE, Chieffo NM, Freeman TD, Carlo DJ, Brostoff SW. Vβ17 T Cell Receptor Peptide Vaccine: Results of a Phase I Dose-Finding Study in Patients with Rheumatoid Arthritis. Ann NY Acad Sci, in press.

27

EVIDENCE THAT IMMUNOSUPPRESSION IS AN INTRINSIC PROPERTY OF THE ALPHA-FETOPROTEIN MOLECULE

D. J. Semeniuk, R. Boismenu, J. Tam, W. Weissenhofer, and R. A. Murgita

McGill University
Department of Microbiology and Immunology
Montreal, Canada, H3A 2B4
Department of Surgery
Rudolfinerhaus
Vienna, Austria

ABSTRACT

Among the proteins that comprise the albumin family, alpha-fetoprotein (AFP) is the only member which exhibits immunoregulatory properties. However, some investigations have argued that AFP-mediated immunosuppression is not an inherent property of the molecule itself, but is instead, hypothesized to be either a function of a low molecular weight inhibitor bound to AFP or to a post-translational modification of the protein. AFP cannot be isolated from natural sources in quantities sufficient for the detailed biochemical and functional analyses required to resolve these issues. We have therefore produced recombinant forms of the protein (rAFP) by cloning the cDNA's for mouse and human AFP in both eukaryotic and prokaryotic expression systems. As described in this report, we were able to abundantly express rAFP's in bacterial, baculovirus and yeast expression systems. Recombinant proteins derived from each expression system were recognized by polyclonal and monoclonal anti-AFP antibodies as determined by immunoblot analysis. Pure recombinant protein samples, as characterized by polyacrylamide gel analyses, N-terminal sequencing and FPLC and HPLC chromatography, were evaluated for their immunoregulatory properties in murine and human *in vitro* immunological assays. The results of these studies establish that rAFP is functionally equivalent to natural fetal derived AFP molecules. Importantly, the data reported here demonstrate that AFP-mediated immunoregulation is an activity intrinsic to the molecule itself and cannot be attributed to either putative non-covalently bound moieties or to post-translational modifications such as glycosylation and sialylation. These studies provide a basis for initiating detailed investigations into the potential clinical usefulness of AFP as an immunotherapeutic agent.

Immunobiology of Proteins and Peptides VIII
Edited by M. Z. Atassi and G. S. Bixler, Jr., Plenum Press, New York, 1995

255

INTRODUCTION

Alpha-fetoprotein (AFP), is a 70 kD glycoprotein produced by fetal tissues, by regenerating hepatocytes and by certain tumours in adults (reviewed in Deutsch, 1991). Several studies have shown that AFP is a member of a multigene family comprising albumin (Gorin et al., 1981), vitamin D-binding protein (Schoentgen et al., 1986), and more recently, afamin (Lichenstein et al., 1994). Mammalian AFP's are known to share many physio-chemical properties with albumin. The homology between albumin and AFP was initially demonstrated by their immunological cross reactivities in the denatured state (Ruoslahti and Engvall, 1976) and by limited sequence homology (Ruoslahti and Terry, 1976). When compared with the known albumin sequence, the overall AFP-albumin homology was determined to be 30 to 40% (Morinaga et al., 1983; Gorin et al.,1981), . These results, in addition to those obtained by electron microscopic studies, image processing and circular dichronic measurements, demonstrated significant similarities of the overall structures of AFP and albumin (Luft and Lorscheider,1983), including the three-domain structure proposed for albumin (Brown, 1976) and subsequently for AFP (Kioussis et al.,1981; Morinaga et al.,1983). Not unexpectedly, AFP and serum albumin, appear to have comparable binding properties and transport roles for a variety of ligands (Savu et al., 1981). However, the transient presence of AFP in the maternal circulation during pregnancy and its reappearance in the course of certain neoplasias, suggest that AFP may be involved in the regulation of immune processes. Indeed, extensive *in vitro* and *in vivo* studies performed by numerous laboratories indicate that one important biological function of AFP is to down regulate select T-lymphocyte responses in order to control potentially harmful autoreactive events during fetal development (Yachnin, 1976; Hooper and Murgita, 1981; Hooper et al., 1989; Hoskin and Murgita, 1989). However, there has been debate in the literature as to whether AFP immunoregulation is an inherent property of the polypeptide chain of the molecule itself, or is instead, related to the ability of AFP to bind a serum-derived low molecular weight inhibitor(s) or is dependent upon post-translational modification of the protein. A number of investigators have proposed that the immunoregulatory properties of AFP depend on the presence of polyunsaturated fatty acids associated with this oncofetal protein, namely arachidonic and docosahexaenoic acids (Deutsch, 1991; Haourigui et al., 1992). More recently, it has been reported that the immunomodulatory behaviour of amniotic fluid can be correlated with TGF-β_2-like (Altman et al., 1990) or to a TGF-β_2 (Lang and Searle, 1994) contamination of AFP. Finally, attempts have been made to determine whether post-translational modifications, including sialylation, play a role in mediating AFP immunoregulatory activity (Zimmerman et al., 1977; Lester et al., 1976, 1977; van Oers et al. 1989).

We have previously described a method for the quantitative isolation of seven molecular variants of murine fetal AFP (van Oers et al.,1989). When each isolated variant was evaluated for immunoregulatory activity, it was determined that a single biologically active AFP isomer exists among the several natural isomeric forms. Moreover, no correlation could be shown to exist between functional activity and degree of sialylation (van Oers et al., 1989) or fatty acid ligand binding patterns (van Oers et al., 1995). These findings put an overall constraint on the ligand-binding theory since it would require that exogenous immunosuppressive ligands would have to bind exclusively to a singular and unique immunoregulatory isomer. In this report, we describe a series of experiments which resolves many previous discrepancies in the literature by showing that pure recombinant AFP exerts immunoregulatory activity in the absence of putative active ligands. Recombinant protein isolated from a bacterial expression system was as effective in down regulating *in vitro* T cell responses as was rAFP derived from eukaryotic expression systems.

MATERIALS AND METHODS

Generation of Mouse and Human AFP cDNA's for Expression Cloning into *E. coli*, Baculovirus and *Pichia pastoris* Expression Systems

The generation of cDNA sequences encoding full-length mouse (Mo) and human (Hu) AFP employed in the E.coli and baculovirus expression system has been previously described (Boismenu et al.,1995a, 1995b). The sequences corresponding to full-length MoAFP and HuAFP that were inserted into the yeast expression vector pPIC9 (Invitrogen, San Diego CA), were produced by PCR amplification, using oligonucleotide primer pairs that contained specific restriction endonuclease sites to facilitate directional cloning. The synthesis of full-length MoAFP was performed by employing pGBAFP (cloning plasmid containing the DNA sequence encoding MoAFP (Boismenu et al.,1995a) as template DNA, in addition to the following oligonucleotide primers: NH_2 (5'-AAA AAA CTC GAG AAA AGA GAG TTG CAC GAA AAT GAG-3') and COOH (5' - AAA AAA GAA TTC TTA AAC GCC CAA AGC ATC-3'). HuAFP was generated using pI18 (cloning vector containing the DNA sequence encoding HuAFP) as the source of template and the following primers: NH_2 (5'-AAA AAA CTC GAG AAA AGA GAG ACA CTG CAT AGA AAT GAA-3') and COOH (5'-AAA AAA GAA TTC TTA AAC TCC CAA AGC AGC ACG-3'). Each PCR reaction contained 34µL H_2O, 10µL 10 X reaction buffer, 20µL 1mM dNTP's, 2 µL DNA template, 10 µL 10 pmol/µL 5'-primer, 10 pmol/µL 3' primer, 1µL glycerol, 10 µL DMSO and 1 µL *Pfu* DNA polymerase (Stratagene, La Jolla CA). Annealing, extension, and denaturation temperatures were 50^0C, 72^0C and 94^0C, respectively, for 30 cycles, using the Gene Amp PCR System 9600 (Perkin Elmer Cetus). The DNA obtained from the PCR reactions were purified by isolating the fragments from a 0.7% TAE agarose gel, followed by gel extraction using the Geneclean kit (Bio101 Inc., La Jolla,CA). The PCR amplification fragments were then digested with XhoI and EcoRI, followed by purification away from the endonucleases by gel extraction as described above.

Cloning of MoAFP and HuAFP into *E.coli*, baculovirus and *Pichia pastoris* Expression Systems

The cloning and expression of the AFP cDNA's in *E.coli*, baculovirus and yeast were performed with plasmids pTrp4 (Boismenu et al, 1995a; 1995b), pVT-PLacZ (Semeniuk et al, 1995a; 1995b), and pPIC9 (Invitrogen), respectively. The vector pTrp4 was digested with ClaI, treated with Klenow enzyme (Pharmacia, Baie d'Urfe, QC) and dephosphorylated employing calf intestine phosphatase (Boehringer Mannheim, Laval, QC). The isolated cDNA's were incubated with Klenow fragment, purified, and inserted into the vector downstream of the trp promoter. DH5 α transformants were screened for the correct orientation of the rAFP cDNAs by restriction endonuclease analysis. Plasmid DNA from positive transformants was isolated and transformed into a variety of protease deficient *E.coli* strains.

The baculovirus vector pVT-PlacZ was modified by replacing the multiple cloning site with oligonucleotide fragments that would reduce the number of non-AFP coding nucleotides between the melittin signal peptide cleavage site and the location of the AFP cDNAs insertion at the EcoRI endonuclease sequence. The hydrolysed cDNA's were ligated into the modified pVT-PLacZ vectors at either the EcoRI endonuclease site for MoAFP cDNA, or between the EcoRI and BamHI DNA sequences for HuAFP cDNA. The DH5 α transformants were verified for the presence and correct orientation of the AFP cDNA's under the control of the polyhedrin promoter by performing restriction enzyme analysis.

Plasmid pPIC9 is an *E.coli - P.pastoris* shuttle vector containing a *S. cerevisiae* alpha mating factor (α-MF) prepro leader sequence. The vector was cleaved with XhoI and EcoRI restriction endonucleases and the PCR products encoding MoAFP and HuAFP were ligated to the linearized vector. The DH5α transformants were screened for the presence and correct orientation of the AFP cDNA's by restriction enzyme analysis.

Expression and Purification of Recombinant (r)AFP's

E. coli Derived (r) AFP. The expression and purification of rAFP's from E.coli has been previously described (Boismenu et al, 1995a; 1995b). Briefly, an overnight culture of LB broth containing rAFP-expressing *E.coli* was diluted 100 fold in antibiotic supplemented (tetracycline-HCl (50μg/ml) and ampicillin-Na (100μg/ml)) M9CA media in order to induce the activity of the trp promoter under tryptophan starvation conditions. The cells were gown to an A_{550} of 0.8 and harvested by centrifugation at 4000 RPM (Beckman J6-MI). The *E.coli* pellets were subjected to chemical lysis, liberating inclusion bodies containing rAFP. The inclusion bodies were washed with a 0.1% Triton-X solution and then dissolved in an alkaline denaturing buffer containing 6M guanidine. Resolubilized rAFP was diluted 50 fold with Tris buffer and incubated overnight to allow the proteins to refold. Following renaturation, the solution of rAFP was concentrated on Amicon YM 30 membranes (Amicon, Oakville, ON) and dialyzed against PBS.

Microaggregates of rAFP were removed by anion exchange chromatography carried out on Q-Sepharose resin. Final purification of rHuAFP was achieved by eluting a pure protein fraction from a Mono Q column during a step gradient run from 0-100% 1M NaCl in 20mM Tris-HCI (pH 8.0). Metal chelating chromatography using Cu^{+2} was performed as a second step in the purification scheme of rMoAFP.

Baculovirus-Derived rAFP. The generation of recombinant baculovirus containing rAFP coding sequences, the expression and purification of rAFP produced in insect cells, were previously described (Semeniuk et al, 1995a; 1995b). Briefly, purified recombinant baculovirus containing the coding sequences of MoAFP and HuAFP were generated by co-transfection of the pVT-PLacZ transfer vectors and wild-type baculovirus, followed by two rounds of plaque purification. Sf9 insect cells seeded at a density of $1x10^6$ cells/ml in 500 ml spinner flasks were infected with recombinant baculovirus in serum-free Grace medium. The supernatant containing secreted rAFP was harvested and cells were removed by centrifuging at 200 x g. The rAFP containing medium was concentrated 10-20 fold by ultrafiltration with a YM30 Amicon membrane, dialysed overnight against PBS and then applied to a Con A lectin column (Pharmacia). Bound recombinant AFP was eluted with 0.4M methyl α-D mannopyranoside and was purified to homogeneity by elution from Mono Q resin during a linear gradient from 0-100% 1M NaCl in 20 mM phosphate buffer, pH 8.0.

Yeast-Derived rAFP. The expression of rAFP polypeptides in *P. pastoris* was performed according to the manufacturer's instructions (Invitrogen). Briefly, recombinant *P. pastoris* yeast cells expressing AFP from the AOX1 (alcohol oxidase gene) locus were generated by transforming spheroblasts of *his4 P. pastoris* strain GS115 with linearized pPIC9 plasmids containing the AFP cDNA's, and plating transformants on histidine-deficient media. Cells in which recombination occurred, possessed the AFP cDNA in addition to the histidinol dehydrogenase gene (HIS4). To screen recombinants containing the AFP cDNA, clones were replica plated on -his/+glycerol and -his/+methanol plates. Colonies which grew normally on -his/+glycerol but poorly on the -his/+met plate were selected because they no longer contained the *AOX1* gene and have a his+/mut- (methanol utilization deficient) phenotype. A number of isolated his+/mut- recombinants were screened for expression levels of rAFP. Those clones which demonstrated the best expression level were then used to produce large amounts of rAFP. An overnight culture of 120 ml of antibiotic-supplemental BMGY media (10g/L yeast extract, 20g/L peptone, 100 ml 1M potassium phosphate buffer, pH6.0, 100ml 10X yeast nitrogen base, 2ml 500X biotin and 100 ml 10X glycerol) containing recombinant *P. pastoris* was used to inoculate 12L of 50 μg/ml ampicillin-Na supplemented BMGY for 48 hours. The cells were spun down at 4000 RPM (Beckman J6-MI) and resuspended in 2L of BMMY (BMGY with 100ml of 10X methanol replacing 100 ml 10X glycerol) to induce the synthesis of rAFP. The inductions were collected after 48 hrs and centrifuged. The supernatant was concentrated to 40 ml with an Amicon S1 Y30 cartridge. The rAFP containing media was equilibrated with 0.5 X PBS, pH 8.0, and filtered through a 0.45 μm filter (Millepore). The protein preparation was loaded onto an FPLC Mono Q column equilibrated with 20 mM Tris, pH8.0 and rAFP was found in the flow-through. This fraction enriched with rAFP was then chromatographed on a copper chelating Superose (Pharmacia) resin equilibrated with 20mM phosphate buffer, 1M NaCl, pH7.0. A fraction containing rAFP at a purity of greater than 90% was eluted during a gradient of 0-100% 50 mM NaAcetate, 1M NH_4Cl, pH4.0.

Cell Cultures

Mitogen Transformations. Adult CBA/J thymocyte suspensions were passed through commercial Mouse CD4 columns which negatively select for $CD4^+$ $CD8^-$ cells (Biotex Laboratories, Edmonton, AB). The selected thymocytes were cultured at a concentration of 2.5×10^5 cells in 96-well round-bottomed microculture plates for 48 h with ConA (1 μg/ml, Pharmacia). Experiments designed to detect MoAFP-mediated anti-proliferative effects were carried out by adding purified rAFP, rTGF-β1 and β2, and anti-TGF-β antibodies at the initiation of the cultures. Mouse albumin (Calbiochem, San Diego, CA) was used as the protein control. The culture medium employed was RPMI 1640 supplemented with 0.5% autologous normal mouse serum, plus 4mM L-glutamine, 20mM Hepes, 100 U/ml penicillin, 100ug/ml streptomycin, 1mM sodium pyruvate (Gibco) and 5×10^{-5}M 2-mercaptoethanol. Total volume of the cell cultures was 200μl. Cells were maintained at 37^0C in 5% CO_2. The cultures were pulsed at 42 hrs with 1μCi ^{3}H-thymidine and harvested 6 hrs later onto glass fibre mats. The amount of ^{3}H-thymidine incorporation was measured with a Packard TR 2500 liquid scintillation counter (Packard Instruments, Inc., Mississauga, ON). The results are expressed as mean cpm ± standard error of the mean (SEM) of triplicate cultures.

Heparinized blood from normal donors was diluted 1:1 with PBS and human peripheral blood lymphocytes (PBL) were separated from red blood cells by density centrifugation on Ficoll-Hypaque (Sigma, St-Louis, MO). They were washed at least 3 times with PBS and verified for cell viability by the Trypan Blue exclusion method. Human PBL (2.5×10^5 cells) were cultured following the same procedure described for the murine mitogen assay with the exception that 2 mg/ml of human serum albumin (HSA) (ICN, Mississauga, ON) was employed to supplement the RPMI 1640 medium instead of normal human serum. The results are expressed as the mean cpm thymidine incorporation ± SEM of triplicate cultures.

Autologous Mixed Lymphocyte Reactions (AMLR). Isolated PBLs were fractionated into a T cell population by passing 1.5×10^8 PBL over a commercial Ig-anti-Ig affinity column (Biotex Laboratories), while non-T lymphocytes were isolated by sheep red blood cell rosetting procedures that have been previously described (Hooper et al., 1989). Cultures consisted of 2.5×10^5 T cells incubated with an equivalent number of autologous irradiated (2500 rads) non-T stimulator cells. The medium employed was RPMI 1640 supplemented as above with the exception that 10% fresh human serum autologous to the responder T cell donor was used. Purified preparations of rAFP and HSA were added at the initiation of cultures. AMLR cultures were incubated for 4 to 7 days at 37^0C in 5% CO_2. DNA synthesis was assayed and results are reported as described above.

RESULTS

Expression and Purification of rAFP from *E.coli*, Baculovirus Infected Insect Cells and Yeast

E. Coli Derived rAFP. Assessing a variety of *E.coli* vectors and bacterial strains, we found a combination that yielded relatively high levels of expression of rAFP (Boismenu et al., 1995a). An *E.coli* vector successfully employed for overexpression of rAFP was pTrp4 (Fig. 1.A). This vector contains an optimized ribosome binding site flanking the Shine-Dalgarno site which was shown to enhance the expression of a variety of heterologous polypeptides under the control of the tryptophan promoter (Olsen et al., 1989). When pTrp4 was transformed into protease deficient bacterial strains, we observed expression of full-length rAFP at levels equal to or greater than 10% of the total amount of cellular protein isolated from the bacteria (Boismenu et al., 1995a).

Employing denaturation and refolding conditions to isolate soluble, monomeric rAFP from inclusion bodies, we found that a significant proportion of AFP molecules were in the form of microaggregates. These multimeric rAFP molecules were efficiently removed during a Q-Sepharose chromatography step as determined by native PAGE analysis (data not shown). Final purification of rHuAFP was achieved by collecting the protein fraction by eluting with 220-230 mM NaCl in 20 mM Tris, pH8.0 during Mono Q chromatography. Soluble, monomeric human rAFP preparations were greater than 98% pure, yielding 1mg of purified protein per litre of bacterial culture (Table 1). Recombinant mouse AFP, on the other hand, could not be as readily purified by Mono Q chromatography. We therefore exploited the metal binding capacity of AFP (Aoyagi et al., 1978) to achieve a purified form of the recombinant mouse protein (Table 1).

Table 1. Summary of the expression and purification of recombinant AFP molecules from *E. coli*, baculovirus, and yeast expression systems

	Expression system	Mode of expression	Purification scheme	Quantity of purified protein per litre culture
Mouse	*E. coli*	Inclusion bodies	1. Q-Sepharose 2. Cu^{+2} Chelating	0.5 mg
	Baculovirus	Secreted	1. Con A resin 2. Mono Q	2 mg
	Yeast	Secreted	1. Mono Q 2. Cu^{+2} Chelating	1 mg
Human	*E. coli*	Inclusion bodies	1. Q-Sepharose 2. Mono Q	1 mg
	Baculovirus	Secreted	1. Con A resin 2. Mono Q	2 mg
	Yeast	Secreted	N.D.	N.D.

N.D. Not Determined.

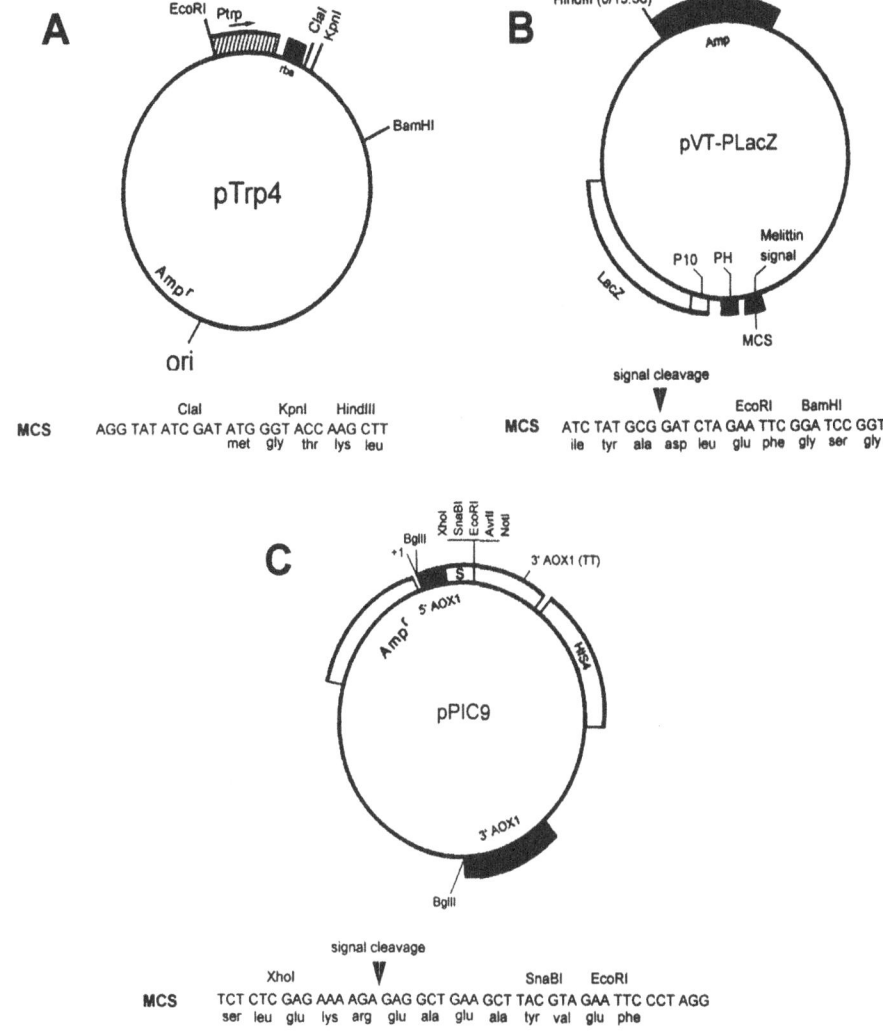

Figure 1. Expression vectors employed in the production of recombinant AFP in (A) *E. coli*, (B) baculovirus, and (C) *Pichia pastoris* expression systems. (A) The cDNAs for MoAFP and HuAFP were inserted by blunt end ligation into the Cla I site in pTrp4. The expression of AFP cDNAs is under the control of the trp promoter. (B) MoAFP and HuAFP coding sequences were inserted in frame with the melittin signal leader sequence at the EcoRl and EcoRl plus BamHl sites, respectively, in plasmid pVT-PLacZ. AFP fusion genes are transcribed by the baculovirus polyhedrin promoter. (C) The PCR amplification products of MoAFP and HuAFP cDNAs were inserted into the Xhol/EcoRl sites of the vector pPIC9, in frame with the α-MF signal peptide sequence. Transcription of the AFP cDNAs are under the control of the AOX1 promoter. Arrows indicate the expected cleavage site between the signal peptide and the mature AFP proteins.

Baculvirus Derived rAFP. The cDNAs containing the coding sequences of MoAFP and HuAFP were cloned into plasmid pVT-PlacZ (Fig. 1B) in frame with the 3'-end of the insect derived melittin signal peptide sequence. An earlier derivative of vector pVT-PLacZ that possessed the melittin signal leader, demonstrated enhanced secretion of a heterologous polypeptide (Tessier et al., 1991). We found that rMoAFP and rHuAFP represented approximately 20% of the total proteins secreted into serum-free medium by the Sf9 insect cells and were monomeric as analyzed by non-reducing alkaline PAGE. Both rAFPs were then purified to homogeneity using similar chromatographic procedures. Most of the baculovirus derived

AFP bound to immobilized Con A. This resulted in effective removal of more than 90% of contaminating proteins which were nonadhered to lectin columns. Final purification of the baculovirus derived rAFP preparations was accomplished by eluting protein with 270-310 mM NaCl from MonoQ beads. Approximately 2mg of purified protein was obtained per litre of growth culture (Table 1).

Yeast Derived rAFP. The cDNA's of MoAFP and HuAFP cloned into the yeast expression vector pPIC9 (Fig 1C), had to be generated by PCR amplification in order to recreate the 3'- terminus of the α-MF signal peptide DNA sequence lost during vector hydrolysis with XhoI. This strategy permitted the AFP cDNA's to be fused to the yeast derived signal sequence so that the translated heterologous gene product would be targeted to the secretory pathway of the host cell for release into the culture medium (Table 1). Five to seven milligrams of recombinant AFP's were secreted into one litre of culture. The majority of the contaminating yeast proteins were bound to a Mono Q anion-exchange resin whereas the fraction specifically enriched for rAFP eluted in the flow through. The final step in purification was accomplished by Cu^{+2} chelating chromatography, yielding 1 mg of protein per litre of growth culture (Table 1).

Demonstration of Immunosuppressive Function by Recombinant MoAFP Produced by Baculovirus and Yeast Expression Systems

Purified natural AFP in amounts necessary for detailed structure/function analyses has up until now been unavailable. We have circumvented this problem by generating relatively abundant quantities of purified soluble, and monomeric recombinant protein preparations. As shown in Table 2, rMoAFP produced in both yeast and insect cells, inhibited the proliferative response of mitogen-stimulated CD4$^+$ CD8$^-$ thymocytes. Since recovery of purified rAFP was much more efficient from the baculovirus system compared with yeast, murine functional assays were subsequently performed using the baculovirus derived product.

Table 2. Anti-lymphoproliferative activity of recombinant mouse AFP produced from baculovirus and yeast

Amount of protein added to cultures (µg/ml)[a]	Con A stimulated CD4$^+$CD8$^-$ thymocyte proliferation [³H] Thymidine incorporation CPM ± SE (% suppression)	
	Baculovirus MoAFP	Yeast MoAFP
100	1,694 ± 123 (99)	1,688 ± 392 (98)
25	4,635 ± 291 (97)	6,224 ± 360 (93)
6.2	6,871 ± 714 (96)	29,498 ± 668 (67)
3.1	13,782 ± 730 (91)	43,994 ± 3,041 (52)
1.5	29,625 ± 3,042 (82)	74,453 ± 9,958 (18)
0.8	47,602 ± 3,096 (71)	86,565 ± 4,337 (5)
0.4	72,982 ± 3,057 (56)	—
0.1	123,247 ± 7,299 (26)	—
0	167,703 ± 14,294	91,196 ± 2,048

[a]CBA/J CD4$^+$CD8$^-$ thymocytes (2.5 x 10^5) were incubated for 48 hr in the presence of 1 µg/ml Con A and 0.5% syngeneic normal mouse serum. Purified protein samples were added at the initiation of the culture.

Immunosuppression by Recombinant AFP Is Independent of Protein Post-Translational Modifications and of the Presence of Serum Co-factors

Experiments were performed to address the issue of whether post-translational modifications play a role in influencing the immunosuppressive properties of AFP. We have previously shown a lack of correlation between functional activity and degree of sialylation of AFP isomers (van Oers et al., 1989). However, such correlative studies could not formally exclude a contributory role for sialic acids. The availability of recombinant AFP derived from eukaryotic expression systems, which do modify proteins, and from a prokaryotic expression system which does not, allowed us to address this question in a more direct fashion. This was accomplished by comparing recombinant human AFP molecules produced in the baculovirus and in the *E. coli* expression systems for their ability to suppress the proliferation of autoreactive T cells in the autologous mixed lymphocyte reaction, an assay which we have previously shown to be sensitive to the anti-lymphoproliferative effects of AFP (Hooper et al., 1989). As shown in Figure 3a, both rHuAFP's added to the assay cultures at 100 µg/ml, are equivalent in their ability to suppress T cell autoproliferation at 144 hours. The addition of an identical amount of human serum albumin failed to diminished lympho-proliferative responses.

Recent reports have indicated that the observed immunosuppressive activity associated with amniotic fluid derived AFP is a result of TGF-β complexing with this onco-fetal protein (Altman et al., 1990; Lang and Searle, 1994). Employing baculovirus derived rAFP, we wanted to determine whether the presence of TGF-β's, that might be derived either from serum additives to the culture medium, or by ConA activated CD4$^+$ CD8$^-$ thymocytes (Kerhl et al., 1986), is required to mediate AFP immunosuppression. In the experiment shown in Figure 2, ConA responses were suppressed by more than 90% upon the addition of either 100 µg/ml rAFP, 2 ng/ml rTGF-β_1 or 2 ng/ml rTGF-β_2. When antibodies specific for either TGF-β_1 or -β_2 are added to the assay cultures, only the suppressive activity of the ligands for which the antibodies are specific is blocked. The proliferative responses of the media controls are unaltered, suggesting that no active TGF-β's are being produced by the mitogen stimulated T lymphocytes during the course of this assay. Finally, and most importantly, rAFP immunosuppression is unaffected by the addition of anti-TGF-β antibodies.

Having eliminated TGF-β's and post-synthetic modifications as factors contributing to the immunoregulatory activity associated with AFP, we wanted to rule out the possibility that other active factors present in the *in vitro* culture assays as a result of exogenous serum additives, do not contribute to the AFP anti-lymphoproliferative function. Therefore, we examined the ability of rHuAFP to suppress the mitogen induced proliferation of peripheral blood lymphocytes (PBL) in RPMI tissue culture media supplemented only with 2 mg/ml purified human albumin (ICN, Mississauga, ON). As can be seen in Figure 3b, marked suppression of ConA stimulated PBL's occurred with 100 µg/ml rHuAFP whereas the same concentration of albumin was ineffective.

DISCUSSION

This investigation demonstrates that immunosuppression is an inherent property of AFP and is not solely dependent on extrinsic co-factor(s) or post-translational modifications as has previously been suggested (Altman et al.,1990; Deutsch,1991; Lang and Searle,1994). These firm conclusions were drawn from the results of experiments performed using purified preparations of recombinant AFP derived from both eukaryotic and prokaryotic expression

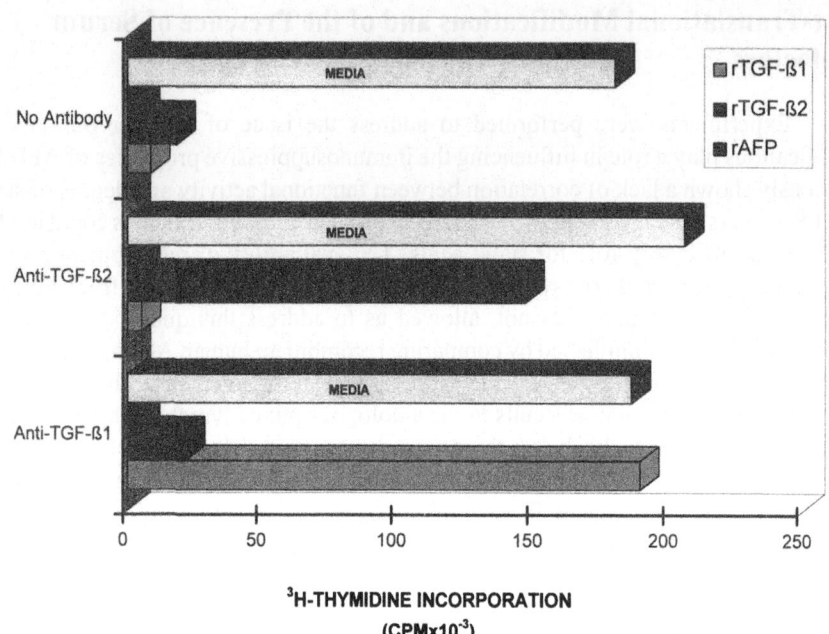

Figure 2. Immunosuppressive activity of rAFP does not require the presence of or interaction with Transforming Growth Factor-βs (TGF-βs). Anti-TGF-β1 polyclonal antibodies (10 μg/ml) and anti-TGF-β$_2$ monoclonal antibodies (20 μg/ml) were added to lymphocyte proliferation cultures containing either 100 μg/ml rAFP, 2 ng/ml rTGF-β$_1$, 2 ng/ml rTGF-β$_2$, or no protein addition whatsoever. Proliferative responses measured are described in Materials and Methods. The standard errors of the mean (SEM) were found to be less than 5% of the mean value.

systems. It was shown here that recombinant AFP molecules derived from bacteria, baculovirus and yeast were equally immunosuppressive. Because the prokaryotic expression system yielded a biologically active recombinant protein, the sophisticated post-translational modifications associated with the synthetic machinery of higher eukaryotes were shown not to be required for AFP-mediated anti-lymphoproliferative activity. Furthermore, additional evidence was presented to help dispel the notion that a putatively biologically active ligand must be bound to AFP in order to impart immunosuppression. This was accomplished by demonstrating the immunoregulatory effects of rAFP in the complete absence of either active TGF-β's or other serum co-factors.

Attempts by others to overexpress rAFP's in *E.coli* and yeast (Nishi et al., 1988; Yamamoto et al., 1990) have met with very little success, resulting in either unsatisfactory yields and/or insoluble product. The presence of 0.2% SDS in protein preparations was reported to be necessary in order to maintain *E. coli* derived recombinant rat AFP in a soluble state (Nishi et al., 1988). In addition, the recombinant protein reacted poorly in radioimmunoassays and did not exhibit estradiol-binding activity. Efforts to produce rHuAFP resulted in very low yields of immunopurified protein from yeast culture media (Yamamoto et al., 1990).

In the present investigation, an *E.coli* expression system was devised that was capable of expressing rAFP's at mg/litre quantities. The recombinant protein remained soluble in phosphate buffered saline (PBS), pH 7.2 (Table 1) without detergent or urea. Milligram quantities of soluble, monomeric rAFP was also readily recoverable from baculovirus and yeast expression systems (Table 1). This was accomplished by employing expression vectors

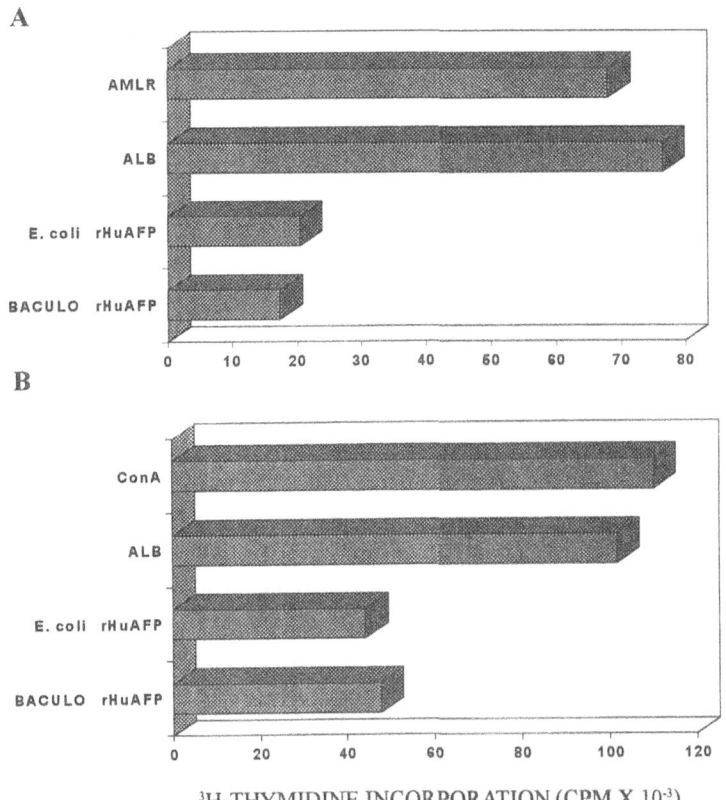

Figure 3. Recombinant AFP-mediated immunosuppression is not dependent upon (a) post-translational modifications or (b) other serum co-factors. a) The autologous mixed lymphocyte reaction (AMLR) was prepared by co-culturing 250,000 T cells with an equivalent amount of autologous irradiated non-T lymphocytes as described in Materials and Methods. Recombinant human AFP preparations derived from *E.coli* and baculovirus and human albumin were added at a concentration of 100 μg/ml at the initiation of culture. Proliferative responses were measured at 144 hours by ^{3}H-thymidine incorporation. (b) PBLs (2 x 10[5] stimulated with 1 μg/ml ConA were cultured in RPMI medium supplemented with only 2 mg/ml human serum albumin for 48 hrs. Human albumin and rHuAFP derived from *E. coli* and baculovirus were added to the initiation of the cultures at a concentration of 100 μg/ml. Proliferative responses were measured as the amount of ^{3}H-thymidine incorporated during DNA synthesis. The SEM were determined to represent less than 5% of the value of the mean.

specific for secretion of foreign protein into the extracellular environment. The obvious advantage of heterologous protein secretion in the two eukaryotic expression systems employed here is that relatively small quantities of host cellular proteins are discharged into the media. This permits isolation of recombinant protein employing relatively straightforward and conventional purification techniques.

Reported variations in the immunosuppressive potency of AFP depending on the source and method of purification combined with observations that certain serum ligands preferentially associate with AFP rather than albumin, has led some researchers to believe that AFP immunoregulation is a function of a complex between co-factors and the protein (Deutsch et al., 1991; Altman et al, 1990; Lang and Searle, 1994). However, the discovery of a single active isomer (van Oers et al., 1989) imposes a considerable constraint on the argument that a noncovalently bound moiety imparts functional activity to AFP. For instance, it would have to be assumed that putative immunosuppressive ligands bind exclusively to the unique suppressive AFP isoform which represents only 6% of the native population of AFP isomers. van Oers et al. (1989, 1995) further demonstrated that immunosuppressive activity did not correlate to either the amount of sialic acids or polyunsaturated fatty acids associated with individual AFP isoforms. To address more directly whether AFP-mediated immunosuppression is an inherent property of the polypeptide portion of the AFP molecule itself or alternatively is somehow dependent on a complex with another active factor or to post-synthetic modifications, we have in the present investigation focused on a structure-function comparison of eukaryotic versus prokaryotic-derived recombinant AFP's. Such recombinant molecules would not, under the growth and recovery conditions employed, have had any prior exposure to exogenous factors of mammalian origin which could complex with the recombinant protein. Therefore, these recombinant molecules allowed us to perform experiments aimed at answering in a definitive manner whether exogenous factors are involved. Furthermore, a direct comparison of functional activity of the unmodified prokaryotic recombinant protein with that of the eukaryotic-derived products permitted us to address the question of whether glycosylation, sialylation, etc. were related to activity. As depicted in Table 2, rMoAFP protein preparations derived from both recombinant baculovirus and yeast demonstrated the anti-lymphoproliferative activity that has been routinely observed with native AFP. The results in Fig. 2 indicate that immunosuppression is not mediated by TGF-β, derived either from serum or from mitogen activated T cells (Kerhl et al., 1986). The anti-TGF-β blocking experiments performed in Fig. 2 further demonstrate that latent TGF-β did not become activated by acidification in culture or by some other as yet unknown mechanism to become available to combine with rAFP during the course of the functional assays.

Unlike TGF-β's isolated from serum, TGF-β_1 and -β_2 and TGF-β_2-like factors in human amniotic fluid (Lang and Searle, 1994) and mouse amniotic fluid (Altman et al., 1990) are apparently physiologically active as determined by their effect in lymphocyte inhibition assays. TGF-βs are hypothesized to be activated (Lawrence et al., 1984) at the maternal-fetal interface by trophoblast-derived proteinases (Lang and Searle, 1994). The immunosuppression associated with AFP in amniotic fluid was blocked upon the addition of anti-TGF-β antibodies (Altman et al.,1990; Lang and Searle,1994), leading these investigators to conclude that AFP plays a passive role in amniotic fluid by acting only as a carrier protein for biologically active TGF-β_2 molecules. While the results of the present investigation renders untenable the conclusion that AFP is by itself inactive, we cannot exclude the possibility that AFP acts in concert with TGF-βs to exert additional patterns of immune regulations. However, it is also possible to hypothesize a TGF-β sequestration role for AFP in analogy with what is known from the TGF-β complexing function of α_2-macroglobulin in serum. It is well documented in the literature that TGF-β in serum exists in a latent form, complexed to α_2-macroglobulin (O'Connor-McCourt and Wakefield, 1987; Philip and O'Connor-MCourt, 1991). Alpha$_2$-macroglobulin may serve an important multifunctional role at sites of inflammation by scavenging active TGF-β at the site of injury and limit the potentially harmful action the TGF-β may induce at this site (Wahl, 1994). In amniotic fluid, AFP may act in a similar manner by actively binding TGF-β and thus neutralizing its biological activities to protect the fetus from the potential deleterious side effects of TGF-β, including excess inflammation and/or fibrosis. According to this reasoning, it would be expected that TGF-β complexed to AFP would be in a state of latency. Other neutralizing functions of AFP have been previously shown by demonstrating that AFP binds to estrogen and protects the developing brain against the effects of maternal estrogens (Attardi and Ruoslahti, 1976). Inhibition of AFP-mediated immunosuppressive activity by anti-TGF-β_2 antibodies could be the consequence of steric hindrance caused by a tertiary complex comprising AFP, TGF-β_2 and anti-TGF-β_2 antibodies, preventing AFP from apparently interacting with target lymphoid cells. An essential role for other possible serum-derived co-factors in AFP-mediated immunosuppression were eliminated when inhibition of mitogen transformed PBL's was demonstrated in serum-free media (Fig. 3b).

Attempts have been made to assign a role for post-translational modification with immunosuppressive potency of AFP (Lester et al, 1977; Zimmerman et al, 1977). In one such study, a variety of human AFP isolates were evaluated for their suppression of mitogen transformed PBL's and their activities appeared to correlate with the amount of the most acidic of three electrophoretic variants (Lester et al., 1976). Quantitation of sialic acid residues and desialylation of AFP molecules revealed that no relationship could be established between biological potency and sialic acid content. Additional studies performed by Yachnin's group (Lester et al, 1977) indicated that a post-synthetic modification of AFP was responsible for the reduction in the immunosuppressive potency of AFP and that this alteration to the protein most probably involved the addition of a charged moiety to AFP other than sialic acid. These results are not in complete agreement with those of Zimmerman et al. (1977) who demonstrated that mouse AFP immunosuppressive activity was associated with a sialylated species of the protein. However, subsequent work to further characterize the relationship between sialylation and anti-lymphoproliferative activity of both mouse (van Oers et al., 1989) and human (Lester et al., 1977) AFP demonstrated that sialic acids play no central role in mediating the immunosuppressive effect of AFP. Because baculovirus rAFP is produced in eukaryotic cells, it undergoes modification following protein translation, which may include sialylation (O'Reilly et al., 1992). Therefore, a comparison of the immunosuppressive potency of rAFP synthesized in eukaryotic and prokaryotic cells, showed that post-synthetic modifications are not directly involved in AFP immunosuppressive action *in vitro*. The results presented here serve to validate the earlier conclusions of van Oers et al.(1989) and Lester et al.(1977).

We conclude on the basis of the data presented here, that immunosuppression is an inherent property of AFP, resident within the polypeptide component of the molecule.

REFERENCES

Altman, J.D., Schneider, S.L., Thompson, D.A., Cheng, H.-L., Tomasi, T.B., 1990, A Transforming Growth Factor β2 (TGF- β2)-like Immunosuppressive Factor in Amniotic Fluid and Localization of TGF-β2 mRNA in the Pregnant Uterus, *J. Exp. Med.* 172:1391-1401.15.

Attardi, B., and Ruosiahti, E.. 1976. Foetoneonatal oestradiol-binding protein in mouse brain cytosol is α foetoprotein, *Nature* 263:685-687.

Aoyagi, Y., Ikenaka, T., and Ichida, F., 1978, Copper(II)-binding ability of human AFP, *Cancer Res.* 38:3483-3486.

Boismenu, R., Semeniuk, D., and Murgita, R. A., 1995a, Expression cloning of mouse alpha-fetoprotein cDNA: Recovery of full-length recombinant AFP molecules and peptide domain fragments in a form suitable for structure-function analysis, Submitted for publication: *Protein Expression and Purification.*

Boismenu, R., Semeniuk, D., Bennett, J.A., Jacobson, H., and Murgita, R.A., 1995b, Expression cloning of Human Alpha-fetoprotein and its growth regulatory properties on autoreactive lymphocytes and estrogen-stimulated tissues, Submitted for publication: *Cancer Research.*

Brown, J.R., 1976, Structural origins of mammalian albumin, *Fed. Proc. 35:2141-2144.*

Deutsch, H.F., 1991, Chemistry and Biology of α-fetoprotein. *Adv. Cancer Res.* 56:253-312.

Gorin, M.B., Cooper, D.L., Eiferman, F., van de Rijn, P., and Tilghman, S.M., 1981, The evolution of alpha-fetoprotein and albumin. A comparison of the primary amino acid sequences of mammalian alpha-fetoprotein and albumin. *J. Biol. Chem.* 256:1954-1959.

Haourigui, M., Thobie, N., Martin, M.E., Benassayag, C., and Nunez, E.A., 1992, In vivo transient rise in plasma free fatty acids alters the functional properties of α-Fetoprotein, *Biochimica et Biophysica Acta.* 25:157-165.

Hooper, D.C., and Murgita, R.A., 1981, Regulation of murine T-cell responses to autologous antigens by alpha-fetoprotein, *Cell. Immunol* 63:417-425.

Hooper, D.C., O'Neill, G., Gronvik, K.-O., Gold, P., and Murgita, R.A., 1989, Human AFP inhibits cell proliferation and NK-like cytotoxic activity generated in autologous, but not in allogeneic mixed lymphocyte reactions, in *Biological Activities of Alpha-Fetoprotein.* Vol ll , CRC Press, Boca Raton, FL, 184-197.

Hoskin, D., and Murgita, R. A., 1989, Specific maternal anti-fetal lymphoproliferative responses and their regulation by natural immunosuppressive factors. *Clin. Exp. Immunol.* 76:262-267.

Kerhl, J.H., Wakefield, L.M., Roberts, A.B., Jakowlew, S., Alvarez-Mon, M., Derynck, R., Sporn, M.B., and Fauci, A. S., 1986, The production of TGF-β by human T lymphocytes and its potential role in the regulation of T cell growth. *J. Exp. Med.* 163:1037-1050.

Kioussis, D., Eiferman, F., van de Rijn, P., Gorin, M.B., Ingram, R.S., and Tilghman, S.M., 1981, The evolution of the alpha-fetoprotein and albumin genes in the mouse. *J. Biol Chem.* 256:1960-1967.

Lang, A.K., and Searle, R.F., 1994, The immunomodulatory activity of human amniotic fluid can be correlated with transforming growth factor (TGF-β1) and β2 activity, *Clin. Exp. Immunol.* 97:158-163.

Lawrence, D.A., Pircher, R., Kryceve-Martiniere, C., and Julien, P., 1984, Normal embryo fibroblasts release transforming growth factors in a latent form. *J. Cell. Physiol.* 121:184-188.

Lester, E.P., Miller, J.B., and Yachnin, S., 1976, Human alphafetoprotein as a modulator of human lymphocyte transformation: Correlation of biological potency with electrophoretic variants. *Proc. Natl. Acad. Sci. USA* 73:4645-4648.

Lester, E.P., Miller, J.B., and Yachnin, S., 1977, A post-synthetic modification of human α-fetoprotein controls its immunosuppressive potency, *Proc. Natl. Acad. Sci. USA* 74:3988-3992.

Lichenstein, H.S., Lyons, D.E., Wurfel, M.M., Johnson, D.A., McGinley, M.D., Leidli, J.C., Trollinger, D.B., Mayer, J.P., Wright, S.D., and Zukowski, M.M., 1994, Afamin is a new member of the albumin, α-fetoprotein, and vitamin D-binding protein gene family, *J. Biol. Chem.* 269(27): 18149-18154.

Luft, A.J. and Lorscheider, F. L., 1983, Structural analysis of human and bovine alpha-fetoprotein by electron microscopy, image processing, and circular dichroism. *Biochemistry.* 22:5978-5981.

Morinaga, T., Sakai, M., Wegmann, T.G., and Tamaoki, T., 1983, Primary structures of human alpha-fetoprotein and its mRNA. *Proc. Natl. Acad. Sci. USA* 80:4604-4608.

Nishi, S., Koyama, Y., Sakamoto, T., Soda, M., and Kairiyama, C.B., 1988, Expression of rat α-fetoprotein cDNA in *Escherichia coli* and in yeast. *J. Biochem.* 104:968-972.

O'Connor-McCourt, M. D., and Wakefield, L.M., 1987, Latent Transforming Growth Factor-β in Serum. A specific complex with α2-Macroglobulin, *J. Biol Chem.* 262 (29): 14090-14099.

Olsen, M.K., Rockenbach, S.K., Curry, K.A., and Tomich, C-S.C., 1989, Enhancement of heterologous polypeptide expression by alterations in the ribosome binding site sequence. *J. Biotechnology* 9:179-180.

O'Reilly, D., Miller, L.K., and Luckow, V. A., 1992, Post-translational modification *in Baculovirus expression vectors.* W.H. Freeman and Company. 216-236.

Philip, A., and O'Connor-McCourt, M. D., 1991, Interaction of transforming growth factor-β1 with a α2-macroglobulin. Role in transforming growth factor-β clearance. *J. Biol Chem.* 266:22290-22296.

Ruoslahti, E. and Engvall E., 1976, Immunological crossreaction between alpha-fetoprotein and albumin. *Proc. Natl. Acad. Sci.* 73:4641-4644.

Ruoslahti, E., and Terry, W.D., 1976, α-fetoprotein and serum albumin show sequence homology, *Nature* 260:804-805.

Savu,L., Benassayag, C., Vallette, G., Christeff, N., and Nunez, E., 1981, Mouse α-Fetoprotein and albumin. A comparison of their binding properties with estrogen and fatty acid ligands. *J. Biol. Chem.* 256:9414-9418.

Semeniuk, D.J., Boismenu, R., and Murgita, R.A., 1995a, Recombinant alpha-fetoprotein produced by a baculovirus expression system exhibits immunoregulatory properties that are functionally equivalent to its authentic fetal- derived counterpart. Submitted for publication: *J. Exp. Med.*

Semeniuk, D. J., Boismenu, R., Bennett, J. A., Jacobson, H.I., and Murgita, R. A., 1995b, Recombinant human alpha-fetoprotein expressed in insect cells and *Escherichia coli* demonstrate equivalent biological activities. Submitted for publication: *Proc. Natl. Acad. Sci. USA.*

Schoentgen, F., Metz-Boutigue, M.-H., Jolies, J., Constans, J., and Joiles, P., 1986, Complete amino acid sequence of human vitamin D-binding protein (group specific component): evidence of a three-fold internal homology as in serum albumin and α-fetoprotein, *Biochimica et Biophysica Acta* 871:189-198.

Tessier, D.C., Thomas, D.Y., Khouri, H.E., Laliberte,F., and Vernet, T., 1991, Enhanced secretion from insect cells of a foreign protein fused to the honeybee melittin signal peptide. *Gene* 98:177-183.

van Oers, N.S.C., Cohen, B.L., and Murgita, R. A., 1989, Isolation and characterization of a distinct immunoregulatory isoform of α-fetoprotein produced by the normal fetus, *J. Exp. Med.* 170:811-825.

van Oers, N.S.C., Powell, W.S., Semeniuk, D.J., Weissenhofer, W., and Murgita, R.A., 1995, The capacity of natural fetal-derived alphafetoprotein to regulate immune responses is independent of ligand-binding by fatty acids. Submitted for publication: *Biochemistry.*

Wahl, S.M., 1994, Transforming growth factor β: The good, the bad, and the ugly, *J. Exp. Med.* 180:1587-1590.

Yachnin, S., 1976, Demonstration of the inhibitory effects of human alpha-fetoprotein on in vitro transformation of human lymphocytes, *Proc. Natl. Acad. Sci.* USA 73:2857-2861.

Yamamoto, R., Sakamoto, T., Nishi, S., Sakai, M., Morinaga, T., and Tamaoki, T., 1990, Expression of human α-fetoprotein in yeast. *Life Sciences* 46:16791686.

Zimmerman, E. F., Voorting-Hawking, M., and Michael, J. G., 1977, Immunosuppression by mouse sialylated alpha-fetoprotein. *Nature.* 26:354-356.

INDEX

The manufacturer's authorised representative in the EU is Springer
Nature Customer Service Centre GmbH, Europaplatz 3, 69115 Heidelberg,
Germany. If you have any concerns regarding our products, please
contact ProductSafety@springernature.com

Printed and bound by CPI Group (UK) Ltd, Croydon, CR0 4YY

26/04/2026

02097341-0001